Cell Biology of Galectins

Cell Biology of Galectins

Editor

Alexander Timoshenko

MDPI • Basel • Beijing • Wuhan • Barcelona • Belgrade • Manchester • Tokyo • Cluj • Tianjin

Editor
Alexander Timoshenko
Department of Biology
The University of
Western Ontario
London
Canada

Editorial Office
MDPI
St. Alban-Anlage 66
4052 Basel, Switzerland

This is a reprint of articles from the Special Issue published online in the open access journal *Biomolecules* (ISSN 2218-273X) (available at: www.mdpi.com/journal/biomolecules/special_issues/ Biology_Galectins).

For citation purposes, cite each article independently as indicated on the article page online and as indicated below:

LastName, A.A.; LastName, B.B.; LastName, C.C. Article Title. *Journal Name* **Year**, *Volume Number*, Page Range.

ISBN 978-3-0365-4448-9 (Hbk)
ISBN 978-3-0365-4447-2 (PDF)

© 2022 by the authors. Articles in this book are Open Access and distributed under the Creative Commons Attribution (CC BY) license, which allows users to download, copy and build upon published articles, as long as the author and publisher are properly credited, which ensures maximum dissemination and a wider impact of our publications.

The book as a whole is distributed by MDPI under the terms and conditions of the Creative Commons license CC BY-NC-ND.

Contents

About the Editor vii

Alexander V. Timoshenko
Cell Biology of Galectins: Novel Aspects and Emerging Challenges
Reprinted from: *Biomolecules* **2022**, *12*, 744, doi:10.3390/biom12060744 1

Mohamed I. Gatie, Danielle M. Spice, Amritpal Garha, Adam McTague, Mariam Ahmer and Alexander V. Timoshenko et al.
O-GlcNAcylation and Regulation of Galectin-3 in Extraembryonic Endoderm Differentiation
Reprinted from: *Biomolecules* **2022**, *12*, 623, doi:10.3390/biom12050623 5

Veronica Ayechu-Muruzabal, Melanie van de Kaa, Reshmi Mukherjee, Johan Garssen, Bernd Stahl and Roland J. Pieters et al.
Modulation of the Epithelial-Immune Cell Crosstalk and Related Galectin Secretion by DP3-5 Galacto-Oligosaccharides and -3′Galactosyllactose
Reprinted from: *Biomolecules* **2022**, *12*, 384, doi:10.3390/biom12030384 19

Brenda Lucila Jofre, Ricardo Javier Eliçabe, Juan Eduardo Silva, Juan Manuel Pérez Sáez, Maria Daniela Paez and Eduardo Callegari et al.
Galectin-1 Cooperates with Yersinia Outer Protein (Yop) P to Thwart Protective Immunity by Repressing Nitric Oxide Production
Reprinted from: *Biomolecules* **2021**, *11*, 1636, doi:10.3390/biom11111636 35

Jennifer D. Kaminker and Alexander V. Timoshenko
Expression, Regulation, and Functions of the Galectin-16 Gene in Human Cells and Tissues
Reprinted from: *Biomolecules* **2021**, *11*, 1909, doi:10.3390/biom11121909 55

Federico M. Ruiz, Francisco J. Medrano, Anna-Kristin Ludwig, Herbert Kaltner, Nadezhda V. Shilova and Nicolai V. Bovin et al.
Structural Characterization of Rat Galectin-5, an N-Tailed Monomeric Proto-Type-like Galectin
Reprinted from: *Biomolecules* **2021**, *11*, 1854, doi:10.3390/biom11121854 71

Victor L. Thijssen
Galectins in Endothelial Cell Biology and Angiogenesis: The Basics
Reprinted from: *Biomolecules* **2021**, *11*, 1386, doi:10.3390/biom11091386 87

Grażyna Sygitowicz, Agata Maciejak-Jastrzebska and Dariusz Sitkiewicz
The Diagnostic and Therapeutic Potential of Galectin-3 in Cardiovascular Diseases
Reprinted from: *Biomolecules* **2021**, *12*, 46, doi:10.3390/biom12010046 103

Lucas de Freitas Pedrosa, Avraham Raz and João Paulo Fabi
The Complex Biological Effects of Pectin: Galectin-3 Targeting as Potential Human Health Improvement?
Reprinted from: *Biomolecules* **2022**, *12*, 289, doi:10.3390/biom12020289 125

Yves St-Pierre
Towards a Better Understanding of the Relationships between Galectin-7, p53 and MMP-9 during Cancer Progression
Reprinted from: *Biomolecules* **2021**, *11*, 879, doi:10.3390/biom11060879 157

Nishant V. Sewgobind, Sanne Albers and Roland J. Pieters
Functions and Inhibition of Galectin-7, an Emerging Target in Cellular Pathophysiology
Reprinted from: *Biomolecules* **2021**, *11*, 1720, doi:10.3390/biom11111720 167

About the Editor

Alexander Timoshenko

Alexander Timoshenko is an Assistant Professor at the University of Western Ontario, London, Canada. He earned all his doctoral degrees (PhD and DSc) from the Belarusian State University (Minsk, Belarus) and completed his postdoctoral and visiting scholar research in Germany (Marburg University, Ludwig-Maximilian University, and Thoraxklinik-Heidelberg) and Canada (the University of Western Ontario). The main areas of his research expertise include the biological activity of plant and animal lectins, cell biology of galectins, cell aggregation, molecular mechanisms of cellular stress responses, reactive oxygen species and redox regulation, cancer biology, lymphangiogenic factors, innate immunity and neutrophils. His current research program explores the role of galectins in regulating cell stemness and differentiation considering O-GlcNAc-dependent mechanisms.

Editorial

Cell Biology of Galectins: Novel Aspects and Emerging Challenges

Alexander V. Timoshenko

Department of Biology, Western University, London, ON N6A 5B7, Canada; atimoshe@uwo.ca

Galectins are a family of soluble β-galactoside-binding proteins with diverse glycan-dependent and glycan-independent functions outside and inside the cell [1–3]. There are sixteen recognized mammalian galectin genes, and their expression profiles are very different between cell types, tissues, and species [4–6]. Galectins are known to be involved in regulating multiple processes in cells under normal, stress, and pathological conditions, which suggest they are potential candidates for biomedical applications. However, current success in this direction is challenging, mostly due to the complex network of interacting galectins in cells, different modes of their action, and association with diverse fundamental cellular functions (such as cell growth, differentiation, stemness, apoptosis, autophagy, phagocytosis, and cellular interactions). Functions of galectins depend on their localization in specific cellular compartments and organelles (cytosol, cytoskeleton, mitochondria, nucleus, lysosomes, and plasma membrane) and related intracellular trafficking and secretion, which occurs through non-classical pathways [7,8]. An integrated vision of the galectin cell biology is still warranted, and advanced studies of galectin post-translational modifications, the transcriptional regulation of galectin gene expression, and galectin-mediated transmembrane signaling are highly regarded. This Special Issue covers recent progress in the field of cell biology of galectins, relevant concepts of galectin regulatory mechanisms, and biomedical aspects of these unique multifunctional proteins.

Five articles in this Special Issue represent original and novel research studies with both well-known galectin family members (galectins-1, -3, -4, and -9) and relatively poorly characterized galectins (rat galectin-5 and human galectin-16). Gatie et al. [9] explore the role of galectins in a model of extraembryonic endoderm differentiation and provide experimental evidence in support of the new concept of the *O*-GlcNAc-mediated regulation of galectin expression and secretion [10,11]. The findings from this study together with an excellent recent report from Hanover group [12] indicate that the secretion of galectin-3 is an *O*-GlcNAc-dependent process, which represents a new mechanism of unconventional secretion, requiring deglycosylation of released molecules. Although bioinformatics analysis suggests that all human galectin molecules have potential sites for *O*-GlcNAcylation [11], whether this mechanism works for other galectins remains to be investigated. Ayechu-Muruzabal et al. [13] demonstrate that secretion of galectin-3, -4, and -9 can be stimulated by specific types of galacto-oligosaccharides and CpG oligodeoxynucleotides in a complex transwell co-culture model of intestinal epithelial cells (human colon adenocarcinoma HT-29 cell line) with primary peripheral blood mononuclear cells, which ultimately supports the role of galectins in the mucosal immune response. In this context, the authors highlight the key immunomodulatory contribution of β-3′galactosyllactose, a component of human milk [14], which reveals an interesting aspect of health benefits associated with secreted galectins. In comparison, Jofre et al. [15] analyze the role of galectin-1 in mechanisms, allowing Gram-negative bacteria *Yersinia enterocolytica*, which causes gastrointestinal infections, to escape immunosurveillance and demonstrate the critical role of carbohydrate-dependent interaction with *Yersinia* outer proteins in this context. As such, the authors show that galectin-1 can bind to and protect the bacterial proteins from trypsin digestion and also contribute to the decrease in nitric oxide production by infected macrophages. Thus, the

Citation: Timoshenko, A.V. Cell Biology of Galectins: Novel Aspects and Emerging Challenges. *Biomolecules* **2022**, *12*, 744. https://doi.org/10.3390/biom12060744

Received: 23 May 2022
Accepted: 24 May 2022
Published: 25 May 2022

Publisher's Note: MDPI stays neutral with regard to jurisdictional claims in published maps and institutional affiliations.

Copyright: © 2022 by the author. Licensee MDPI, Basel, Switzerland. This article is an open access article distributed under the terms and conditions of the Creative Commons Attribution (CC BY) license (https://creativecommons.org/licenses/by/4.0/).

multifunctional role of galectins remains a challenging topic for health and disease control. The knowledge of tissue-specific expression of several galectins, such as galectin-12 in adipocytes and leukocytes [16] and galectin-16 in placenta [17], raises further insights into an inherent galectin-mediated regulation of cellular responses, including cellular differentiation. Currently, galectin-16 is at a very early stage of its elaboration and the article in this Special Issue introduces a model of trophoblastic differentiation of BeWo and JEG-3 cells associated with the upregulation of *LGALS16* gene expression [18]. In addition, bioinformatics analyses highlight possible transcriptional and post-transcriptional regulation of *LGALS16* as well as its associations with other tissues and human diseases. To wrap up the experimental part of this Special Issue, the article by Ruiz et al. [19] reports a comprehensive crystallographic characterization and ligand-binding specificity of rat galectin-5, a lectin with a unique N-terminal extension that is likely serving as a molecular binding switch.

Five review articles provide a comprehensive summary of the cell biology of galectins, as well as their functions and applications in normal and pathophysiological context. A review article by Thijssen [20] elegantly describes all the studied functions of endothelial galectins (galectin-1, -3, -8, and -9), making it easy for readers to understand the versatile functions of individual galectins in endothelial cell biology and angiogenesis. The role of commonly expressed galectin-3 is addressed in two reviews highlighting several important themes: (1) the diagnostic and therapeutic potential in cardiovascular diseases by Sygitowicz et al. [21] and (2) the complexity and controversy of galectin-3/pectin interactions in the context of dietary interference by Pedrosa et al. [22]. Finally, two reviews focus on galectin-7, an epithelial cell galectin with pro-apoptotic and pro-carcinogenic properties. The review by St-Pierre [23] provides exceptional information and clarifications on the complex relationship between galectin-7, p53, and MMP-9 molecules, which helps to clarify the role of galectin-7 in cancer cells. Sewgobind et al. [24] summarize the pathophysiological role of galectin-7 in multiple human diseases, justify a rationale to design small-molecule inhibitors of galectin-7, and present structure and examples of novel carbohydrate and non-carbohydrate compounds to be tested against cellular disorders including cancer.

To conclude, the collection of research and review articles in this Special Issue addresses novel and challenging aspects of cell biology of galectins that might catalyze further experimental investigations into the complex network of galectin molecules in cells.

Acknowledgments: I would like to thank all of the authors for their cooperation and valuable contributions to this Special Issue and all of the reviewers for their helpful comments during the peer review process.

Conflicts of Interest: The author declares no conflict of interest.

References

1. Compagno, D.; Jaworski, F.M.; Gentilini, L.; Contrufo, G.; Perez, I.; Elola, M.T.; Pregi, N.; Rabinovich, G.A.; Laderach, D.J. Galectins: Major signaling modulators inside and outside the cell. *Curr. Mol. Med.* **2014**, *14*, 630–651. [CrossRef] [PubMed]
2. Vladoiu, M.C.; Labrie, M.; St-Pierre, Y. Intracellular galectins in cancer cells: Potential new targets for therapy. *Int. J. Oncol.* **2014**, *44*, 1001–1014. [CrossRef] [PubMed]
3. Verkerke, H.; Dias-Baruffi, M.; Cummings, R.D.; Arthur, C.M.; Stowell, S.R. Galectins: An ancient family of carbohydrate binding proteins with modern functions. In *Methods in Molecular Biology*; Humana: New York, NY, USA, 2022; Volume 2442, pp. 1–40. [CrossRef]
4. Timoshenko, A.V. Towards molecular mechanisms regulating the expression of galectins in cancer cells under microenvironmental stress conditions. *Cell. Mol. Life Sci.* **2015**, *72*, 4327–4340. [CrossRef] [PubMed]
5. Nio-Kobayashi, J. Tissue- and cell-specific localization of galectins, β-galactose-binding animal lectins, and their potential functions in health and disease. *Anat. Sci. Int.* **2017**, *92*, 25–36. [CrossRef]
6. Johannes, L.; Jacob, R.; Leffler, H. Galectins at a glance. *J. Cell Sci.* **2018**, *131*, jcs208884. [CrossRef]
7. Hughes, R.C. Secretion of the galectin family of mammalian carbohydrate-binding proteins. *Biochim. Biophys. Acta* **1999**, *1473*, 172–185. [CrossRef]
8. Popa, S.J.; Stewart, S.E.; Moreau, K. Unconventional secretion of annexins and galectins. *Semin. Cell Dev. Biol.* **2018**, *83*, 42–50. [CrossRef]
9. Gatie, M.I.; Spice, D.M.; Garha, A.; McTague, A.; Ahmer, M.; Timoshenko, A.V.; Kelly, G.M. O-GlcNAcylation and regulation of galectin-3 in extraembryonic endoderm differentiation. *Biomolecules* **2022**, *12*, 623. [CrossRef]

10. Timoshenko, A.V. The role of galectins and O-GlcNAc in regulating promyelocytic cell stemness and differentiation. *Mol. Biol. Cell* **2019**, *30*, 3075 (abstract #P1664/B802).
11. Tazhitdinova, R.; Timoshenko, A.V. The emerging role of galectins and O-GlcNAc homeostasis in processes of cellular differentiation. *Cells* **2020**, *9*, 1792. [CrossRef]
12. Mathew, M.P.; Abramowitz, L.K.; Donaldson, J.G.; Hanover, J.A. Nutrient-responsive O-GlcNAcylation dynamically modulates the secretion of glycan-binding protein galectin 3. *J. Biol. Chem.* **2022**, *298*, 101743. [CrossRef] [PubMed]
13. Ayechu-Muruzabal, V.; van de Kaa, M.; Mukherjee, R.; Garssen, J.; Stahl, B.; Pieters, R.J.; van't Land, B.; Kraneveld, A.D.; Willemsen, L.E.M. Modulation of the epithelial-immune cell crosstalk and related galectin secretion by DP3-5 galacto-oligosaccharides and β-3′galactosyllactose. *Biomolecules* **2022**, *12*, 384. [CrossRef] [PubMed]
14. Eussen, S.R.B.M.; Mank, M.; Kottler, R.; Hoffmann, X.-K.; Behne, A.; Rapp, E.; Stahl, B.; Mearin, M.L.; Koletzko, B. Presence and levels of galactosyllactoses and other oligosaccharides in human milk and their variation during lactation and according to maternal phenotype. *Nutrients* **2021**, *13*, 2324. [CrossRef] [PubMed]
15. Jofre, B.L.; Eliçabe, R.J.; Silva, J.E.; Pérez Sáez, J.M.; Paez, M.D.; Callegari, E.; Mariño, K.V.; Di Genaro, M.S.; Rabinovich, G.A.; Davicino, R.C. Galectin-1 cooperates with Yersinia outer protein (Yop) P to thwart protective immunity by repressing nitric oxide production. *Biomolecules* **2021**, *11*, 1636. [CrossRef]
16. Wan, L.; Yang, R.-Y.; Liu, F.-T. Galectin-12 in cellular differentiation, apoptosis and polarization. *Int. J. Mol. Sci.* **2018**, *19*, 176. [CrossRef]
17. Than, N.G.; Romero, R.; Goodman, M.; Weckle, A.; Xing, J.; Dong, Z.; Xu, Y.; Tarquini, F.; Szilagyi, A.; Gal, P.; et al. A primate subfamily of galectins expressed at the maternal-fetal interface that promote immune cell death. *Proc. Natl. Acad. Sci. USA* **2009**, *106*, 9731–9736. [CrossRef]
18. Kaminker, J.D.; Timoshenko, A.V. Expression, regulation, and functions of the galectin-16 gene in human cells and tissues. *Biomolecules* **2021**, *11*, 1909. [CrossRef]
19. Ruiz, F.M.; Medrano, F.J.; Ludwig, A.-K.; Kaltner, H.; Shilova, N.V.; Bovin, N.V.; Gabius, H.-J.; Romero, A. Structural characterization of rat galectin-5, an N-tailed monomeric proto-type-like galectin. *Biomolecules* **2021**, *11*, 1854. [CrossRef]
20. Thijssen, V.L. Galectins in endothelial cell biology and angiogenesis: The basics. *Biomolecules* **2021**, *11*, 1386. [CrossRef]
21. Sygitowicz, G.; Maciejak-Jastrzębska, A.; Sitkiewicz, D. The diagnostic and therapeutic potential of galectin-3 in cardiovascular diseases. *Biomolecules* **2022**, *12*, 46. [CrossRef]
22. Pedrosa, L.d.F.; Raz, A.; Fabi, J.P. The complex biological effects of pectin: Galectin-3 targeting as potential human health improvement? *Biomolecules* **2022**, *12*, 289. [CrossRef] [PubMed]
23. St-Pierre, Y. Towards a better understanding of the relationships between galectin-7, p53 and MMP-9 during cancer progression. *Biomolecules* **2021**, *11*, 879. [CrossRef] [PubMed]
24. Sewgobind, N.V.; Albers, S.; Pieters, R.J. Functions and inhibition of galectin-7, an emerging target in cellular pathophysiology. *Biomolecules* **2021**, *11*, 1720. [CrossRef] [PubMed]

Article

O-GlcNAcylation and Regulation of Galectin-3 in Extraembryonic Endoderm Differentiation

Mohamed I. Gatie [1,2,†], Danielle M. Spice [1,2,†], Amritpal Garha [1], Adam McTague [1], Mariam Ahmer [1], Alexander V. Timoshenko [1] and Gregory M. Kelly [1,2,3,4,5,*]

1. Department of Biology, Western University, London, ON N6A 5B7, Canada; mgatie@uwo.ca (M.I.G.); dspice@uwo.ca (D.M.S.); agarha@uwo.ca (A.G.); amctagu@uwo.ca (A.M.); mahmer@uwo.ca (M.A.); atimoshe@uwo.ca (A.V.T.)
2. Collaborative Specialization in Developmental Biology, Western University, London, ON N6A 5B7, Canada
3. Department of Physiology and Pharmacology, Western University, London, ON N6A 5B7, Canada
4. Lawson Health Research Institute, London, ON N6A 5B7, Canada
5. Children's Health Research Institute, London, ON N6A 5B7, Canada
* Correspondence: gkelly@uwo.ca
† These authors contributed equally to this work.

Abstract: The regulation of proteins through the addition and removal of *O*-linked β-*N*-acetylglucosamine (*O*-GlcNAc) plays a role in many signaling events, specifically in stem cell pluripotency and the regulation of differentiation. However, these post-translational modifications have not been explored in extraembryonic endoderm (XEN) differentiation. Of the plethora of proteins regulated through *O*-GlcNAc, we explored galectin-3 as a candidate protein known to have various intracellular and extracellular functions. Based on other studies, we predicted a reduction in global *O*-GlcNAcylation levels and a distinct galectin expression profile in XEN cells relative to embryonic stem (ES) cells. By conducting dot blot analysis, XEN cells had decreased levels of global *O*-GlcNAc than ES cells, which reflected a disbalance in the expression of genes encoding *O*-GlcNAc cycle enzymes. Immunoassays (Western blot and ELISA) revealed that although XEN cells (low *O*-GlcNAc) had lower concentrations of both intracellular and extracellular galectin-3 than ES cells (high *O*-GlcNAc), the relative secretion of galectin-3 was significantly increased by XEN cells. Inducing ES cells toward XEN in the presence of an *O*-GlcNAcase inhibitor was not sufficient to inhibit XEN differentiation. However, global *O*-GlcNAcylation was found to decrease in differentiated cells and the extracellular localization of galectin-3 accompanies these changes. Inhibiting global *O*-GlcNAcylation status does not, however, impact pluripotency and the ability of ES cells to differentiate to the XEN lineage.

Keywords: *O*-GlcNAc; galectin-3; unconventional secretion; extraembryonic endoderm; cell differentiation

1. Introduction

Many intracellular proteins are post-translationally modified by the addition of β-*N*-acetylglucosamine (GlcNAc) to the hydroxyl moiety of serine or threonine residues [1,2]. This *O*-GlcNAcylation acts in competition with phosphorylation at the same or nearby amino acids [3–5], demonstrating its integral function in protein regulation. Unlike protein phosphorylation and dephosphorylation, which are carried out by a variety of kinases and phosphatases, respectively, *O*-GlcNAcylation is regulated by only two enzymes: *O*-GlcNAc transferase (OGT), which transfers GlcNAc to proteins from a donor UDP-GlcNAc derived from the hexosamine biosynthesis pathway [6], and N-acetyl-β-D-glucosaminidase (*O*-GlcNAcase, OGA), which removes GlcNAc from proteins [7].

O-GlcNAcylation is a dynamic process that is regulated by multiple mechanisms. As UDP-GlcNAc is derived through the hexosamine biosynthesis pathway, the availability of GlcNAc is highly regulated by glucose metabolism [8,9] and insulin [10]. *O*-GlcNAcylation

Citation: Gatie, M.I.; Spice, D.M.; Garha, A.; McTague, A.; Ahmer, M.; Timoshenko, A.V.; Kelly, G.M. *O*-GlcNAcylation and Regulation of Galectin-3 in Extraembryonic Endoderm Differentiation. *Biomolecules* **2022**, *12*, 623. https://doi.org/10.3390/biom12050623

Academic Editors: Davide Vigetti and Vladimir N. Uversky

Received: 26 February 2022
Accepted: 19 April 2022
Published: 22 April 2022

Publisher's Note: MDPI stays neutral with regard to jurisdictional claims in published maps and institutional affiliations.

Copyright: © 2022 by the authors. Licensee MDPI, Basel, Switzerland. This article is an open access article distributed under the terms and conditions of the Creative Commons Attribution (CC BY) license (https://creativecommons.org/licenses/by/4.0/).

is also sensitive to free fatty acids [11] and thermal and oxidative stress [12–14]. The addition or removal of O-GlcNAc in response to nutrients and stress is not universal as the O-GlcNAcylated proteome is dynamic, and evidence exists showing many proteins with increased O-GlcNAcylation, while others have less O-GlcNAc [12]. O-GlcNAcylation is also differentially regulated during development and differentiation. The high levels of O-GlcNAc on specific proteins play important roles in maintaining pluripotency in mouse embryonic stem (ES) cells as it can regulate the activity of OCT4 and SOX2 [15] or inhibit the differentiation of ectoderm in human ES cells [16]. In contrast, global O-GlcNAcylation decreases in response to the induced differentiation of neutrophils [17] and cardiomyocytes [18].

Galectins is a family of proteins that binds β-galactoside-containing glycans and confers the ability to influence apoptosis, autophagy, transcriptional regulation, Wnt signaling and cell division, or interact with soluble and membrane-bound glycoproteins in the extracellular milieux [19]. Changes in the expression of multiple galectin family members occur in response to a variety of stress conditions [20]. Galectins are also involved with the differentiation of many cell lineages, which has been proposed to rely on O-GlcNAc-dependent mechanisms controlling the secretion and localization of galectins in cells [21]. In fact, while intracellular galectin-3 is O-GlcNAcylated, a preferentially deglycosylated protein is secreted from transfected HeLa cells [22]. Both O-GlcNAcylation of galectins and interaction of galectins with O-GlcNAcylated effector molecules are considered as potential molecular patterns in the context of cellular differentiation and proliferation [21–24]. However, little is known about the role of galectins and O-GlcNAc in extraembryonic endoderm (XEN) formation. XEN cells differentiate from the inner embryonic cell mass in mice at E4.5, and they are essential for later patterning and segmentation [25]. We have demonstrated a role of oxidative molecules in the differentiation of the F9 teratocarcinoma cell line, a model of XEN differentiation [26,27], and although much has been reported on their formation, there are few studies directly investigating the role of O-GlcNAc in this lineage [28,29].

To address this shortcoming, the goal of this study was to understand the role of O-GlcNAcylation, and specifically the secretion of O-GlcNAc-sensitive galectin-3 in processes of XEN differentiation. Toward that end, we used the relevant models of mouse embryonic and teratocarcinoma cells to demonstrate that O-GlcNAc levels decreased in response to XEN differentiation and that galectin-3 altered its intracellular to extracellular localization depending on cellular O-GlcNAc homeostasis. We showed that the lowering global O-GlcNAc levels in XEN cells led to the secretion of galectin-3 and had no effects on pluripotency.

2. Materials and Methods

2.1. Cell Culture

Three different mouse cell lines were used in this study representing ES cells (E14TG2a, referred to as E14), extraembryonic endoderm XEN cells (known as E4 cells), and F9 embryonal carcinoma cells inducible to endoderm differentiation by all-trans retinoic acid.

Mouse embryonic stem cell line ES-E14TG2a (E14) was obtained from the University of California, Davis, and they were maintained on 0.1% gelatin-coated tissue culture dishes as previously described [30]. Briefly, ES cells were cultured in 50/50 KnockOut™ DMEM/F-12 and neurobasal™ medium (Gibco) supplemented with N-2™, vitamin A-free B-27™ (Gibco), 1 µM MEK inhibitor PD0325901, and 3 µM GSK3 inhibitor CHIR99021 (both from APExBIO), and 1000 U/mL LIF (Millipore-Sigma, Burlington, MA, USA). E4 mouse extraembryonic endoderm [31] cells were maintained on 0.1% gelatin-coated tissue culture dishes in RPMI 1640 medium (cat. # R5886-500 mL) supplemented with 15% fetal bovine serum (FBS, cat.# 12483020) (both from ThermoFisherScientific). Both media were supplemented with GlutaMAX™ and 2-mercaptoethanol (cat.# 21985023, ThermoFisher-Scientific, Waltham, MA, USA), and all cells were maintained at 37 °C and 5% CO_2 and assessed for mycoplasma and karyotypic abnormalities at the beginning of the project. The chemical induction of E14 cells toward XEN lineage was performed as previously reported [32]. Briefly, ES cells were seeded onto 0.1% gelatin-coated tissue culture plates in XEN medium (RPMI1640 media, 0.5X B-27 Supplement without insulin, 2 mM GlutaMAX™, and 0.1 mM 2-mercaptoethanol) for 2 days. Base XEN medium was supplemented with 3 µM CHIR99021 and 20 ng/mL activin A (R&D Systems), and the induction medium was changed every 2 days for 10 days.

F9 mouse teratocarcinoma cells (ATCC, cat. #CRL-1720) were cultured in DMEM (Corning, cat. # 10-013-CV), supplemented with 10% FBS and 1% penicillin-streptomycin (Gibco), and cultured on adherent tissue culture dishes (Sarstedt). For primitive endoderm differentiation, cells were treated with 0.1 µM all-trans retinoic acid (Millipore-Sigma, cat. # R2625) or vehicle control, dimethyl sulfoxide (DMSO; BioShop), for 4 days. To differentiate cells to the parietal endoderm, F9 cells were treated with 0.1 µM all-trans retinoic acid, and after two days, they were subsequently treated with 1 mM of dibutyryl-cAMP (Millipore-Sigma, cat.# D0627) for 2 days. Cells were cultured at 37 °C and 5% CO_2 and passaged every 4 days or when cultures reached 70% confluency.

2.2. Western and Immunodot Blot Assays

Cells were lysed in RIPA buffer (150 mM NaCl, 1% Triton-X-100, 0.5% deoxycholate, 0.1% SDS, and 50 mM Tris, pH 8.0, with 1:100 Halt protease inhibitor cocktail from ThermoFisherScientific, cat. # PI87786) before being quantified by a DC Protein Assay (Bio-Rad, cat. # 5000116, Hercules, CA, USA). Approximately 10–30 µg of proteins was added to the 5X SDS loading buffer (30% glycerol, 0.02% bromphenol blue, 10% SDS, 250 mM Tris, pH 6.8) containing 10% 2-mercaptoethanol. Proteins were separated by electrophoresis on 10% polyacrylamide gels for 1.5 h at 120 V at room temperature. After separation, proteins were transferred overnight at 20 V at 4 °C to Western blot PVDF membranes (Bio-Rad, cat. # 1620177) in 20% methanol/Tris-glycine transfer buffer. Following transfer, membranes were washed in TBS-T (19.8 mM Tris, 150 mM NaCl, 0.1% Tween-20, pH 7.6), and then they were incubated in blocking solution (5% skim milk in TBS-T) for 2 h at room temperature. Membranes were then washed in TBS-T before incubation overnight at 4 °C with primary antibody solution containing 5% bovine serum albumin (BSA) in TBS-T. Primary antibodies were mouse monoclonal pan-specific antibody to O-GlcNAc, RL2 (1:1000; Thermo Scientific, cat. # MA1-072), rabbit anti-galectin-3 (1:1000, Santa Cruz Biotechnology, cat. # sc-20157), and mouse anti-β-actin (1:1000, Santa Cruz Biotechnology, cat. # sc-47778). Membranes were washed extensively in TBS-T, followed by incubation for 2 h at room temperature with a secondary antibody, either goat anti-mouse (1:10,000, Millipore-Sigma, cat/# AP130P) or goat anti-rabbit (1:10,000; Millipore-Sigma, cat. # AP156P). Finally, membranes were washed in TBS-T, exposed to Luminata™ Classico Western HRP Substrate (Millipore-Sigma) and imaged using a ChemiDoc™ Touch Imaging System (Bio-Rad). Densitometry was performed using ImageLab version 5.2 software (Bio-Rad).

Samples for O-GlcNAc immunodot blot analysis were prepared using a Bio-Dot® Microfiltration apparatus (Bio-Rad), as described previously [17]. Briefly, 4 µg of protein was loaded into each well of the dot blot apparatus and then transferred to the nitrocellulose membrane by gravity filtration for 1.5 h. All subsequent steps were identical to the Western blot procedure with the RL2 primary antibody to detect global levels of O-GlcNAc in cells.

2.3. Quantitative RT-PCR

Total RNA was collected from cells using the Qiashredder (Qiagen, cat. # 79654) and RNeasy Mini kit (Qiagen, cat. # 74104) and was reverse transcribed according to manufacturer's instructions into cDNA using a High-Capacity cDNA Reverse Transcription Kit (ThermoFisherScientific, cat. # 4368814). PCR reactions containing 500 nM of each forward and reverse primers (Table 1), SensiFAST SYBR Mix (FroggaBio, cat. # BIO-98005), and 1 µL of cDNA were carried out using the CFX Connect Real-Time PCR Detection System (Bio-Rad). The comparative cycle threshold method ($2^{-\Delta\Delta Ct}$) was used to determine fold changes in gene expression. The cycle threshold value was normalized to the housekeeping gene *L14* and was subsequently normalized to DMSO-treated controls for F9 cell experiments or to untreated E14 cells for ES and XEN experiments.

Table 1. Primers used for RT-qPCR.

Gene	Forward Primer (5′→3′)	Reverse Primer (5′→3′)
Lgals1	TCTCAAACCTGGGGAATGTCTC	CTCAAAGGCCACGCACTTAATC
Lgals2	AACATGAAACCAGGGATGTCC	CGAGGGTTAAAATGCAGGTTGAG
Lgals3	AGGAGAGGGAATGATGTTGC	TAGCGCTGGTGAGGGTTATG
Lgals4	GGTCGTGGTGAACGGAAATTC	GTGGAGGGTTGTACCCAGGA
Lgals7	GTGAGGAGCAAGGAGCAGAT	CGGTGGTGGAAGTGGAGATA
Lgals8	CCGATAATCCCCTATGTTGG	GTTCACTTTGCCGTAGATGC
Lgals9	TTGAGGAAGGAGGGTATGTG	AACTGGACTGGCTGAGAGAA
Lgals12	TATGGCACAACAATTTTTGGTGG	GCTTGACAGTGTAGAATCGAGGG
Ogt	TTGGCAATTAAACAGAATCCCCT	GGCATGTCGATAATGCTCGAT
Oga	TGGTGCCAGTTTGGTTCCAG	TGCTCTGAGGTCGGGTTCA
Gfpt1	GAAGCCAACGCCTGCAAAATC	CCAACGGGTATGAGCTATTCC
Gfpt2	CCAACGGGTATGAGCTATTCC	GACTCTTTCGACCAATGTGGAA
Oct4	CCCAATGCCGTGAAGTTGGA	GCTTTCATGTCCTGGGACTCCT
Nanog	TCTTCCTGGTCCCCACAGTTT	GCAAGAATAGTTCTCGGGATGAA
FoxA2	CCCTACGCCAACATGAACTCG	GTTCTGCCGGTAGAAAGGGA
Dab2	GGAGCATGTAGACCATGATG	AAAGGATTTCCGAAAGGGCT
Gata6	ATGGCGTAGAAATGCTGAGG	TGAGGTGGTCGCTTGTGTAG
L14	GGGAGAGGTGGCCTCGGACGC	GGCTGGCTTTCACTCAAAGGCC

2.4. RNA Sequencing

RNA was extracted from ESC and XEN cells are described above using Qiashredder and RNeasy kits (both from Qiagen). Samples were quantified, and RNA purity was evaluated using both a Nanodrop spectrophotometer (ThermoFisherScientific) and an Agilent 2100 bioanalyzer (Agilent Technologies). The BGISEQ-500 platform was used for library construction and sequencing, where reads were mapped to reference genome (GRCm38-mm10) using Bowtie [33]. Gene expression levels were then calculated using RSEM [34] and DEseq2 [35] was using to determine differentially expressed genes with a fold change greater than or equal to 2.0 and a significance value less than or equal to 0.0001.

2.5. Galectin-3 ELISA

Extracellular galectin-3 was detected using a Mouse SimpleStep ELISA® kit (Abcam, cat. # ab203369). Briefly, supernatants were collected after 3 days of cell culture and centrifuged at 2000× *g* for 10 min, and subsequent steps were carried out according to manufacturer's instructions. Data analysis was performed after obtaining a standard curve. The secretion levels of galectin-3 from E14 and E4 cells were normalized to intracellular protein concentrations, which were determined using a *DC Protein Assay* kit (Bio-Rad).

2.6. Statistical Analysis

All figures are representative of at least 3 independent biological replicates. An unpaired Student's *t*-test was performed for data comparisons between two groups. One sample *t*-test was performed for data comparisons between two groups where one group was set as a relative value of 1. One-way ANOVA was performed with Tukey's Honest Significant Difference post-hoc analysis for data comparisons between 3 or more groups. All statistical analyses were performed using GraphPad Prism version 8.4.2, (GraphPad Software, San Diego, CA, USA). P-values were considered statistically significant at * $p < 0.05$, ** $p < 0.01$, *** $p < 0.001$, and **** $p < 0.0001$.

3. Results

3.1. Global Reduction in O-GlcNAcylation Levels in Feeder-Free XEN Cells

Global O-GlcNAcylation levels were determined in feeder-free mouse E14 embryonic stem (ES) cells and terminally differentiated mouse E4 extraembryonic endoderm (XEN) cells. Western blot analysis showed multiple O-GlcNAcylated proteins in both cell lines with a significant decrease in global levels of O-GlcNAc in the XEN cells as quantified by immunodot blot technique (Figure 1a). O-GlcNAcylation status was also examined by measuring the expression of *Ogt*, which encodes the enzyme that catalyzes the addition of GlcNAc to serine/threonine residues, and the expression of *Oga*, which encodes the enzyme responsible for removing the modification. *Ogt* and *Oga* expression levels were found to be significantly decreased in XEN cells, but the ratio of *Ogt/Oga* was found to be significantly higher in XEN cells (Figure 1b). At first glance, this increased *Ogt/Oga* ratio would suggest more O-GlcNAc accompanied differentiation; however, it contradicts the dot blot analysis used to examine global O-GlcNAc levels (Figure 1a). Further evidence supports the global decrease in O-GlcNAc observed in Figure 1a, as the expressions of the rate-limiting enzyme of the hexosamine biosynthesis pathway, *Gfpt1* and *Gfpt2*, both show significantly reduced expression in XEN (Figure 1c). This decrease in O-GlcNAcylation with differentiation was also representative in the F9 cell model (Supplementary Materials, Figure S1a,b) and by previous reports from this lab [17] and others [15,18]. Incidentally, F9 cells differentiated toward XEN showed increased expression in *Ogt* and *Oga*, but there was no apparent change in the *Ogt/Oga* ratio (Supplementary Materials, Figure S1b).

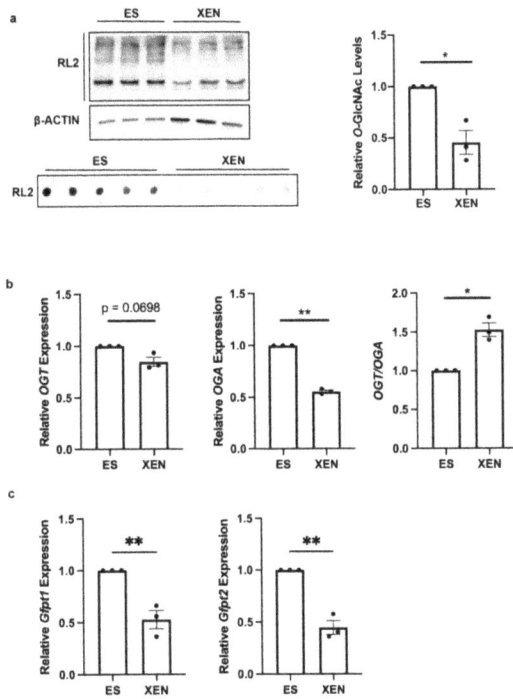

Figure 1. Global O-GlcNAcylation is decreased in XEN compared to ES cells: (**a**) Western blot, immunodot blot, and immunodot blot densitometry of O-GlcNAcylated proteins from ESC and XEN cells detected by the RL2 antibody; (**b**) expression levels and ratio of *Ogt* and *Oga* transcripts and (**c**) expression levels of *Gfpt1* and *Gfpt2* in ES and XEN cells as detected by RT-qPCR. Bars represent mean ± SEM, N = 3. * $p < 0.05$, ** $p < 0.01$.

3.2. OGA Inhibition Alters Pluripotency and XEN Marker Expression Patterns

To determine the effects of global O-GlcNAcylation levels on pluripotency and endodermal markers, ES and XEN cells were treated with 10 µM Thiamet G (TG), a well-established OGA inhibitor elevating O-GlcNAc levels in different types of cells [17,36]. As an example, the efficiency of this treatment was evident in XEN cells, which originally had relatively low homeostatic level of O-GlcNAc (Figure 2a). The inhibition of OGA for 24 h in ES cells resulted in no significant change in Oct4 expression (Figure 2b), but there was decreased Nanog expression (Figure 2b). Exploring the expression of XEN markers FoxA2, Gata6, and Dab2 [31,37] in TG-treated ES cells all showed no significant changes (Figure 2b). Furthermore, OGA inhibition in XEN cells resulted in no significant change in Gata6 expression, but caused a decrease in FoxA2 and Dab2 levels (Figure 2c). Surprisingly pluripotency marker Nanog had reduced expression in TG-treated XEN cells; however, there was no change in Oct4 (Figure 2c). Although the decrease in FoxA2 and Dab2 in XEN cells would suggest that the high global O-GlcNAc might positively regulate pluripotency, the simultaneous decrease in Nanog expression in ES cells and XEN following the chemically induced inhibition of OGA contradicts this idea.

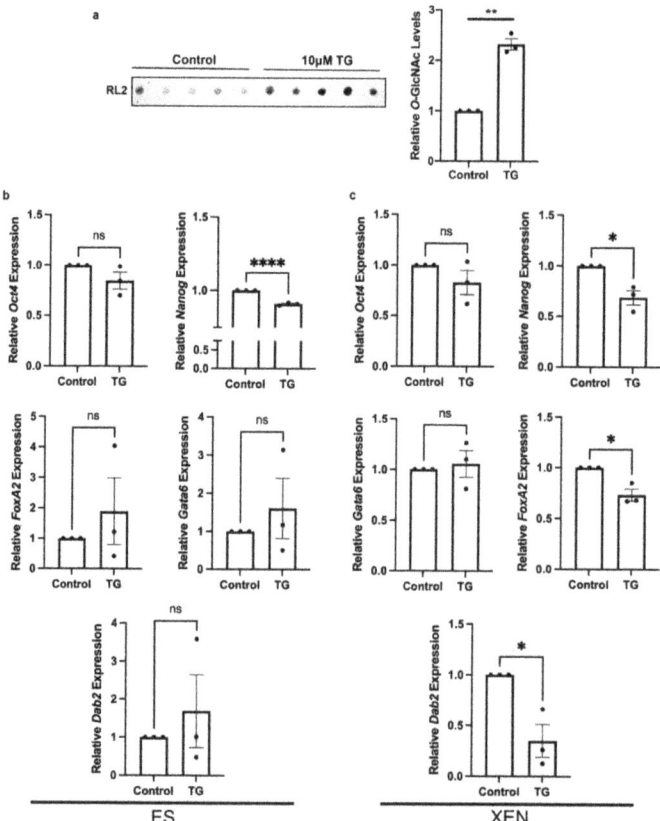

Figure 2. OGA inhibition alters pluripotency marker expression in ES cells and XEN marker expression in XEN cells. The cells were treated with 10 µM TG for 24 h: (**a**) immunodot blot and densitometry of O-GlcNAcylated proteins detected by the RL2 antibody in XEN cells; (**b,c**) expression of pluripotency gene expression Oct4 and Nanog and XEN marker expression FoxA2, Gata6, and Dab2 in ES cells (left panel) and XEN cells (right panel). Bars represent mean values ± SEM, N = 3. * $p < 0.05$, ** $p < 0.01$, **** $p < 0.0001$.

3.3. Lgals3 Expression Is Decreased in XEN Compared to ES Cells

Since we and others have shown that global O-GlcNAcylation decreased in differentiated cells [15,17,18] and our observed increase in O-GlcNAcylation with TG treatment would be expected to favor pluripotency, the overall levels of O-GlcNAcylated proteins should also decrease with differentiation. We previously demonstrated altered oxidative stress during XEN-like cell differentiation [26] and have shown that the expression and levels of galectins are affected in response to differentiation and oxidative stress in HL-60 cells [38]. Galectins are proteins presumably regulated by O-GlcNAcylation, which impacts their intracellular and extracellular functions [21]. RNA sequencing analysis of transcripts from ES and XEN cells was employed to explore general trends in the expression of galectin genes and others involved in the O-GlcNAcylation process (*Ogt*, *Mgea5*, *Gfpt1* and *Gfpt2*). Our sequencing analysis showed changes in many galectin genes, with some such as *Lgasl3* showing decreased transcript levels in XEN cells and others such as *Lgals2* showing increased transcript levels (Figure 3a). Although RNA sequencing analysis showed altered expression levels of various galectin genes, validation by RT-qPCR revealed that all but *Lgals3* (Figure 3b), encoding galectin-3 protein, showed no difference in expressions between ES and XEN cells or had extremely variable expression between replicates, making the statistical analysis of these genes less reliable (Figure 3b). In contrast, F9 cells showed a significant decrease in *Lgals1*, *Lgals2*, *Lgals4*, *Lgals7*, *Lgals8*, *Lgals9*, and *Lgals12*, but no significant change in *Lgals3* expression with differentiation (Supplementary Materials, Figure S2) was observed, which again underscores the differences between E14 and F9 cell types. As the functional role of galectins, which are soluble proteins, may depend not only on their expression levels but also on their O-GlcNAc-mediated localization in cells [21–24], we next analyzed intracellular levels of galectin-3 and galectin-3 secretion in the context of embryonic stem cell differentiation.

3.4. Elevated Extracellular to Intracellular Ratio of Galectin-3 in XEN Cells

Galectin-3 is a candidate protein, the secretion of which is regulated by O-GlcNAcylation [22]. Our Western blot analysis showed a significant decrease in intracellular galectin-3 in XEN compared to ES cells (Figure 4a), as did an ELISA examining the extracellular levels of galectin-3 (Figure 4b). However, the ratio of extracellular to intracellular levels of galectin-3 significantly increased in XEN versus ES cells (Figure 4c), indicating that although there is less intracellular galectin-3 overall, XEN cells were prone to secrete this galectin. Thus, XEN cells had less *Lgals3* expression, less overall galectin-3 levels, and more intensive galectin-3 secretory activity, which accompanied extraembryonic endoderm differentiation. Therefore, since the localization of galectins influences their function [19], this extracellular shuttling or secretion of galectin-3 may be due to reduced global O-GlcNAcylation. The changing localization of galectin-3 highlights its highly likely role during differentiation and highlights an intracellular function when cells are in the pluripotent state.

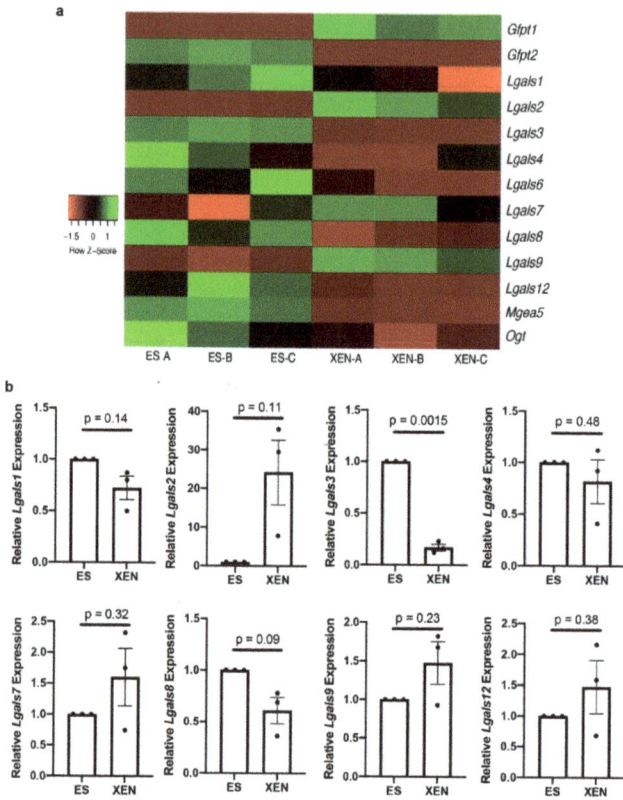

Figure 3. RNA sequencing and gene expression analysis of galectin genes in ES cells and XEN: (a) heat map of RNA sequencing of transcripts in ES cells and XEN showing *Lgals* genes 1–4, 7–9, and 12 and genes involved in O-GlcNAcylation *Ogt*, *Mgea5*, *Gfpt1*, and *Gfpt2*; data are shown for 3 biological replicates denoted as -A, -B, and -C; (b) expression levels of *Lgals1*, *Lgals2*, *Lgals3*, *Lgals4*, *Lgals7*, *Lgals8*, *Lgals9*, and *Lgals12* as detected by RT-qPCR. Bars represent mean ± SEM, N = 3.

3.5. Inhibiting OGT in ES Cells Increases Galectin-3 Secretion

To explore whether low O-GlcNAcylation is a driving force of galectin-3 secretion, we tested the effects of an OGT inhibitor alloxan [39] on ES and XEN cells. As expected, alloxan at a concentration of 5 mM was effective in reducing global O-GlcNAcylation levels in ES cells (Figure 5a). This treatment led to a significant increase in galectin-3 secretion by ES cells as measured by ELISA, while no significant change was evident following OGA inhibition with TG (Figure 5b). Interestingly, no change was seen when XEN cells were treated with either alloxan or TG (Figure 5b), suggesting an irreversibly upregulated galectin-3 secretory system in this case. Thus, these results would suggest that galectin-3 localization in ES cells is dependent on O-GlcNAc homeostasis, as noted earlier [22], and this localization may serve as an indicator to discriminate pluripotency from differentiation.

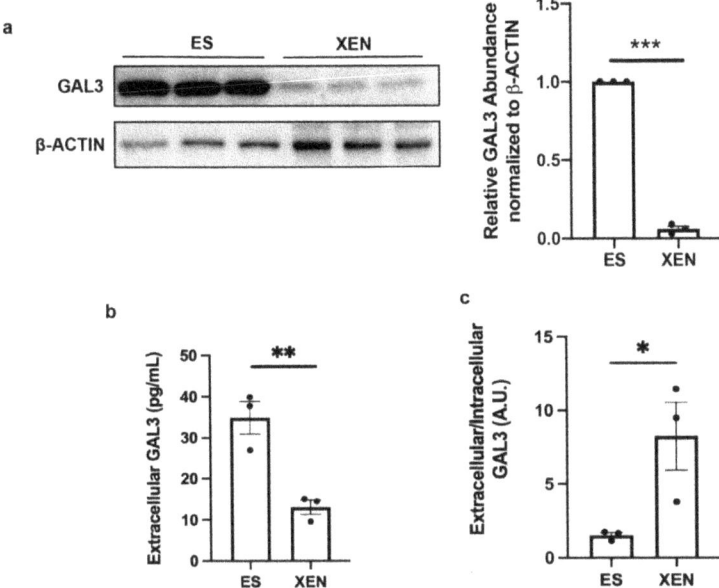

Figure 4. ES cells increased intracellular galectin-3 levels while XEN cells increased extracellular galectin-3 levels compared to their own intracellular protein levels: (**a**) Western blot and densitometry of intracellular galectin-3 in ES and XEN cells; (**b**) ELISA of extracellular galectin-3 in the cell culture media of ES and XEN cells after 8 days normalized to total intracellular protein levels (mg/mL); (**c**) extracellular to intracellular ratio of galectin-3 in ES and XEN cell cultures both normalized to total intracellular protein levels (mg/mL) represented as arbitrary units (A.U.). Bars represent mean ± SEM, N = 3. * $p < 0.05$, ** $p < 0.01$, *** $p < 0.001$.

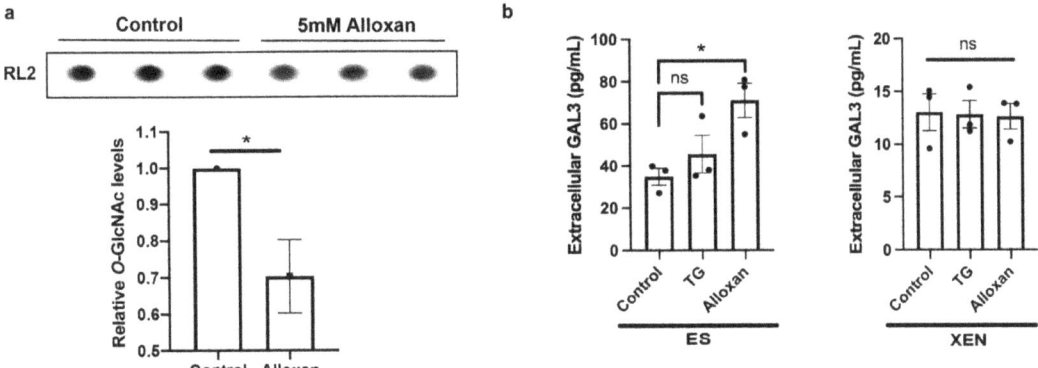

Figure 5. OGT inhibition in ES cells increases extracellular galectin-3. (**a**) Immunodot blot of O-GlcNAc detecting antibody RL2 in ES cells treated with 5 mM alloxan, an OGT inhibitor, and densitometry; (**b**) ELISA measuring secretion of galectin-3 from ES and XEN cells treated with 10 µM TG or 5 mM alloxan for 8 days normalized to total intracellular protein (mg/mL). Bars represent mean values ± SEM, N = 3. * $p < 0.05$.

3.6. OGA Inhibition Is Not Sufficient to Inhibit XEN Differentiation

Decreased global O-GlcNAc levels in XEN cells (Figure 1a) coincided with the decrease in intracellular galectin-3 levels (Figure 4a) and increased the secretion of galectin-3 (Figure 4c), which was also stimulated by an OGT inhibitor alloxan in ES cells (Figure 5b). Thus, it is possible that O-GlcNAc and the localization of galectin-3 could influence pluripotency. If so, inhibiting OGT activity would promote differentiation while inhibiting OGA, which we showed previously (Figure 2), would promote pluripotency. To test this further and examine over time what effect O-GlcNAcylation had on pluripotency and differentiation, ES cells were differentiated toward XEN lineage for 8 days in the presence or absence of TG to disrupt OGA activity (Figure 6). The regular differentiation of ES cells toward a XEN lineage was confirmed through morphology as ES cells form tight colonies, while XEN cells and ES cells were induced to differentiate show a cobblestone-like morphology with some rounded cells (Figure 6a). Cell differentiation was also confirmed by conducting gene expression analysis demonstrating that pluripotency markers *Oct4* and *Nanog* have reduced expression, while differentiation markers *Gata6* and *Dab2* increased in expression over time (Figure 6b). Our results from testing differentiation in the presence of TG revealed that the relative expression of the pluripotency marker *Nanog* decreased similarly to differentiation conditions in the absence of TG (Figure 6c). In addition, *Oct4* expression in the presence of TG also decreased to the same extent as differentiation conditions in the absence of TG (Figure 6c). This suggests that the inhibition of OGA did not promote pluripotency. Furthermore, blocking OGA did not alter the expression of differentiation markers *Gata6* and *Dab2* (Figure 6c), which again contradicts the predictions. Together, this evidence would suggest that the inhibition of OGA, which increases global O-GlcNAcylation in stem cells (Figure 2), neither promotes pluripotency nor inhibits differentiation.

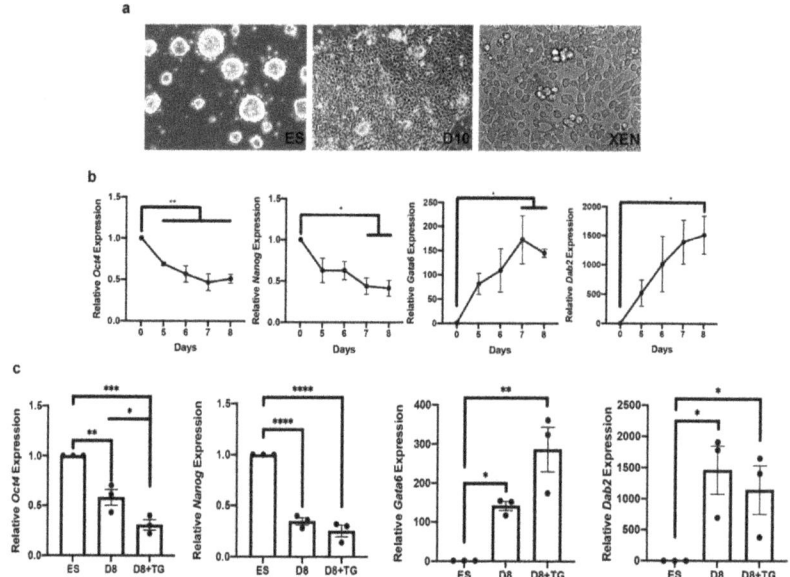

Figure 6. OGA inhibition is not sufficient to inhibit ESC differentiation toward XEN phenotype: (**a**) phase contrast microscopy of undifferentiated ES, differentiated ES toward XEN after 10 days in differentiation media, and XEN cells; (**b**) time-dependent changes in the expression of genetic biomarkers of pluripotency (*Oct4* and *Nanog*) and differentiation (*Gata6* and *Dab2*) in ES cells differentiated toward the XEN phenotype; (**c**) effects of TG (10 µM) on the expression of same biomarkers of pluripotency/differentiation as in (**b**) on day 8 of ES cell differentiation toward XEN. Bars represent mean ± SEM, N = 3. * $p < 0.05$, ** $p < 0.01$, *** $p < 0.001$, **** $p < 0.0001$.

4. Discussion

The objective of this study was to understand the role of O-GlcNAcylation during XEN differentiation and to explore if this specific post-translational modification can regulate galectin expression and localization. Here, we report that global O-GlcNAc decreased in XEN cells, although inhibiting OGA had no effect on differentiation. Galectin-3 localization and secretion, however, depend on both cell differentiation status and O-GlcNAc homeostasis.

Although the reduction in global O-GlcNAcylation is associated with the differentiation programs of many cell lineages, there are a few exceptions [21]. In our model, a decrease in global O-GlcNAc in XEN compared to ES cells (Figure 1) is consistent with previous work on neutrophilic and enterocytic differentiation [17] and supports another study that showed decreased global O-GlcNAcylation when ES cells were differentiated toward cardiomyocytes [18]. The global decrease in O-GlcNAcylation, as seen during XEN and cardiomyocyte differentiation, coincided with the decreased expression of *Ogt*, the gene encoding the enzyme responsible for adding O-GlcNAc to proteins (Figure 1) [18]. This overall decrease in O-GlcNAcylation is likely in response to the exit from pluripotency, and this was observed by Jang et al. who demonstrated the pluripotency markers OCT4 and SOX2 are both modified by O-GlcNAc, which is later removed during differentiation [15].

Since the trend of a high degree of the O-GlcNAcylation is consistent with pluripotency and decreases occur in response to differentiation, we explored how chemical inhibition of the modification-removing enzyme, OGA, would affect both stem cells and those induced to form XEN. Blocking OGA was expected to increase the amount of global O-GlcNAc, and this was expected to maintain or enhance pluripotency in ES cells, while decreasing the differentiation to XEN. In fact, OGA inhibition in XEN cells did decrease XEN marker *FoxA2* expression (Figure 2e); however, it did not inhibit the differentiation of ES cells induced toward XEN lineage (Figure 6). In contrast, OGA inhibition in ES cells did inhibit the expression of the pluripotency marker *Nanog* (Figure 2c); however, it did not inhibit the differentiation-induced decrease in pluripotency markers in ES cells induced toward XEN (Figure 6).

Previous work has provided evidence that proteins within the galectin family can be regulated by O-GlcNAcylation [17,21,22]. Few studies have examined galectins in F9 cells [40,41], where we saw no changes in *Lgal3* expression in F9 cell XEN differentiation (Supplementary Materials, Figure S2); however, we observed a decrease in *Lgals3* expression in terminally differentiated XEN cells compared to ES cells (Figure 3). Unlike the changes in O-GlcNAc levels, which are consistent in many previous reports [15,17,18,21], galectins appear to have context-specific changes in expression, as evident from our own studies [17,20,38].

As *Lgals3* was the only galectin for which its expression decreased in XEN cells compared to ES cells (Figure 3b), we further explored the regulation of this protein in pluripotency and differentiation. Similarly to other members of the galectin family, galectin-3 is a multifunctional molecule with different modes of action inside and outside the cell [19]. In addition to confirming previous findings of low intracellular galectin-3 expression in XEN cells [37], ELISA analysis revealed a significant increase in galectin-3 secretion in this cell model (Figure 4). Although it is not explored in this study, others have reported on the role of galectin-3 in clathrin-independent endocytosis by interacting with secreted and membrane bound glycoproteins [19,22], which could alter detectable galectin-3 through ELISA. In a developmental context, Iacobini et al. discussed the role of galectin-3 in osteogenic lineages, highlighting its localization as enhancing or inhibiting specific cell types [42]. Given our results showing increased intracellular galectin-3 in ES cells, it is tempting to speculate that extracellular galectin-3 is playing a role in enhancing XEN differentiation. Firstly, the alloxan studies, through the inhibition of OGT, reduced the extent of O-GlcNAcylation (Figure 5a) and increased extracellular galectin-3 levels in ES cell culture (Figure 5b), which is in line with the study of Mathew et al. who showed that the enhanced secretion of galectin-3 occurs as a result of it not being tagged with

O-GlcNAc [22]. Although used in the present study as an OGT inhibitor, a limitation of alloxan is its unspecific inhibition of OGT that can also, at high concentrations, inhibit OGA [43]. Given this previous study, we chose an alloxan concentration that was unlikely to reduce OGA activity from controls.

Our findings support the new concept of the O-GlcNAc-mediated regulation of galectin localization and secretion in cells [21]. Galectin-3 secretion is considered to occur through unconventional mechanisms, meaning that it is not released from the cell through the classical endoplasmic reticulum/Golgi secretory pathway [19]. Instead, other means include direct translocation, export via the lysosome or endosome, export via microvesicles, and/or release in exosomes [44–46]. O-GlcNAcylation is known to regulate protein sorting exosomes; however, it is still unclear if its regulation is binary or a more dynamic process [47]. To our knowledge, no studies have investigated the composition or role of exosomes in XEN differentiation, so galectin-3 may be exiting the cell regardless of its O-GlcNAc status. While this requires further study, our evidence indicates that the decrease in global O-GlcNAc status occurs concurrently with increased galectin-3 secretion and the exit from pluripotency.

Given the differences in the two populations, ES and XEN cells, we wanted to explore the role of OGA inhibition when ES cells were induced toward the XEN lineage. Given that OGA inhibition increased global O-GlcNAc and decreased *FoxA2* expression in XEN cells (Figure 2), it was predicted OGA inhibition should inhibit XEN differentiation. However, when ES cells were induced toward XEN in the presence of the OGA inhibitor, there was no apparent change in pluripotency and XEN marker expression compared to differentiating cells in the absence of the inhibitor (Figure 6). As others have demonstrated the importance of increased O-GlcNAc in ES cell self-renewal and the maintenance of pluripotency transcription factor function [15], this result was unexpected. For instance, Maury et al. demonstrated that excess O-GlcNAc inhibited human pluripotent stem cell differentiation without inhibiting cell exits from pluripotency [16]. Taken together, this suggests that although there are changes in O-GlcNAc globally in XEN cells compared to ES cells, inhibiting only the enzyme involved in removing O-GlcNAc modifications may not be sufficient to inhibit XEN differentiation. The ultimate role of O-GlcNAc in ES/XEN model is still undergoing evaluation, and other cell models and experimental design may help to further elaborate this notion.

Despite some conflicting information, which is likely cell type-dependent, we were the first to explore the role of O-GlcNAcylation in XEN maintenance and differentiation. Our work highlighted the differences in global O-GlcNAc between ES and XEN cells and showed that galectin-3 intracellular levels decreased in XEN; more importantly, the elevation of galectin-3 secretion was observed as the result of O-GlcNAcylation occurring concomitantly with XEN differentiation.

Supplementary Materials: The following are available online at https://www.mdpi.com/article/10.3390/biom12050623/s1, Figure S1: Global O-GlcNAcylation decreases with F9 cell differentiation toward a XEN-like lineage; Figure S2: Expressions of many galectins are decreased in F9 cells differentiated toward a XEN-like lineage.

Author Contributions: Conceptualization, M.I.G., D.M.S., A.V.T. and G.M.K.; methodology, M.I.G. and D.M.S.; formal analysis, M.I.G. and D.M.S.; investigation, M.I.G., D.M.S., A.G., A.M. and M.A.; writing—original draft preparation, M.I.G. and D.M.S.; writing—review and editing, D.M.S., A.V.T. and G.M.K.; supervision, G.M.K.; funding acquisition, G.M.K. and A.V.T. All authors have read and agreed to the published version of the manuscript.

Funding: This research was funded by the Natural Sciences and Engineering Research Council of Canada to GMK (R2615A02) and to AVT (R5082A05).

Institutional Review Board Statement: Not applicable.

Informed Consent Statement: Not applicable.

Data Availability Statement: A data appears in this manuscript.

Acknowledgments: The authors would like to acknowledge current and past members of the Timoshenko and Kelly labs for their support.

Conflicts of Interest: The authors declare no conflict of interest.

References

1. Holt, G.D.; Hart, G.W. The subcellular distribution of terminal N-acetylglucosamine moieties: Localization of a novel protein-saccharide linkage, O-linked GlcNAc. *J. Biol. Chem.* **1986**, *261*, 8049–8057. [CrossRef]
2. Bond, M.R.; Hanover, J.A. A little sugar goes a long way: The cell biology of O-GlcNAc. *J. Cell Biol.* **2015**, *208*, 869–880. [CrossRef] [PubMed]
3. Comer, F.I.; Hart, G.W. Reciprocity between O-GlcNAc and O-phosphate on the carboxyl terminal domain of RNA Polymerase II. *Biochemistry* **2001**, *40*, 7845–7852. [CrossRef] [PubMed]
4. Hart, G.W.; Housley, M.P.; Slawson, C. Cycling of O-linked β-N-acetylglucosamine on nucleocytoplasmic proteins. *Nature* **2007**, *446*, 1017–1022. [CrossRef]
5. Zeidan, Q.; Hart, G.W. The intersections between O-GlcNAcylation and phosphorylation: Implications for multiple signaling pathways. *J. Cell Sci.* **2010**, *123*, 13–22. [CrossRef]
6. Haltiwangers, R.S.; Blomberg, M.A.; Hart, G.W. Glycosylation of nuclear and cytoplasmic proteins: Purification and characterization of a uridine diphospho-N-acetylglucosamine:polypeptide b-N-acetylglucosaminyltransferase. *J. Biol. Chem.* **1992**, *267*, 9005–9013. [CrossRef]
7. Dong, D.L.-Y.; Hart, G.W. Purification and characterization of an O-GlcNAc selective N-acetyl-b-D-glucosaminidase from rat spleen cytosol. *J. Biol. Chem.* **1994**, *269*, 19321–19330. [CrossRef]
8. McClain, D.A. Hexosamines as mediators of nutrient sensing and regulation in diabetes. *J. Diabetes Complicat.* **2002**, *16*, 72–80. [CrossRef]
9. Love, D.C.; Hanover, J.A. The hexosamine signaling pathway: Deciphering the "O-GlcNAc code". *Sci. STKE* **2005**, *2005*, re13. [CrossRef]
10. Walgren, J.L.E.; Vincent, T.S.; Schey, K.L.; Buse, M.G. High glucose and insulin promote O-GlcNAc modification of proteins, including alpha-tubulin. *Am. J. Physiol. Endocrinol. Metab.* **2003**, *284*, E424–E434. [CrossRef]
11. Hawkins, M.; Barzilai, N.; Liu, R.; Hu, M.; Chen, W.; Rossetti, L. Role of the glucosamine pathway in fat-induced insulin resistance. *J. Clin. Investig.* **1997**, *99*, 2173–2182. [CrossRef] [PubMed]
12. Lee, A.; Miller, D.; Henry, R.; Paruchuri, V.D.P.; O'Meally, R.N.; Boronina, T.; Cole, R.N.; Zachara, N.E. Combined antibody/lectin enrichment identifies extensive changes in the O-GlcNAc sub-proteome upon oxidative stress. *J. Proteome Res.* **2016**, *15*, 4318–4336. [CrossRef] [PubMed]
13. Groves, J.A.; Maduka, A.O.; O'Meally, R.N.; Cole, R.N.; Zachara, N.E. Fatty acid synthase inhibits the O-GlcNAcase during oxidative stress. *J. Biol. Chem.* **2017**, *292*, 6493–6511. [CrossRef] [PubMed]
14. Zachara, N.E.; O'Donnell, N.; Cheung, W.D.; Mercer, J.J.; Marth, J.D.; Hart, G.W. Dynamic O-GlcNAc modification of nucleocytoplasmic proteins in response to stress: A survival response of mammalian cells. *J. Biol. Chem.* **2004**, *279*, 30133–30142. [CrossRef]
15. Jang, H.; Kim, T.W.; Yoon, S.; Choi, S.-Y.; Kang, T.-W.; Kim, S.-Y.; Kwon, Y.-W.; Cho, E.-J.; Youn, H.-D. O-GlcNAc regulates pluripotency and reprogramming by directly acting on core components of the pluripotency network. *Cell Stem Cell* **2012**, *11*, 62–74. [CrossRef] [PubMed]
16. Maury, J.J.P.; Chan, K.K.-K.; Zheng, L.; Bardor, M.; Choo, A.B.-H. Excess of O-linked N-acetylglucosamine modifies human pluripotent stem cell differentiation. *Stem Cell Res.* **2013**, *11*, 926–937. [CrossRef]
17. Sherazi, A.A.; Jariwala, K.A.; Cybulski, A.N.; Lewis, J.W.; Karagiannis, J.; Cumming, R.C.; Timoshenko, A.V. Effects of global O-GlcNAcylation on galectin gene-expression profiles in human cancer cell lines. *Anticancer Res.* **2018**, *38*, 6691–6697. [CrossRef]
18. Kim, H.-S.; Park, S.Y.; Choi, Y.R.; Kang, J.G.; Joo, H.J.; Moon, W.K.; Cho, J.W. Excessive O-GlcNAcylation of proteins suppresses spontaneous cardiogenesis in ES cells. *FEBS Lett.* **2009**, *583*, 2474–2478. [CrossRef]
19. Johannes, L.; Jacob, R.; Leffler, H. Galectins at a glance. *J. Cell Sci.* **2018**, *131*, jcs208884. [CrossRef]
20. Timoshenko, A.V. Towards molecular mechanisms regulating the expression of galectins in cancer cells under microenvironmental stress conditions. *Cell. Mol. Life Sci.* **2015**, *72*, 4327–4340. [CrossRef]
21. Tazhitdinova, R.; Timoshenko, A.V. The emerging role of galectins and O-GlcNAc homeostasis in processes of cellular differentiation. *Cells* **2020**, *9*, 1792. [CrossRef] [PubMed]
22. Mathew, M.P.; Abramowitz, L.K.; Donaldson, J.G.; Hanover, J.A. Nutrient-responsive O-GlcNAcylation dynamically modulates the secretion of glycan-binding protein galectin 3. *J. Biol. Chem.* **2022**, *298*, 101743. [CrossRef] [PubMed]
23. Magescas, J.; Sengmanivong, L.; Viau, A.; Mayeux, A.; Dang, T.; Burtin, M.; Nilsson, U.J.; Leffler, H.; Poirier, F.; Terzi, F.; et al. Spindle pole cohesion requires glycosylation-mediated localization of NuMA. *Sci. Rep.* **2017**, *7*, 1474. [CrossRef] [PubMed]
24. Hart, C.; Chase, L.G.; Hajivandi, M.; Agnew, B. Metabolic labeling and click chemistry detection of glycoprotein markers of mesenchymal stem cell differentiation. *Methods Mol. Biol.* **2011**, *698*, 459–484. [CrossRef] [PubMed]
25. Rossant, J.; Tam, P.P.L. Emerging asymmetry and embryonic patterning in early mouse development. *Dev. Cell* **2004**, *7*, 155–164. [CrossRef] [PubMed]

26. Dickson, B.J.; Gatie, M.I.; Spice, D.M.; Kelly, G.M. NOX1 and NOX4 are required for the differentiation of mouse F9 cells into extraembryonic endoderm. *PLoS ONE* **2017**, *12*, e0170812. [CrossRef]
27. Gatie, M.I.; Kelly, G.M. Metabolic profile and differentiation potential of extraembryonic endoderm-like cells. *Cell Death Discov.* **2018**, *4*, 42. [CrossRef]
28. Nairn, A.V.; Aoki, K.; dela Rosa, M.; Porterfield, M.; Lim, J.-M.; Kulik, M.; Pierce, J.M.; Wells, L.; Dalton, S.; Tiemeyer, M.; et al. Regulation of glycan structures in murine embryonic stem cells: Combined transcript profiling of glycan-related genes and glycan structural analysis. *J. Biol. Chem.* **2012**, *287*, 37835–37856. [CrossRef]
29. Speakman, C.M.; Domke, T.C.E.; Wongpaiboonwattana, W.; Sanders, K.; Mudalair, M.; van Aalten, D.M.F.; Barton, G.J.; Stavridis, M.P. Elevated O-GlcNAc levels activate epigenetically repressed genes and delay mouse ESC differentiation without affecting naive to primed cell transition. *Stem Cells* **2014**, *32*, 2605–2615. [CrossRef]
30. Ying, Q.-L.; Wray, J.; Nichols, J.; Batlle-Morera, L.; Doble, B.; Woodgett, J.; Cohen, P.; Smith, A. The ground state of embryonic stem cell self-renewal. *Nature* **2008**, *453*, 519–523. [CrossRef]
31. Kunath, T.; Arnaud, D.; Uy, G.D.; Okamoto, I.; Chureau, C.; Yamanaka, Y.; Heard, E.; Gardner, R.L.; Avner, P.; Rossant, J. Imprinted X-inactivation in extra-embryonic endoderm cell lines from mouse blastocysts. *Development* **2005**, *132*, 1649–1661. [CrossRef]
32. Anderson, K.G.V.; Hamilton, W.B.; Roske, F.V.; Azad, A.; Knudsen, T.E.; Canham, M.A.; Forrester, L.M.; Brickman, J.M. Insulin fine-tunes self-renewal pathways governing naive pluripotency and extra-embryonic endoderm. *Nat. Cell Biol.* **2017**, *19*, 1164–1177. [CrossRef] [PubMed]
33. Langmead, B.; Salzberg, S.L. Fast gapped-read alignment with Bowtie 2. *Nat. Methods* **2012**, *9*, 357–359. [CrossRef] [PubMed]
34. Parrish, N.; Hormozdiari, F.; Eskin, E. Assembly of non-unique insertion content using next-generation sequencing. *BMC Bioinform.* **2011**, *12*, 21–40. [CrossRef] [PubMed]
35. Love, M.I.; Huber, W.; Anders, S. Moderated estimation of fold change and dispersion for RNA-seq data with DESeq2. *Genome Biol.* **2014**, *15*, 550. [CrossRef] [PubMed]
36. Cecioni, S.; Vocadlo, D.J. Tools for probing and perturbing O-GlcNAc in cells and in vivo. *Curr. Opin. Chem. Biol.* **2013**, *17*, 719–728. [CrossRef]
37. Gatie, M.I.; Cooper, T.T.; Khazaee, R.; Lajoie, G.A.; Kelly, G.M. Lactate enhances mouse ES cell differentiation towards XEN cells in vitro. *Stem Cells* **2022**, *40*, 239–259. [CrossRef]
38. Vinnai, J.R.; Cumming, R.C.; Thompson, G.J.; Timoshenko, A.V. The association between oxidative stress-induced galectins and differentiation of human promyelocytic HL-60 cells. *Exp. Cell Res.* **2017**, *355*, 113–123. [CrossRef]
39. Konrad, R.J.; Zhang, F.; Hale, J.E.; Knierman, M.D.; Becker, G.W.; Kudlow, J.E. Alloxan is an inhibitor of the enzyme O-linked N-acetylglucosamine transferase. *Biochem. Biophys. Res. Commun.* **2002**, *293*, 207–212. [CrossRef]
40. Lu, Y.; Amos, B.; Cruise, E.; Lotan, D.; Lotan, R. A parallel association between differentiation and induction of galectin-1, and inhibition of galectin-3 by retinoic acid in mouse embryonal carcinoma F9 cells. *Biol. Chem.* **1998**, *379*, 1323–1331. [CrossRef]
41. Lu, Y.; Lotan, D.; Lotan, R. Differential regulation of constitutive and retinoic acid-induced galectin-1 gene transcription in murine embryonal carcinoma and myoblastic cells. *Biochim. Biophys. Acta Gene Struct. Expr.* **2000**, *1491*, 13–19. [CrossRef]
42. Iacobini, C.; Fantauzzi, C.B.; Pugliese, G.; Menini, S. Role of galectin-3 in bone cell differentiation, bone pathophysiology and vascular osteogenesis. *Int. J. Mol. Sci.* **2017**, *18*, 2481. [CrossRef] [PubMed]
43. Lee, T.N.; Alborn, W.E.; Knierman, M.D.; Konrad, R.J. Alloxan is an inhibitor of O-GlcNAc-selective N-acetyl-β-D-glucosaminidase. *Biochem. Biophys. Res. Commun.* **2006**, *350*, 1038–1043. [CrossRef]
44. Bänfer, S.; Schneider, D.; Dewes, J.; Strauss, M.T.; Freibert, S.-A.; Heimerl, T.; Maier, U.G.; Elsässer, H.P.; Jungmann, R.; Jacob, R. Molecular mechanism to recruit galectin-3 into multivesicular bodies for polarized exosomal secretion. *Proc. Natl. Acad. Sci. USA* **2018**, *115*, E4396–E4405. [CrossRef]
45. Hughes, R.C. Secretion of the galectin family of mammalian carbohydrate-binding proteins. *Biochim. Biophys. Acta* **1999**, *1473*, 172–185. [CrossRef]
46. Popa, S.J.; Stewart, S.E.; Moreau, K. Unconventional secretion of annexins and galectins. *Semin. Cell Dev. Biol.* **2018**, *83*, 42–50. [CrossRef] [PubMed]
47. Chaiyawat, P.; Weeraphan, C.; Netsirisawan, P.; Chokchaichamnankit, D.; Srisomsap, C.; Svasti, J.; Champattanachai, V. Elevated O-GlcNAcylation of extracellular vesicle proteins derived from metastatic colorectal cancer cells. *Cancer Genom. Proteom.* **2016**, *13*, 387–398.

Article

Modulation of the Epithelial-Immune Cell Crosstalk and Related Galectin Secretion by DP3-5 Galacto-Oligosaccharides and β-3′Galactosyllactose

Veronica Ayechu-Muruzabal [1], Melanie van de Kaa [1], Reshmi Mukherjee [2], Johan Garssen [1,3], Bernd Stahl [2,3], Roland J. Pieters [2], Belinda van't Land [3,4], Aletta D. Kraneveld [1] and Linette E. M. Willemsen [1,*]

[1] Division of Pharmacology, Utrecht Institute for Pharmaceutical Sciences (UIPS), Utrecht University, 3584 CG Utrecht, The Netherlands; v.ayechumuruzabal@uu.nl (V.A.-M.); mel_kaa@live.nl (M.v.d.K.); j.garssen@uu.nl (J.G.); a.d.kraneveld@uu.nl (A.D.K.)
[2] Department of Chemical Biology and Drug Discovery, Utrecht Institute for Pharmaceutical Sciences (UIPS), Utrecht University, 3584 CG Utrecht, The Netherlands; reshmi.mukherjee@gmail.com (R.M.); bernd.stahl@danone.com (B.S.); r.j.pieters@uu.nl (R.J.P.)
[3] Danone Nutricia Research, 3584 CT Utrecht, The Netherlands; belinda.vantland@danone.com
[4] Center for Translational Immunology, The Wilhelmina Children's Hospital, University Medical Center Utrecht, 3584 EA Utrecht, The Netherlands
* Correspondence: l.e.m.willemsen@uu.nl

Abstract: Prebiotic galacto-oligosaccharides (GOS) were shown to support mucosal immune development by enhancing regulatory-type Th1 immune polarization induced by synthetic CpG oligodeoxynucleotides (TLR9 agonist mimicking a bacterial DNA trigger). Epithelial-derived galectin-9 was associated with these immunomodulatory effects. We aimed to identify the most active fractions within GOS based on the degree of polymerization (DP), and to study the immunomodulatory capacities of DP3-sized β-3′galactosyllactose (β-3′GL) using a transwell co-culture model of human intestinal epithelial cells (IEC) and activated peripheral blood mononuclear cells (PBMC). IEC were apically exposed to different DP fractions of GOS or β-3′GL in the presence of CpG, and basolaterally co-cultured with αCD3/CD28-activated PBMC, washed, and incubated in fresh medium for IEC-derived galectin analysis. Only DP3-5 in the presence of CpG enhanced galectin-9 secretion. DP3-sized β-3′GL promoted a regulatory-type Th1 response by increasing IFNγ and IL-10 or galectin-9 concentrations as compared to CpG alone. In addition, IEC-derived galectin-3, -4, and -9 secretion was increased by β-3′GL when combined with CpG. Therefore, the GOS DP3-5 and most effectively DP3-sized β-3′GL supported the immunomodulatory properties induced by CpG by enhancing epithelial-derived galectin secretion, which, in turn, could support mucosal immunity.

Keywords: galacto-oligosaccharides; galectins; intestinal epithelial cells; β-3′galactosyllactose; immunomodulation; mucosal immunity

1. Introduction

Non-digestible oligosaccharides (NDO) are the third major component in human milk [1]. Based on the amount and structure diversity of NDO in human milk, a 9:1 mixture of short-chain galacto- and long-chain fructo-oligosaccharides (GOS/FOS) was studied for its effects on the microbiota and the intestinal mucosa [2]. Various clinical studies have shown that this GOS/FOS mixture promoted the growth of commensal bacteria, induced stool softening, reduced the incidence of infections and the incidence of atopic dermatitis, as well as modulated the antibody profile in infants at high risk of allergy [3–8]. Furthermore, when combined with *Bifidobacterium breve* M-16V, it effectively lowered allergic symptoms in a murine model for cow's milk or hen's egg allergy in association with increased intestinal and/or serum galectin-9 levels [9–12]. In addition, in children affected

with atopic dermatitis, this synbiotic mixture was found to enhance serum galectin-9 levels after 12 weeks of intervention, in association with reduced atopic dermatitis symptom scores and lower risks of developing asthma [10,13,14].

The major component of the GOS/FOS (9:1) mixture is GOS, which is composed of galactose units coupled to a terminal glucose with a degree of polymerization (DP) ranging between 2 and 8 [13,14]. Upon ingestion, GOS reach the lower parts of the gastrointestinal tract intact where fermentation by the gut microbiota occurs. Consumption of GOS promotes the growth of beneficial commensal bacteria and provides health benefits to the host [15,16]. Besides the microbiota-dependent effects, GOS was also shown to have direct effects on the epithelial barrier. Regarding the effect of GOS on intestinal epithelial cells (IEC), it was shown that GOS can inhibit the adherence of pathogenic bacteria to IEC [17,18], enhance the barrier function by preventing the disruption of gut barrier integrity [19], and promote goblet cell function [20]. Furthermore, GOS supported the absorption of minerals such as iron and calcium in young infants [21,22], as well as lowered the incidence and severity of travelers' diarrhea in humans travelling to high-risk countries [23].

In addition to the effects on the microbiota and the IEC, NDO such as GOS have been shown to promote direct immunomodulatory effects. GOS was shown to interact with T-cells and dendritic cells, and to selectively promote the release of regulatory IL-10 in vitro [24,25]. The increased IL-10 was also observed in a study performed in suckling piglets [26]. Furthermore, in a double-blind, placebo-controlled study performed in healthy elderly, GOS supplementation positively influenced immune parameters by increasing the production of IL-10, increasing NK cell activity, and reducing pro-inflammatory IL-6 and IL-1β measured in peripheral blood mononuclear cells (PBMC) [27,28].

Although previous studies have shown effects of GOS either on epithelial or on immune cells, we have identified immunomodulatory properties of the NDO mixture GOS/FOS in a transwell model developed to study the crosstalk between IECs and immune cells [29–32]. The crosstalk between IEC and underlying immune cells is key to maintain the intestinal mucosal homeostasis and to develop appropriate immune responses [33]. Previous studies using this well-established in vitro IEC/PBMC co-culture model reported regulatory-type Th1 responses upon exposure to NDO in association with bacterial DNA or synthetic CpG oligodeoxynucleotides, known to be TLR9 ligands. The immunomodulatory effects observed upon exposure to NDO and CpG in the IEC/PBMC model could support mucosal immune development and were shown to be mediated by epithelial-derived galectin-9, which was found to be a key mediator contributing to the effects observed in vitro [29–31,34,35]. These studies also showed that the immunomodulatory properties of CpG and NDO occurred only when the PBMC underlying the IEC were activated, mimicking inflammatory conditions [29].

GOS as the main component in the GOS/FOS mixture contains multiple oligomers (DP2-8), out of which the active immunomodulatory component has not yet been identified. Due to the variety of structures present in GOS, this study aimed to investigate the most active oligomers within GOS by investigating their immunomodulatory capacity using a transwell co-culture model combining IEC and activated PBMC. Therefore, specific GOS DP fractions were isolated by size-exclusion chromatography and exposed to IEC in the presence of CpG oligodeoxynucleotides to study the crosstalk with the underlying immune cells and their effect on IEC-derived galectins. Additionally, we studied the immunomodulatory effects of a specific NDO present in the GOS mixture and found in human milk, namely β-3′galactosyllactose (β-3′GL) [14,36], using the IEC/PBMC co-culture model. Studying the immunomodulatory properties of specific NDO structures will provide further insights regarding their potential role in mucosal immune development.

2. Materials and Methods

2.1. GOS DP Separation by Size Exclusion Chromatography

Vivinal GOS syrup (derived from lactose, 45% pure) produced by the elongation of galactose catalyzed by β-galactosidases (Friesland Campina, Amersfoort, The Netherlands) was diluted in Milli-Q water (1:1) and fractionated using a Bio-Gel P-2 column. Milli-Q water was used as eluent. The flow rate used was 0.2 mL/min. The fractions collected (6–12 mL) were freeze-dried (Christ, Osterode am Harz, Germany) and fractions containing DP4-7, DP3-7, or DP3-5 were pooled upon analysis by electrospray ionization-mass spectrometry (ESI-MS) (microTOF-Q-II Bruker, Billerica, MA, USA) using a HILIC column (X-Bridge^{TM} HILIC, Waters, Milford, MA, USA). An amount of 0.1% ammonia was used in the running buffer (acetonitrile:water; gradient of acetonitrile going from 5% to 50% aqueous solution in 10 min) (Figure 1). Additionally, the DP3-5 fraction was further separated into DP3, DP4, and DP5 (Figure 2) using the same experimental conditions as described above.

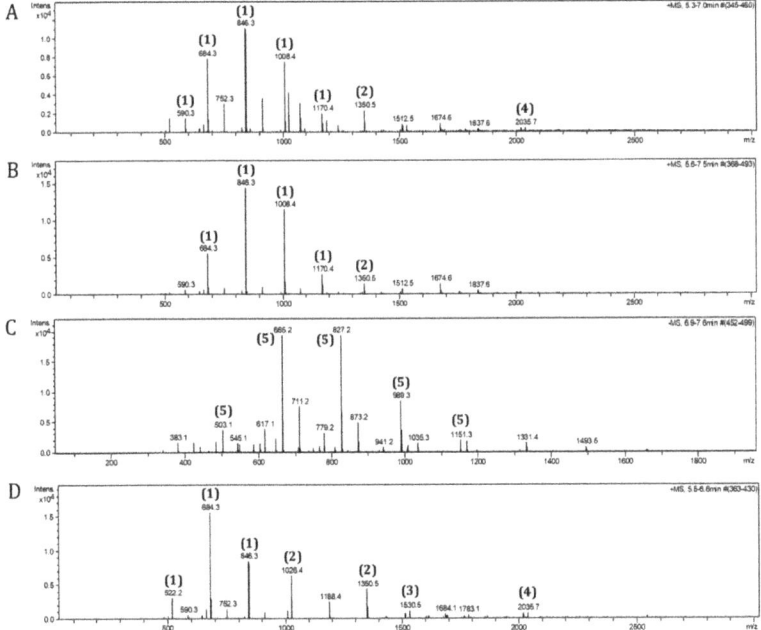

Figure 1. Electrospray ionization-mass spectrometry (ESI-MS) spectra of DP fractions. GOS DP oligomers were separated by size-exclusion chromatography from GOS mixture. The profiles of GOS (**A**), GOS DP4-7 (**B**), GOS DP3-7 (**C**), and GOS DP3-5 (**D**) are shown as ammonia adducts (1) = [M+NH$_4$]$^+$, (2) = [2M+NH$_4$]$^+$, (3) = [3M+NH$_4$]$^+$, (4) = [M+NH$_4$]$^+$, and (5) = [M, no ammonia adduct].

2.2. Culture of IEC

The human colon adenocarcinoma HT-29 cell line (ATCC, HTB-38, Manassas, VA, USA) was used as a model for IEC. The HT-29 cell line was cultured in McCoy 5A medium (Gibco, Invitrogen, Carlsbad, CA, USA) supplemented with 10% fetal calf serum (FCS), penicillin (100 U/mL), and streptomycin (100 µg/mL) (Sigma-Aldrich, St. Louis, MO, USA). IEC were grown in 75 cm^2 flasks (Greiner Bio-One, Alphen aan den Rijn, The Netherlands) and maintained at 37 °C, 5% CO$_2$. Medium was refreshed every 2–3 days. One week before the experiments, IEC were diluted 8–10 times based on surface area and seeded in 12-well transwell inserts (Costar Corning Incorporated, New York, NY, USA). When confluency was reached, the IEC monolayers were used to perform co-culture experiments.

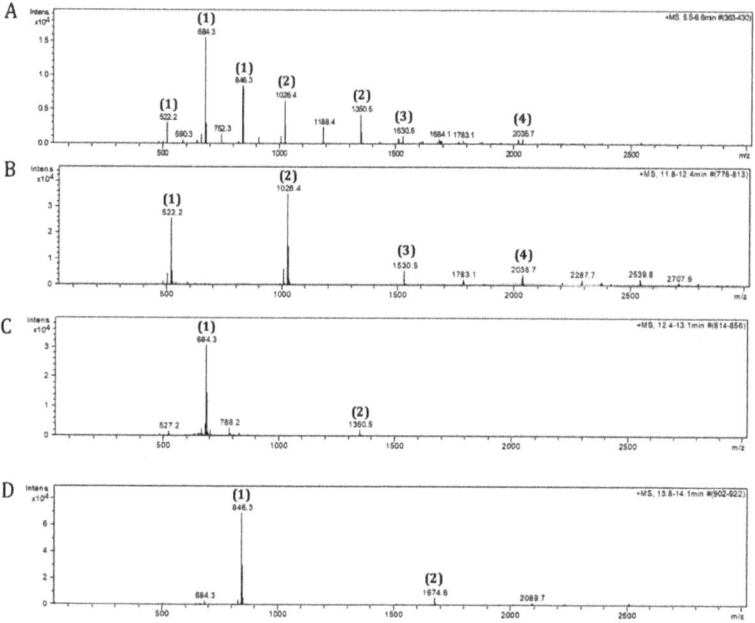

Figure 2. Electrospray ionization-mass spectrometry (ESI-MS) spectra of DP fractions. GOS DP oligomers were separated by size-exclusion chromatography from GOS mixture. The profiles of GOS DP3-5 (**A**), GOS DP3 (**B**), GOS DP4 (**C**), and GOS DP5 (**D**) are shown as ammonia adducts (1) = $[M+NH_4]^+$, (2) = $[2M+NH_4]^+$, (3) = $[3M+NH_4]^+$, and (4) = $[M+NH_4]^+$.

2.3. Isolation of PBMC

Buffy coats from healthy donors were purchased (Blood bank, Amsterdam, The Netherlands) and used to isolate human PBMC. Buffy coats were diluted (1:1) using PBS supplemented with 2% FCS. PBMC were isolated by density gradient centrifugation (1000× g, 13 min) using Leucosep™ tubes (20 mL per tube) (Greiner Bio-One). After washing, the remaining red blood cells were lysed (4.14 g of NH_4Cl, 0.5 g of $KHCO_3$, 18.6 mg of Na_2EDTA in 500 mL of demineralized water, sterile-filtered, pH = 7.4). Isolated PBMC were resuspended in RPMI 1640 supplemented with 2.5% FCS, penicillin (100 U/mL), and streptomycin (100 µg/mL).

2.4. IEC/PBMC Co-Culture Model

IEC grown in transwell filters were apically exposed to 0.1 or 0.5% (w/v) GOS DP fractions or β-3′GL (Carbosynth, Berkshire, UK) in the presence or absence of synthetic CpG oligodeoxynucleotides (ODN M362, 0.1, 0.5, or 5.0 µM) (Invivogen, San Diego, CA, USA). In the basolateral compartment, αCD3 and αCD28-activated PBMC were added (2 × 10^6 cells/mL) (0.15 µg/mL and 0.2 µg/mL, respectively, from Sanquin or BD Biosciences, San Jose, CA, USA) and incubated for 24 h (37 °C, 5% CO_2), after which the basolateral supernatant was collected and stored at −20 °C for cytokine analysis.

Subsequent to IEC/PBMC co-culture, transwell inserts containing IEC monolayers were collected for quantitative Polymerase Chain Reaction (qPCR) analysis or transferred into a new plate separated from the PBMC and washed with PBS. Then, fresh medium was added and IEC were incubated in fresh medium for an additional 24 h (total 48 h; 24 h in IEC/PBMC co-culture and 24 h in culture of IEC in fresh medium) to determine the IEC-derived basolateral mediator release. In addition, IEC were collected and stored for qPCR analysis.

2.5. Enzyme-Linked Immunosorbent Assay (ELISA)

The basolateral supernatants were used to analyze the cytokine and galectin-9 secretion in the IEC/PBMC co-culture, as well as the IEC-derived galectin-3, -4, and -9. Commercially available kits were used to determine IFNγ, TNFα, IL-13 (all from Thermo Fischer Scientific, Waltham, MA, USA), IL-10 (U-Cytech, Utrecht, The Netherlands), and galectin-3 (R&D systems, Minneapolis, MN, USA) following the manufacturer's protocol. Human galectin-4 and galectin-9 were measured using antibody pairs (both from R&D), as described before [29].

2.6. Gene Expression Analysis by qPCR

RNA was isolated from IEC samples using the Nucleospin® RNA Plus kit (Macherey-Nagel, Düren, Germany). Contaminating DNA was removed by incubating with DNAse for 15 min on ice (Qiagen, Hilden, Germany). Complementary DNA (cDNA) was obtained using the iScript™ cDNA synthesis kit (Bio-Rad, Veenendaal, The Netherlands) following the manufacturer's protocol. The IQ SYBR Green Supermix and CFX96 real-time PCR detection system (both from Bio-Rad) were used for the quantification of gene expression. Commercially available primers for galectin-3, -4, and -9 were used and compared to $RPS13$, as a reference gene (all from Qiagen). Relative mRNA expression was calculated as $100 \times 2^{(Ct\ reference\ -\ Ct\ gene\ of\ interest)}$ [37].

2.7. Statistical Analysis

All statistical analyses were performed using GraphPad Prism 8 software (GraphPad Software, San Diego, CA, USA). When the data did not fit a normal distribution, transformation was applied prior to ANOVA analysis. One-way repeated measures ANOVA followed by Bonferroni's post hoc test with selected pairs were used for the statistical analysis. The conditions with and without CpG were analyzed separately as represented by the dotted line. Within the analysis of CpG-exposed conditions, a comparison between the medium control group and CpG alone was included. Probability values of $p < 0.05$ were considered significant.

3. Results

3.1. Immunomodulatory Effects of GOS DP Fractions in IEC/PBMC Co-Culture Model

The immunomodulatory effects of DP fractions isolated from GOS were studied using a transwell IEC/PBMC co-culture model used to investigate the crosstalk between epithelial cells and innate, as well as adaptive, immune cells. IEC were apically exposed to the GOS DP fractions DP4-7, DP3-7, and DP3-5 (0.5% w/v) in combination with 5 µM CpG. In the basolateral compartment, αCD3/CD28-activated PBMC were added and incubated for 24 h.

Exposure of GOS DP4-7, DP3-7, or DP3-5 alone did not affect IFNγ, IL-10, or galectin-9 secretion, but CpG alone enhanced IL-10 secretion by activated PBMC in the IEC/PBMC co-culture (Figure 3A–C). Only combined exposure to GOS DP4-7 and CpG resulted in significantly increased IFNγ concentrations as compared to CpG alone (for DP3-7 and DP3-5; $p = 0.053$) (Figure 3A). Upon combined exposure to DP3-5 and CpG, significantly increased galectin-9 concentrations were observed as compared to CpG alone. Meanwhile for IL-10, this did not reach significance ($p = 0.06$) (Figure 3B,C).

Th1-type IFNγ secretion was increased by GOS DP4-7 when combined with CpG as compared to CpG alone. However, as GOS DP3-5 and CpG enhanced galectin-9 secretion, while showing a similar pattern for IFNγ and IL-10 secretion, the following studies were performed using GOS DP3-5.

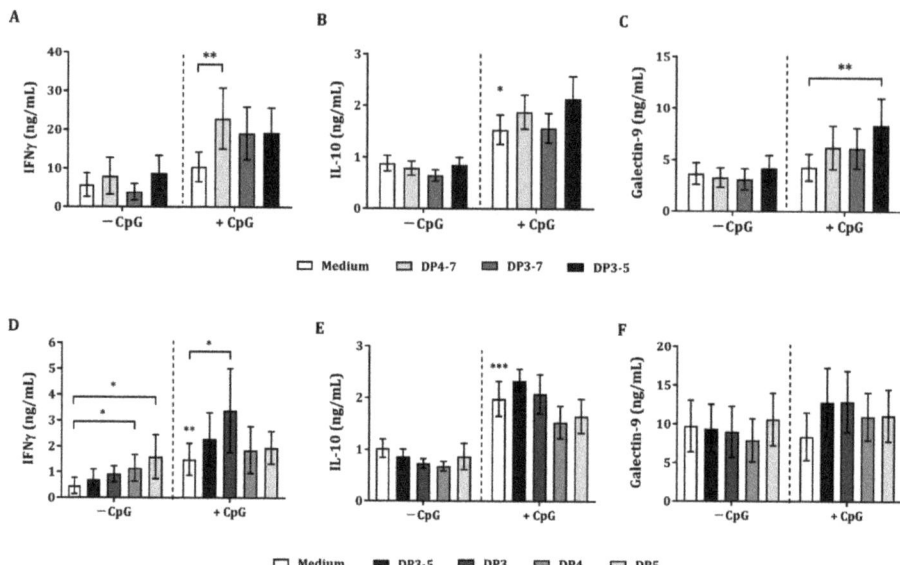

Figure 3. Cytokine and galectin-9 secretion in IEC/PBMC co-culture. IECs were apically exposed to GOS DP4-7, DP3-7, and DP3-5 in combination with 5 µM CpG and basolaterally to αCD3/CD28-activated PBMC. Additionally, GOS DP3-5 was further separated into DP3, DP4, and DP5 fractions. IEC were apically exposed to GOS DP3, DP4, and DP5 in combination with 5 µM CpG and basolaterally to αCD3/CD28-activated PBMC. After 24 h of incubation, IFNγ (**A,D**), IL-10 (**B,E**), and galectin-9 (**C,F**) concentrations were measured in the basolateral supernatant. Data represent mean ± SEM of $n = 6$ independent PBMC donors (except for $n = 4$ in (**A**) and $n = 5$ in (**C,D**)). Statistical analysis was performed separately for conditions with and without CpG (represented as dotted line). However, a comparison between the medium control and CpG-exposed condition was included (* $p < 0.05$, ** $p < 0.01$, *** $p < 0.001$).

Additionally, the GOS DP3-5 fraction was further separated into DP3, DP4, and DP5 fractions to study the most active oligomer/s within GOS DP3-5. Therefore, IEC were apically exposed to 0.5% (w/v) DP3, DP4, and DP5 in the presence or absence of 5 µM CpG. In the basolateral compartment, αCD3/CD28-activated PBMC were added and incubated for 24 h.

There was no effect on IFNγ, IL-10, and galectin-9 concentrations upon exposure to GOS DP3-5, DP3, DP4, and DP5 in the absence of CpG, except for DP4 and DP5, which significantly increased IFNγ concentrations (Figure 3D–F). CpG alone significantly increased IFNγ and IL-10 concentrations but did not affect galectin-9 secretion (Figure 3D–F). Only exposure to DP3 in combination with CpG further increased IFNγ concentrations compared to CpG alone. DP3-5 did not enhance IFNγ but, similar to DP3, showed a similar pattern in IL-10 and galectin-9 concentrations, although these did not reach significance (Figure 3D–F). No effect was observed on IFNγ, IL-10, and galectin-9 concentrations upon exposure to DP4 and DP5 in combination with CpG (Figure 3D–F).

3.2. Galectin-9 mRNA Expression and IEC-Derived Galectin-9 Secretion by Apical Exposure to GOS DP3-5 and CpG

To study the involvement of galectin-9 secretion resulting from apical GOS DP3-5 and 5 µM CpG exposure, IEC were collected after IEC/PBMC co-culture. In addition to this, IEC were washed and incubated in fresh medium for up to 48 h (24 h in IEC/PBMC co-culture and an additional 24 h in culture of IEC in fresh medium), after which galectin-9 mRNA expression and IEC-derived galectin-9 secretion were studied.

In IEC, the relative galectin-9 mRNA abundance was not significantly affected by GOS DP3-5, CpG, or the combination after 24 h in IEC/PBMC co-culture (Figure 4A). However, after 48 h (24 h in IEC/PBMC co-culture and an additional 24 h in IEC culture in fresh medium) of exposure to CpG alone in the presence or absence of GOS DP3-5, galectin-9 mRNA expression was significantly increased as compared to the medium control and/or GOS DP3-5 exposure (Figure 4A).

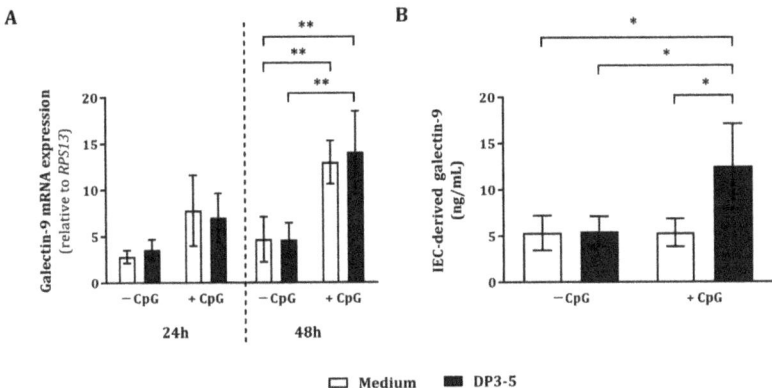

Figure 4. Galectin-9 mRNA expression and IEC-derived galectin-9 release. After IEC/PBMC co-culture, IEC were collected and galectin-9 mRNA expression was measured (**A**). Alternatively, IEC were washed and incubated in fresh medium for an additional 24 h (48 h in total; 24 h in IEC/PBMC co-culture and an additional 24 h in IEC culture in fresh medium), after which IEC and the basolateral supernatant were collected. The relative galectin-9 mRNA abundance at 24 h and 48 h (**A**) and the IEC-derived galectin-9 secretion at 48 h (**B**) were analyzed. Data represent mean ± SEM of $n = 3$ and $n = 6$ independent PBMC donors for (**A**,**B**), respectively (* $p < 0.05$, ** $p < 0.01$).

IEC-derived galectin-9 secretion at 48 h was not affected by GOS DP3-5 or CpG alone (Figure 4B). Combined exposure to GOS DP3-5 and CpG significantly increased IEC-derived galectin-9 as compared to the medium, GOS DP3-5, and CpG alone (Figure 4B).

Galectin-9 mRNA expression, but not IEC-derived galectin-9 secretion, was upregulated by CpG. Combined exposure to GOS DP3-5 and CpG upregulated both galectin-9 mRNA expression, as well as IEC-derived galectin-9 secretion.

3.3. Increased Th1-Type IFNγ and Regulatory Galectin-9 by β-3′GL and CpG in the IEC/PBMC Co-Culture Model

GOS DP3-5 and GOS DP3 showed immunomodulatory capacities when combined with CpG. Therefore, we aimed to study a specific DP3-sized NDO present in the GOS mixture and found in human milk, namely β-3′GL. IEC were apically exposed to β-3′GL (0.1 and 0.5% w/v) alone or in combination with 0.1 µM CpG and basolaterally to activated PBMC after which IFNγ, IL-10, and galectin-9 secretion was studied. Additionally, IEC-derived galectins were measured after an additional 24 h of incubation of IEC with fresh medium (48 h in total). Instead of 5 µM CpG, we used 0.1 µM in the following experiments in order to better identify the additional effects of NDO on top of the CpG effect, as was shown for GOS DP3-5 in Figure S1.

Exposure to β-3′GL, in either concentration, or GOS DP3-5 alone did not have an effect on IFNγ, IL-10, and galectin-9 concentrations (Figure 5A–C). Exposure to CpG alone significantly increased IL-10 concentrations as compared to the medium control but did not affect IFNγ and galectin-9 concentrations (Figure 5A–C). There was no effect on the cytokine and galectin secretion upon combined exposure to CpG and 0.1% β-3′GL or DP3-5 (Figure 5A–C). When IEC were exposed to 0.5% β-3′GL and CpG, significantly increased IFNγ and galectin-9 concentrations were observed, as compared to CpG alone or to 0.1%

β-3'GL and CpG (Figure 5A,C). CpG-induced IL-10 concentrations were not further enhanced by 0.5% β-3'GL or GOS DP3-5 (Figure 5B).

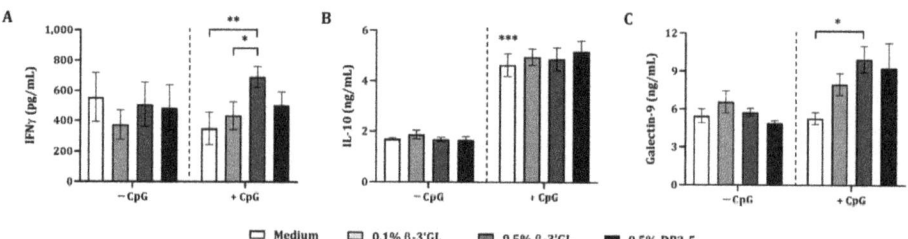

Figure 5. Cytokine and galectin-9 secretion by β-3'GL and CpG in the IEC/PBMC co-culture model. IEC were apically exposed to β-3'GL (0.1% and 0.5% w/v) or GOS DP3-5 (0.5% w/v) in combination with 0.1 μM CpG and basolaterally to αCD3/CD28-activated PBMC. After 24 h of incubation, IFNγ (**A**), IL-10 (**B**), and galectin-9 (**C**) were measured in the basolateral supernatant. The data shown are represented as mean ± SEM of $n = 6$ independent PBMC donors (except for IFNγ $n = 5$) (* $p < 0.05$, ** $p < 0.01$, *** $p < 0.001$).

Apical exposure of IEC to GOS DP3-5 or β-3'GL alone did not have an effect on the cytokine secretion in the IEC/PBMC co-culture model. Exposure to 0.5% β-3'GL and CpG more strongly promoted the secretion of Th1-type IFNγ and regulatory IL-10 compared to 0.1% β-3'GL or GOS DP3-5 and CpG. For the following studies, 0.5% β-3'GL was used.

3.4. Increased Epithelial-Derived Galectin Secretion by β-3'GL and CpG

To further study the immunomodulatory effects of β-3'GL and CpG and the involvement of galectins in the IEC/PBMC co-culture model, IEC were exposed to GOS DP3-5 or β-3'GL (0.5% w/v) and 0.1 μM CpG.

Exposure to CpG alone did not affect IFNγ or galectin-9 concentrations but significantly increased IL-10 concentrations compared to medium control levels (Figure 6A–C). Combined exposure to CpG and β-3'GL or GOS DP3-5 significantly increased IL-10 concentrations as compared to the medium control and/or CpG alone, but no effect was observed on galectin-9 concentrations (Figure 6B,C). Increased IFNγ concentrations were observed upon β-3'GL and CpG as compared to CpG alone (Figure 6A). Significantly decreased IL-13 and TNFα concentrations were observed upon exposure to CpG alone as compared to the medium control (Figure S2). These remained reduced and were not further affected by combined exposure to CpG and GOS DP3-5 or β-3'GL.

To study the involvement of epithelial-derived mediators in the immunomodulatory effects promoted by NDO and CpG, IEC-derived galectin-3, -4, and -9 secretion was measured after 48 h (24 h in IEC/PBMC co-culture and an additional 24 h in IEC culture in fresh medium) and correlated to the cytokine secretion in the IEC/PBMC co-culture. Additionally, galectin-3, -4, and -9 mRNA expression was measured at 48 h.

There was no effect on the IEC-derived galectin-3 and -4 secretion by CpG alone, but increased IEC-derived galectin-9 was observed as compared to the medium control (Figure 6D–F). The increased IEC-derived galectin-9 was also observed upon exposure to GOS DP3-5 or β-3'GL in combination with CpG as compared to the medium control (Figure 6F). Only upon combined exposure to β-3'GL and CpG significant increases were observed in IEC-derived galectin-3 and -4 concentrations as compared to the medium control, CpG alone, and/or GOS DP3-5 and CpG (Figure 6D,E).

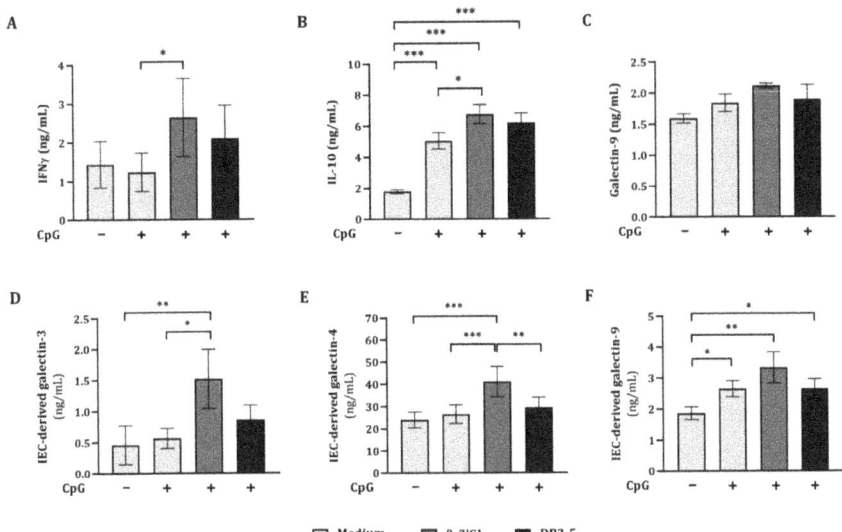

Figure 6. Cytokine and galectin-9 secretion in IEC/PBMC co-culture and IEC-derived galectin secretion at 48 h. IEC were apically exposed to GOS DP3-5 or β-3′GL (0.5% w/v) in combination with 0.1 µM CpG, and basolaterally to αCD3/CD28-activated PBMC. After 24 h of incubation, IFNγ (**A**), IL-10 (**B**), and galectin-9 (**C**) were measured in the basolateral compartment. Additionally, after IEC/PBMC co-culture, IEC were washed and incubated in fresh medium for an additional 24 h (48 h in total; 24 h in IEC/PBMC culture and an additional 24 h in IEC culture in medium), after which IEC-derived galectin-3 (**D**), -4 (**E**), and -9 (**F**) were measured in the basolateral supernatant. The data shown are represented as mean ± SEM of $n = 6$ independent PBMC donors (except for galectin-9 $n = 4$) (* $p < 0.05$. ** $p < 0.01$, *** $p < 0.001$).

In order to link the IEC-derived galectins with the outcome of the immune response, correlations were calculated. These results are summarized in Table 1. A positive correlation was observed between IFNγ and IEC-derived galectin-3, -4, and -9. There was no correlation between IL-10 and IEC-derived galectin-3 and -4. However, a positive correlation was observed between IL-10 and IEC-derived-9. Only IEC-derived galectin-9 was positively correlated to galectin-9, but not to IEC-derived galectin-3 or -4. A negative correlation was observed between IL-13 and IEC-derived galectin-3 and -9, but not with galectin-4 (Figure S2). Meanwhile, TNFα concentrations were not correlated to epithelial-derived galectins (Figure S2).

IEC-derived galectin-3, -4, and -9 were upregulated upon the exposure of IEC to β-3′GL and CpG in the IEC/PBMC co-culture model. IEC-derived galectin-9 secretion was correlated positively to IFNγ, IL-10, and galectin-9 secretion and negatively to IL-13 in the IEC/PBMC co-culture. However, IEC-derived galectin-3 was correlated positively to IFNγ and negatively to IL-13, while IEC-derived galectin-4 was correlated only to IFNγ concentrations.

Additionally, the galectin gene expression was measured at 48 h. There was no effect on the galectin-3, -4, and -9 mRNA expression upon exposure to 0.1 µM CpG alone or in the presence of DP3-5 (Figure S3). Combined exposure to β-3′GL and CpG significantly decreased galectin-4 mRNA expression as compared to the medium control or GOS DP3-5 in combination with CpG (Figure S3). Galectin-9 mRNA expression was significantly increased by exposure to β-3′GL and CpG as compared to GOS DP3-5 and CpG (Figure S3).

Table 1. Correlations of cytokines and galectin-9 in the IEC/PBMC co-culture and IEC-derived galectins. Cytokine and galectin-9 secretions measured in the IEC/PBMC co-culture were correlated to IEC-derived galectin concentrations (48 h) using Pearson correlations.

	IEC-Derived Galectin-3	IEC-Derived Galectin-4	IEC-Derived Galectin-9
IFNγ	+ r = 0.65 p < 0.0001	+ r = 0.49 p = 0.003	+ r = 0.38 p = 0.02
IL-10	n.s. r = 0.06 p = 0.7	n.s. r = 0.07 p = 0.7	+ r = 0.37 p = 0.03
Galectin-9	n.s. r = 0.02 p = 0.9	n.s. r = 0.3 p = 0.1	+ r = 0.44 p = 0.03

+ Positive correlation; n.s. nonsignificant correlation.

4. Discussion

Human milk is highly abundant in NDO for which diverse immune regulatory functions have been described [1]. GOS comprise the main component in the prebiotic mixture resembling the amount and structure diversity of NDO in human milk [2,4] and are composed of a complex variety of NDO with DPs ranging between 2 and 8 [14]. The purpose of this study was to evaluate the most active oligomers within the GOS mixture regarding the immunomodulatory capacity using the IEC/PBMC co-culture model in which the crosstalk between IEC and immune cells was studied.

Up to 60% of GOS DP structures have a DP size of DP2 or DP3 [13,14]. Within the DP2 fraction, various NDO structures have been characterized also including lactose residues [14]. Previous studies did not observe immunomodulatory effects upon exposure to lactose in the IEC/PBMC co-culture model [35]. However, due to the presence of lactose and the inability to separate this from other possible active NDO within the DP2 fraction, this study focused on analyzing GOS fractions with size of DP3 and longer fractions, namely DP4-7, DP3-7, and DP3-5 fractions, which were isolated by size-exclusion chromatography.

In the current study, Th1-type IFNγ concentrations were increased upon exposure to GOS DP4-7 in association with CpG. However, only GOS DP3-5 showed increased CpG-induced regulatory-type galectin-9 secretion with a similar Th1 secretion, suggesting a pattern of Th1 and regulatory cytokine secretion. Therefore, GOS DP3-5 was used for further studies, based on previous findings identifying galectin-9 as a key mediator in driving a regulatory-type Th1 response [31,33,36]. The current study showed the presence of structures with immunomodulatory properties within the GOS DP3-5 fraction, and, also within this fraction, GOS DP3 was identified for being able to upregulate IFNγ concentrations only when combined with CpG. This suggests that the most active oligomers in terms of Th1-type regulatory immunomodulation might be DP3-sized.

Previous studies investigated the involvement of galectins in general and galectin-9 in particular in supporting the regulatory-type Th1 immunomodulatory effects by the blocking of galectins, which resulted in the suppression of regulatory-type Th1 immune effects [31]. Furthermore, stimulation of αCD3/CD28-activated PBMC with recombinant galectin-9 enhanced IFNγ and IL-10 secretion and increased the percentage of Th1 and regulatory T-cells [10], which reinforces the role of galectin-9 as a key immune regulator.

The TLR9 agonist CpG used to mimic a bacterial trigger (bacterial DNA) in the IEC/PBMC co-culture model was required to support the immunomodulatory effects of NDO. Although TLR9 is mostly described as an endosomal receptor, previous studies have observed a surface expression of TLR9 in IEC [32,38–40]. Even though combined exposure to CpG and GOS DP3-5 resulted in increased galectin-9 secretion, the addition of GOS DP3-5 did not result in the further upregulation in galectin-9 mRNA in the IEC at 24 h or at 48 h after apical CpG exposure. This suggests that CpG can regulate galectin-9 expression at the level of gene transcription, while the oligosaccharides facilitate the basolateral release of galectin-9, and thereby support mucosal immunomodulation and/or development.

As GOS DP3-5 showed the most potent immunomodulatory activity out of the DP fractions tested, and this effect was mimicked to some extent by GOS DP3, we further investigated the capacity of a DP3-sized NDO, namely β-3′GL, in promoting immunomodulatory effects. A recent study determined the presence of low concentrations of β-3′GL in human milk samples [36]. Furthermore, β-3′GL is present in the GOS DP3 mixture (Figure 7A) [14]. Thus, we further studied the structure-specific effects of β-3′GL compared to GOS DP3-5 in combination with CpG in the IEC/PBMC model. Increasing the concentration of β-3′GL to 0.5% (w/v) supported a Th1-type regulatory immune response, as shown by increased IFNγ and galectin-9 or IL-10 secretion on top of CpG alone, which is in line with previous studies describing the immunomodulatory effects of other NDO [29–31,34]. In those studies, similar immune polarization profiles for GOS/FOS and 2′fucosyllactose (2′FL), a NDO abundantly present in human milk, were also shown in the presence of CpG [29,31], suggesting that this type of immunomodulation may be relevant for immune development [1]. Similar to 2′FL, β-3′GL may be an important NDO structure present in human milk, capable of supporting mucosal immune development driven by microbial signals (such as bacterial CpG DNA) in early life [29].

Due to the influence of IEC-derived galectins in supporting the immunomodulatory effects boosted by NDO and CpG described in previous studies [29–31,34], we studied IEC-derived galectin secretion. Only β-3′GL in the presence of CpG boosted the secretion of IEC-derived galectin-3, -4, and -9. However, only IEC-derived galectin-9 secretion was upregulated by DP3-5 and CpG. Besides, only IEC-derived galectin-9, but not IEC-derived galectin-3 or -4, correlated to IL-10 and galectin-9 secretion in the IEC/PBMC co-culture, which indicates that IEC-derived galectin-9 supports the regulatory-type immunity. Other studies have confirmed that incubation of activated PBMC with recombinant human galectin-9 was able to enhance not only IFN-γ secretion, but also regulatory IL-10 release [10]. However, beyond galectin-9, epithelial-derived galectin-3 and -4 were also found to significantly correlate positively with IFNγ concentrations. This suggests that all IEC-derived galectins might have been involved in increasing the IFNγ release by the activated PBMC under β-3′GL and CpG-exposed conditions. These results emphasize the need to better understand the complexity involved within the mucosal interactions between IEC and immune cells, as well as the role of galectins in these processes.

The IEC/PBMC model used was set-up to mimic the epithelial cell and immune cell crosstalk representing the intestinal mucosa. The model makes use of PBMC instead of lamina propria mononuclear cells (LPMC), because LPMC isolation requires access to clinical bowel samples and a laborious isolation procedure [41]. LPMC do show similarities as well as differences in the composition and function when compared to PBMC [42,43]. However, mitogen stimulation or activation via the T-cell receptor in both PBMC and LPMC results in induced levels of IFNγ, IL-10, and TNFα [41–43]. Furthermore, the results obtained from the IEC/PBMC model used were validated using in vivo animal models for food allergy [10,11], and in clinical samples of a NDO dietary intervention study, serum galectin-9 levels were shown to be enhanced [10].

The high dose of CpG was found to enhance the transcription of galectin-9; however, this was not further modified by GOS DP3-5, while it was capable of enhancing galectin-9 secretion. In addition, the effects of β-3′GL on epithelial galectin release could not be explained by increased galectin mRNA expression. Little is known about the factors inducing the transcription of galectins, as well as their intracellular storage. Galectins are widely known for their ability to recognize and bind extracellular carbohydrates with high affinity for β-galactoside structures such as NDO. Several studies have described the affinity of specific NDO structures to galectins according to their chemical structure and observed how chemical modifications such as fucosylation or galactosylation, among others, can improve the binding affinity of galectins to specific NDO structures. The increased binding affinity might, in turn, result in improved biological functions of NDO such as raft formation or attachment to pathogenic bacteria [44–47]. Although further research is needed, recent studies have proposed that changes in O-GlcNAcylation might be involved in regulating galectin expression [48].

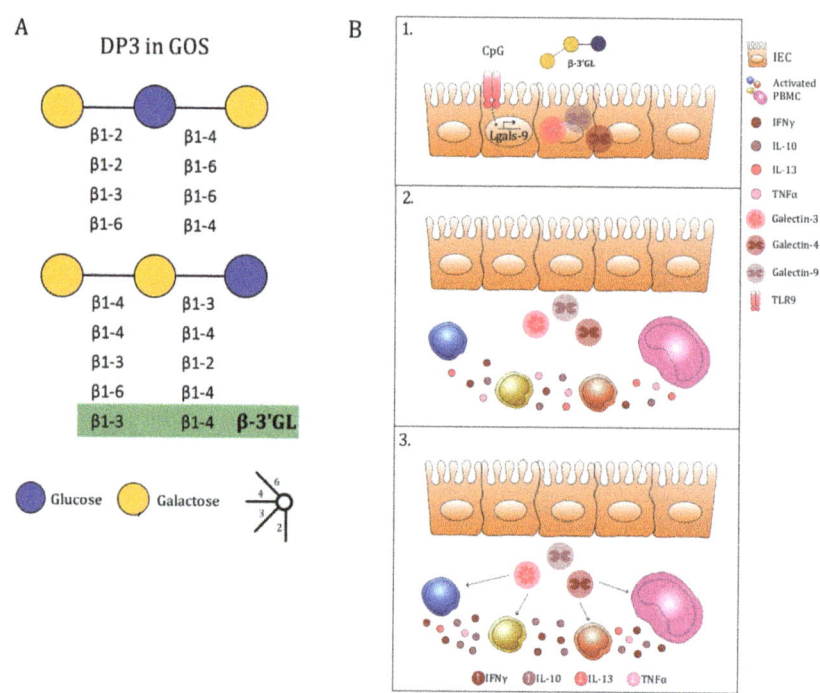

Figure 7. DP3-sized oligomers in GOS and description of the effects by β-3′GL and CpG in the IEC/PBMC co-culture model. DP3-sized structures present in GOS including β-3′GL are shown in (**A**). A summary of the effects observed in the IEC/PBMC co-culture model upon combined exposure to β-3′GL and CpG is shown in (**B**). Galectins are synthesized in the cytosol of IEC upon exposure to β-3′GL and CpG. Additionally, an upregulation of galectin-9 mRNA expression might be promoted by CpG (1). Galectins are then pushed out from the IEC toward the underlying immune compartment (2). Upon inflammatory conditions, defined by the activation of PBMC using αCD3/CD28, the immune cells in the basolateral compartment representing the lamina propria produce cytokines (2). Combined exposure to β-3′GL and CpG enhances IEC-derived galectin-3, -4, and -9 secretion. The IEC-derived galectins modulate the cytokine secretion by upregulating IFNγ and IL-10 while downregulating IL-13 and TNFα secretion (3).

To summarize, GOS DP3-5 in the presence of CpG, a TLR9 agonist mimicking a bacterial trigger, was able to promote immunomodulatory effects by enhancing immune responses in association with modified epithelial-derived galectin secretion. Moreover, exposure to DP3-sized β-3′GL, a NDO present in human milk, was most effective in enhancing CpG-induced galectin-9 release, while also enhancing galectin-3 and -4 secretion, which correlated with the instruction of a regulatory-type Th1 response. The most relevant findings described in this manuscript are summarized in Figure 7B. The use of PBMC might not fully resemble the immune cell populations present in the lamina propria and, therefore, the immune responses might differ from those shown in this manuscript. However, previous studies have shown the predictive value of the IEC/PBMC model in vivo in a murine model for food allergy, by identifying dietary interventions with immunomodulatory properties in which galectin-9 had a key role [10]. This emphasizes the translational value of the IEC/PBMC co-culture model. Our aim is to use the IEC/PBMC co-culture model as a first step to identify relevant bioactive components that should later be studied using more complex models. In addition to this, by the use of these models combining epithelial and immune cells, we aim to better understand the complex interactions occurring between these types of cells, which further contribute to our understanding.

In conclusion, epithelial-derived galectins were demonstrated to be key players in the mucosal immune development supported by GOS of which DP3-size β-3′GL showed relevant immunomodulatory properties.

Supplementary Materials: The following supporting information can be downloaded at: https://www.mdpi.com/article/10.3390/biom12030384/s1, Figure S1: Lower CpG concentrations effectively supported immunomodulatory effects by GOS DP3-5; Figure S2: Cytokine secretion in IEC/PBMC co-culture model; Figure S3: Galectin mRNA expression.

Author Contributions: Conceptualization, B.S., B.v.L., A.D.K. and L.E.M.W.; investigation, V.A.-M., M.v.d.K. and R.M.; writing—original draft preparation, V.A.-M., B.v.L. and L.E.M.W.; writing—review and editing, V.A.-M., M.v.d.K., R.M., J.G., B.S., R.J.P., B.v.L., A.D.K. and L.E.M.W.; supervision, B.v.L., A.D.K. and L.E.M.W.; project administration, B.v.L., A.D.K. and L.E.M.W. All authors have read and agreed to the published version of the manuscript.

Funding: This research was funded by a seed grant from the Utrecht Institute for Pharmaceutical Sciences (UIPS) and Danone Nutricia Research B.V.

Institutional Review Board Statement: Not applicable.

Informed Consent Statement: Not applicable.

Data Availability Statement: The data presented in this study are available upon request from the corresponding author.

Acknowledgments: The authors would like to thank Anke Muijsers and Erik Zubiria for helping with the illustrations.

Conflicts of Interest: None of the authors have a competing financial interest in relation to the presented work; J.G. is head of the division of Pharmacology, Utrecht Institute for Pharmaceutical Sciences, Faculty of Science at Utrecht University, and partly employed by Danone Nutricia Research B.V. B.v.L. and B.S. are employed by Danone Nutricia Research B.V. B.v.L. is affiliated at and leading a strategic alliance between University Medical Centre Utrecht/Wilhelmina Children's Hospital and Nutricia Research B.V. B.S. has an associated position at Utrecht Institute for Pharmaceutical Sciences, CBDD, Faculty of Science at Utrecht University.

Abbreviations

β-3′galactosyllactose (β-3′GL); 2′fucosyllactose (2′FL); complementary DNA (cDNA), electrospray ionization-mass spectrometry (ESI-MS); enzyme-linked immunosorbent assay (ELISA); fetal calf serum (FCS); fructo-oligosaccharides (FOS); galacto-oligosaccharides (GOS); intestinal epithelial cells (IEC); lamina propria mononuclear cells (LPMC); non-digestible oligosaccharides (NDO); oligodeoxynucleotides (ODN); peripheral blood mononuclear cells (PBMC); quantitative Polymerase Chain Reaction (qPCR).

References

1. Ayechu-Muruzabal, V.; van Stigt, A.H.; Mank, M.; Willemsen, L.E.M.; Stahl, B.; Garssen, J.; van't Land, B. Diversity of human milk oligosaccharides and effects on early life immune development. *Front. Pediatr.* **2018**, *6*, 239. [CrossRef] [PubMed]
2. Salminen, S.; Stahl, B.; Vinderola, G.; Szajewska, H. Infant Formula Supplemented with Biotics: Current Knowledge and Future Perspectives. *Nutrients* **2020**, *12*, 1952. [CrossRef] [PubMed]
3. Boehm, G.; Fanaro, S.; Jelinek, J.; Stahl, B.; Marini, A. Prebiotic concept for infant nutrition. *Acta Paediatr. Int. J. Paediatr. Suppl.* **2003**, *91*, 64–67. [CrossRef] [PubMed]
4. Boehm, G.; Moro, G. Structural and functional aspects of prebiotics used in infant nutrition. *J. Nutr.* **2008**, *138*, 1818S–1828S. [CrossRef] [PubMed]

5. Miqdady, M.; Al Mistarihi, J.; Azaz, A.; Rawat, D. Prebiotics in the infant microbiome: The past, present, and future. *Pediatr. Gastroenterol. Hepatol. Nutr.* **2020**, *23*, 1–14. [CrossRef]
6. Moro, G.; Arslanoglu, S.; Stahl, B.; Jelinek, J.; Wahn, U.; Boehm, G. A mixture of prebiotic oligosaccharides reduces the incidence of atopic dermatitis during the first six months of age. *Arch. Dis. Child.* **2006**, *91*, 814–819. [CrossRef] [PubMed]
7. Moro, G.; Minoli, I.; Mosca, M.; Fanaro, S.; Jelinek, J.; Stahl, B.; Boehm, G. Dosage-related bifidogenic effects of galacto- and fructooligosaccharides in formula-fed term infants. *J. Pediatr. Gastroenterol. Nutr.* **2002**, *34*, 291–295. [CrossRef]
8. Van Hoffen, E.; Ruiter, B.; Faber, J.; M'Rabet, L.; Knol, E.F.; Stahl, B.; Arslanoglu, S.; Moro, G.; Boehm, G.; Garssen, J. A specific mixture of short-chain galacto-oligosaccharides and long-chain fructo-oligosaccharides induces a beneficial immunoglobulin profile in infants at high risk for allergy. *Allergy Eur. J. Allergy Clin. Immunol.* **2009**, *64*, 484–487. [CrossRef]
9. Schouten, B.; Van Esch, B.C.A.M.; Hofman, G.A.; Van Doorn, S.A.C.M.; Knol, J.; Nauta, A.J.; Garssen, J.; Willemsen, L.E.M.; Knippels, L.M.J. Cow milk allergy symptoms are reduced in mice fed dietary synbiotics during oral sensitization with whey. *J. Nutr.* **2009**, *139*, 1398–1403. [CrossRef] [PubMed]
10. De Kivit, S.; Saeland, E.; Kraneveld, A.D.; Van De Kant, H.J.G.; Schouten, B.; Van Esch, B.C.A.M.; Knol, J.; Sprikkelman, A.B.; Van Der Aa, L.B.; Knippels, L.M.J.; et al. Galectin-9 induced by dietary synbiotics is involved in suppression of allergic symptoms in mice and humans. *Allergy Eur. J. Allergy Clin. Immunol.* **2012**, *67*, 343–352. [CrossRef]
11. De Kivit, S.; Kostadinova, A.I.; Kerperien, J.; Morgan, M.E.; Muruzabal, V.A.; Hofman, G.A.; Knippels, L.M.J.; Kraneveld, A.D.; Garssen, J.; Willemsen, L.E.M. Dietary, nondigestible oligosaccharides and Bifidobacterium breve M-16V suppress allergic inflammation in intestine via targeting dendritic cell maturation. *J. Leukoc. Biol.* **2017**, *102*, 105–115. [CrossRef]
12. Van Esch, B.C.A.M.; Abbring, S.; Diks, M.A.P.; Dingjan, G.M.; Harthoorn, L.F.; Vos, A.P.; Garssen, J. Post-sensitization administration of non-digestible oligosaccharides and Bifidobacterium breve M-16V reduces allergic symptoms in mice. *Immun. Inflamm. Dis.* **2016**, *4*, 155–165. [CrossRef] [PubMed]
13. Coulier, L.; Timmermans, J.; Richard, B.; Van Den Dool, R.; Haaksman, I.; Klarenbeek, B.; Slaghek, T.; Van Dongen, W. In-depth characterization of prebiotic galactooligosaccharides by a combination of analytical techniques. *J. Agric. Food Chem.* **2009**, *57*, 8488–8495. [CrossRef]
14. Van Leeuwen, S.S.; Kuipers, B.J.H.; Dijkhuizen, L.; Kamerling, J.P. 1H NMR analysis of the lactose/β-galactosidase-derived galacto-oligosaccharide components of Vivinal®GOS up to DP5. *Carbohydr. Res.* **2014**, *400*, 59–73. [CrossRef] [PubMed]
15. Matsuki, T.; Tajima, S.; Hara, T.; Yahagi, K.; Ogawa, E.; Kodama, H. Infant formula with galacto-oligosaccharides (OM55N) stimulates the growth of indigenous bifidobacteria in healthy term infants. *Benef. Microbes* **2016**, *7*, 453–461. [CrossRef] [PubMed]
16. Sierra, C.; Bernal, M.J.; Blasco, J.; Martínez, R.; Dalmau, J.; Ortuño, I.; Espín, B.; Vasallo, M.I.; Gil, D.; Vidal, M.L.; et al. Prebiotic effect during the first year of life in healthy infants fed formula containing GOS as the only prebiotic: A multicentre, randomised, double-blind and placebo-controlled trial. *Eur. J. Nutr.* **2015**, *54*, 89–99. [CrossRef]
17. Shoaf, K.; Mulvey, G.L.; Armstrong, G.D.; Hutkins, R.W. Prebiotic galactooligosaccharides reduce adherence of enteropathogenic Escherichia coli to tissue culture cells. *Infect. Immun.* **2006**, *74*, 6920–6928. [CrossRef] [PubMed]
18. Sarabia-Sainz, H.M.; Armenta-Ruiz, C.; Sarabia-Sainz, J.A.I.; Guzmán-Partida, A.M.; Ledesma-Osuna, A.I.; Vázquez-Moreno, L.; Montfort, G.R.C. Adhesion of enterotoxigenic Escherichia coli strains to neoglycans synthesised with prebiotic galactooligosaccharides. *Food Chem.* **2013**, *141*, 2727–2734. [CrossRef]
19. Akbari, P.; Fink-Gremmels, J.; Willems, R.H.A.M.; Difilippo, E.; Schols, H.A.; Schoterman, M.H.C.; Garssen, J.; Braber, S. Characterizing microbiota-independent effects of oligosaccharides on intestinal epithelial cells: Insight into the role of structure and size: Structure–activity relationships of non-digestible oligosaccharides. *Eur. J. Nutr.* **2017**, *56*, 1919–1930. [CrossRef]
20. Bhatia, S.; Prabhu, P.N.; Benefiel, A.C.; Miller, M.J.; Chow, J.; Davis, S.R.; Gaskins, H.R. Galacto-oligosaccharides may directly enhance intestinal barrier function through the modulation of goblet cells. *Mol. Nutr. Food Res.* **2015**, *59*, 566–573. [CrossRef]
21. Whisner, C.M.; Martin, B.R.; Schoterman, M.H.C.; Nakatsu, C.H.; McCabe, L.D.; McCabe, G.P.; Wastney, M.E.; Van Den Heuvel, E.G.H.M.; Weaver, C.M. Galacto-oligosaccharides increase calcium absorption and gut bifidobacteria in young girls: A double-blind cross-over trial. *Br. J. Nutr.* **2013**, *110*, 1292–1303. [CrossRef] [PubMed]
22. Paganini, D.; Uyoga, M.A.; Cercamondi, C.I.; Moretti, D.; Mwasi, E.; Schwab, C.; Bechtler, S.; Mutuku, F.M.; Galetti, V.; Lacroix, C.; et al. Consumption of galacto-oligosaccharides increases iron absorption from a micronutrient powder containing ferrous fumarate and sodium iron EDTA: A stable-isotope study in Kenyan infants. *Am. J. Clin. Nutr.* **2017**, *106*, 1020–1031. [CrossRef] [PubMed]
23. Drakoularakou, A.; Tzortzis, G.; Rastall, R.A.; Gibson, G.R. A double-blind, placebo-controlled, randomized human study assessing the capacity of a novel galacto-oligosaccharide mixture in reducing travellers' diarrhoea. *Eur. J. Clin. Nutr.* **2010**, *64*, 146–152. [CrossRef]
24. Lehmann, S.; Hiller, J.; Van Bergenhenegouwen, J.; Knippels, L.M.J.; Garssen, J.; Traidl-Hoffmann, C. In vitro evidence for immune-modulatory properties of non-digestible oligosaccharides: Direct effect on human monocyte derived dendritic cells. *PLoS ONE* **2015**, *10*, e0132304. [CrossRef]
25. Bermudez-Brito, M.; Sahasrabudhe, N.M.; Rösch, C.; Schols, H.A.; Faas, M.M.; De Vos, P. The impact of dietary fibers on dendritic cell responses in vitro is dependent on the differential effects of the fibers on intestinal epithelial cells. *Mol. Nutr. Food Res.* **2015**, *59*, 698–710. [CrossRef]

26. Wang, J.; Tian, S.; Yu, H.; Wang, J.; Zhu, W. Response of Colonic Mucosa-Associated Microbiota Composition, Mucosal Immune Homeostasis, and Barrier Function to Early Life Galactooligosaccharides Intervention in Suckling Piglets. *J. Agric. Food Chem.* **2019**, *67*, 578–588. [CrossRef]
27. Vulevic, J.; Drakoularakou, A.; Yaqoob, P.; Tzortzis, G.; Gibson, G.R. Modulation of the fecal microflora profile and immune function by a novel trans-galactooligosaccharide mixture (B-GOS) in healthy elderly volunteers. *Am. J. Clin. Nutr.* **2008**, *88*, 1438–1446. [CrossRef] [PubMed]
28. Vulevic, J.; Juric, A.; Walton, G.E.; Claus, S.P.; Tzortzis, G.; Toward, R.E.; Gibson, G.R. Influence of galacto-oligosaccharide mixture (B-GOS) on gut microbiota, immune parameters and metabonomics in elderly persons. *Br. J. Nutr.* **2015**, *114*, 586–595. [CrossRef] [PubMed]
29. Ayechu-Muruzabal, V.; Overbeek, S.A.; Kostadinova, A.I.; Stahl, B.; Garssen, J.; Van't Land, B.; Willemsen, L.E.M. Exposure of intestinal epithelial cells to 2′-fucosyllactose and CpG enhances galectin release and instructs dendritic cells to drive Th1 and regulatory-type immune development. *Biomolecules* **2020**, *10*, 784. [CrossRef]
30. Hayen, S.M.; Otten, H.G.; Overbeek, S.A.; Knulst, A.C.; Garssen, J.; Willemsen, L.E.M. Exposure of intestinal epithelial cells to short- and long-chain fructo-oligosaccharides and CpG oligodeoxynucleotides enhances peanut-specific T Helper 1 polarization. *Front. Immunol.* **2018**, *9*, 923. [CrossRef]
31. De Kivit, S.; Kraneveld, A.D.; Knippels, L.M.J.; Van Kooyk, Y.; Garssen, J.; Willemsen, L.E.M. Intestinal epithelium-derived galectin-9 is involved in the immunomodulating effects of nondigestible oligosaccharides. *J. Innate Immun.* **2013**, *5*, 625–638. [CrossRef] [PubMed]
32. De Kivit, S.; van Hoffen, E.; Korthagen, N.; Garssen, J.; Willemsen, L.E.M. Apical TLR ligation of intestinal epithelial cells drives a Th1-polarized regulatory or inflammatory type effector response in vitro. *Immunobiology* **2011**, *216*, 518–527. [CrossRef] [PubMed]
33. Peterson, L.W.; Artis, D. Intestinal epithelial cells: Regulators of barrier function and immune homeostasis. *Nat. Rev. Immunol.* **2014**, *14*, 141–153. [CrossRef] [PubMed]
34. Overbeek, S.A.; Kostadinova, A.I.; Boks, M.A.; Hayen, S.M.; De Jager, W.; Van'T Land, B.; Knippels, L.M.; Garssen, J.; Willemsen, L.E.M. Combined Exposure of Activated Intestinal Epithelial Cells to Nondigestible Oligosaccharides and CpG-ODN Suppresses Th2-Associated CCL22 Release while Enhancing Galectin-9, TGF β, and Th1 Polarization. *Mediat. Inflamm.* **2019**, *2019*, 2019. [CrossRef]
35. Mukherjee, R.; Van De Kaa, M.; Garssen, J.; Pieters, R.J.; Kraneveld, A.D.; Willemsen, L.E.M. Lactulose synergizes with CpG-ODN to modulate epithelial and immune cells cross talk. *Food Funct.* **2019**, *10*, 33–37. [CrossRef] [PubMed]
36. Eussen, S.R.B.M.; Mank, M.; Kottler, R.; Hoffmann, X.K.; Behne, A.; Rapp, E.; Stahl, B.; Mearin, M.L.; Koletzko, B. Presence and levels of galactosyllactoses and other oligosaccharides in human milk and their variation during lactation and according to maternal phenotype. *Nutrients* **2021**, *13*, 2324. [CrossRef]
37. García-Vallejo, J.J.; Van Het Hof, B.; Robben, J.; Van Wijk, J.A.E.; Van Die, I.; Joziasse, D.H.; Van Dijk, W. Approach for defining endogenous reference genes in gene expression experiments. *Anal. Biochem.* **2004**, *329*, 293–299. [CrossRef]
38. Ewaschuk, J.B.; Backer, J.L.; Churchill, T.A.; Obermeier, F.; Krause, D.O.; Madsen, K.L. Surface expression of toll-like receptor 9 is upregulated on intestinal epithelial cells in response to pathogenic bacterial DNA. *Infect. Immun.* **2007**, *75*, 2572–2579. [CrossRef]
39. Pedersen, G.; Andresen, L.; Matthiessen, M.W.; Rask-Madsen, J.; Brynskov, J.; Pedersen, G. Expression of toll-like receptor 9 and response to bacterial cpg oligodeoxynucleotides in human intestinal epithelium. *Clin. Exp. Immunol.* **2005**, *141*, 298–306. [CrossRef]
40. Lee, J.; Mo, J.H.; Katakura, K.; Alkalay, I.; Rucker, A.N.; Liu, Y.T.; Lee, H.K.; Shen, C.; Cojocaru, G.; Shenouda, S.; et al. Maintenance of colonic homeostasis by distinctive apical TLR9 signalling in intestinal epithelial cells. *Nat. Cell Biol.* **2006**, *8*, 1327–1336. [CrossRef]
41. Willemsen, L.E.M.; Schreurs, C.C.H.M.; Kroes, H.; Spillenaar Bilgen, E.J.; Van Deventer, S.J.H.; Van Tol, E.A.F. A coculture model mimicking the intestinal mucosa reveals a regulatory role for myofibroblasts in immune-mediated barrier disruption. *Dig. Dis. Sci.* **2002**, *47*, 2316–2324. [CrossRef] [PubMed]
42. Gotteland, M.; Lopez, M.; Muñoz, C.; Saez, R.; Altshiller, H.; Llorens, P.; Brunser, O. Local and systemic liberation of proinflammatory cytokines in ulcerative colitis. *Dig. Dis. Sci.* **1999**, *44*, 830–835. [CrossRef] [PubMed]
43. Guzy, C.; Schirbel, A.; Paclik, D.; Wiedenmann, B.; Dignass, A.; Sturm, A. Enteral and parenteral nutrition distinctively modulate intestinal permeability and T cell function in vitro. *Eur. J. Nutr.* **2009**, *48*, 12–21. [CrossRef]
44. Hirabayashi, J.; Hashidate, T.; Arata, Y.; Nishi, N.; Nakamura, T.; Hirashima, M.; Urashima, T.; Oka, T.; Futai, M.; Muller, W.E.G.; et al. Oligosaccharide specificity of galectins: A search by frontal affinity chromatography. *Biochim. Biophys. Acta Gen. Subj.* **2002**, *1572*, 232–254. [CrossRef]
45. Horlacher, T.; Oberli, M.A.; Werz, D.B.; Kröck, L.; Bufali, S.; Mishra, R.; Sobek, J.; Simons, K.; Hirashima, M.; Niki, T.; et al. Determination of carbohydrate-binding preferences of human galectins with carbohydrate microarrays. *ChemBioChem* **2010**, *11*, 1563–1573. [CrossRef] [PubMed]
46. Iwaki, J.; Hirabayashi, J. Carbohydrate-binding specificity of human galectins: An overview by frontal affinity chromatography. *Trends Glycosci. Glycotechnol.* **2018**, *30*, S137–S153. [CrossRef]
47. Urashima, T.; Hirabayashi, J.; Sato, S.; Kobata, A. Human milk oligosaccharides as essential tools for basic and application studies on galectins. *Trends Glycosci. Glycotechnol.* **2018**, *30*, SE51–SE65. [CrossRef]
48. Tazhitdinova, R.; Timoshenko, A.V. The Emerging Role of Galectins and O-GlcNAc Homeostasis in Processes of Cellular Differentiation. *Cells* **2020**, *9*, 1792. [CrossRef] [PubMed]

Article

Galectin-1 Cooperates with Yersinia Outer Protein (Yop) P to Thwart Protective Immunity by Repressing Nitric Oxide Production

Brenda Lucila Jofre [1,2], Ricardo Javier Eliçabe [1,2], Juan Eduardo Silva [1,2], Juan Manuel Pérez Sáez [3], Maria Daniela Paez [4], Eduardo Callegari [4], Karina Valeria Mariño [5], María Silvia Di Genaro [1,2], Gabriel Adrián Rabinovich [3,6] and Roberto Carlos Davicino [1,2,7,*]

[1] División de Inmunología, Facultad de Química, Bioquímica y Farmacia, Universidad Nacional de San Luis, San Luis CP5700, Argentina; brendalucila.jofre@gmail.com (B.L.J.); javielicabe@gmail.com (R.J.E.); jesilva9@hotmail.com (J.E.S.); sdigena@gmail.com (M.S.D.G.)
[2] Instituto Multidisciplinario de Investigaciones Biológicas (IMIBIO), Consejo Nacional de Investigaciones Científicas y Técnicas (CONICET), San Luis C5700, Argentina
[3] Laboratorio de Glicomedicina, Instituto de Biología y Medicina Experimental (IBYME), Consejo Nacional de Investigaciones Científicas y Técnicas (IBYME-CONICET), Buenos Aires C1428ADN, Argentina; juanmanuelperezsaez@gmail.com (J.M.P.S.); gabyrabi@gmail.com (G.A.R.)
[4] Division of Basic Biomedical Sciences, Sanford School of Medicine, University of South Dakota, Vermillion, SD 66544, USA; Daniela.Paez@usd.edu (M.D.P.); Eduardo.Callegari@usd.edu (E.C.)
[5] Laboratorio de Glicómica Funcional y Molecular, Instituto de Biología y Medicina Experimental, Consejo Nacional de Investigaciones Científicas y Técnicas (IBYME-CONICET), Buenos Aires C1428ADN, Argentina; kmarino@ibyme.conicet.gov.ar
[6] Departamento de Química Biológica, Facultad de Ciencias Exactas y Naturales, Universidad de Buenos Aires, Buenos Aires C1428, Argentina
[7] Roberto Davicino, División de Inmunología, Facultad de Química, Bioquímica y Farmacia, Universidad Nacional de San Luis, Ejercito de los Andes 950, San Luis CP5700, Argentina
* Correspondence: rcdavici@unsl.edu.ar

Abstract: *Yersinia enterocolitica* (Ye) inserts outer proteins (Yops) into cytoplasm to infect host cells. However, in spite of considerable progress, the mechanisms implicated in this process, including the association of Yops with host proteins, remain unclear. Here, we evaluated the functional role of Galectin-1 (Gal1), an endogenous β-galactoside-binding protein, in modulating Yop interactions with host cells. Our results showed that Gal1 binds to Yops in a carbohydrate-dependent manner. Interestingly, Gal1 binding to Yops protects these virulence factors from trypsin digestion. Given that early control of Ye infection involves activation of macrophages, we evaluated the role of Gal1 and YopP in the modulation of macrophage function. Although Gal1 and YopP did not influence production of superoxide anion and/or TNF by Ye-infected macrophages, they coordinately inhibited nitric oxide (NO) production. Notably, recombinant Gal1 (rGal1) did not rescue NO increase observed in $Lgals1^{-/-}$ macrophages infected with the YopP mutant Ye $\Delta yopP$. Whereas NO induced apoptosis in macrophages, no significant differences in cell death were detected between Gal1-deficient macrophages infected with Ye $\Delta yopP$, and WT macrophages infected with Ye wt. Strikingly, increased NO production was found in WT macrophages treated with MAPK inhibitors and infected with Ye wt. Finally, rGal1 administration did not reverse the protective effect in Peyer Patches (PPs) of $Lgals1^{-/-}$ mice infected with Ye $\Delta yopP$. Our study reveals a cooperative role of YopP and endogenous Gal1 during Ye infection.

Keywords: *Yersinia enterocolitica*; YopP; Galectin-1; nitric oxide; macrophages

1. Introduction

Yersinia enterocolitica (Ye), *Yersinia pseudotuberculosis,* and *Yersinia pestis* are the three human pathogenic bacteria in the genus *Yersinia* [1]. Ye causes food-borne self-limiting severe diarrhea, enteritis, and mesenteric lymphadenitis. In addition to gastrointestinal effects, Ye gradually spreads across the body, causing symptoms in the liver and spleen [2,3]. Ye uses a type III protein secretion machinery to deliver into host cells bacterial effector proteins encoded in the 70-kb *Yersinia* virulence plasmid (pYV). This plasmid includes a set of six effector Yersinia outer proteins (Yops): YopE, YopH, YopM, YopO/YpkA, YopP/YopJ, YopT [4]. YopH counteracts phagocytosis and T-cell activation [5,6], while YopE, YopT, and YopO disrupt actin cytoskeleton [7,8]. In addition, YopP/J inhibits nuclear factor kappa B (NF-kB) signaling, suppresses pro-inflammatory cytokines, modulates antigen uptake, and induces apoptosis in macrophages and dendritic cells [9–13]. Moreover, YopP inhibits the activation of MAPKs inactivating c-Jun-N-terminal kinase (JNK), p38, and extracellular signal-regulated 1/2 kinase (ERK1/2) [14–16]. In this context, YopP can interact directly or indirectly with specific kinases, acting as a "poison kinase" [16]. In this regard, YopP is an acetyltransferase, which uses acetyl-coenzymeA(acetyl-CoA) as a cofactor to acetylate critical serine and threonine residues in the activation loop of MAPKKs and IKK-I3 [12,17]. Surprisingly, MAPK as well as NF-kB, are constrained in scaffolds and the recruitment of YopP to such a scaffold would allow faster inhibition of signaling events compared to a free diffusion of YopP in the cell [18]. In addition, YopP is activated by the host cell factor inositolhexakisphosphate (IP6), which could also explain how YopP is kept in a quiescent state in the bacterium, since bacteria lack the capacity to synthesize IP6 [19]. In activated macrophages however, *Yersiniae* cause pyroptosis, a cell death program independent of YopP, which involves inflammasome activation and processing of caspase-1, release of pro-inflammatory cytokines IL-1β and IL-18, and eventually lysis of macrophages and release of pro-inflammatory intracellular content [20,21]. The prevention of pyroptosis and suppression of inflammatory response by YopP could be crucial for *Yersiniae* ability to colonize the Peyer's patches without an initial immune response [22–24]. In this context, the early control of Ye infection is mediated by innate immune mechanisms, involving natural killer (NK) cells, neutrophils and macrophages [25–27].

Interestingly, M1 and M2 macrophages refer to the two extremes of a spectrum of potential macrophage activation states; however the term M2 has been traditionally used for any macrophage activation states other than M1. The use of M2 as a generic term for macrophage activation is justified by the fact that they share a number of functional characteristics and are involved in immunoregulation and tissue remodeling. In this regard, three subclasses of M2 macrophages have been identified: M2a, triggered by IL-4 or IL-13; M2b, induced by exposure to Toll-like receptor (TLR) agonists and IL-1R; and M2c, induced by IL-10 and glucocorticoids [28]. On the other hand, M1 macrophage activation is defined by high production of toxic intermediates, such as reactive oxygen species (ROS) and NO [28]. However, few reports are available on the role of NO in Ye infection [29,30]. We have previously shown increased NO synthesis and enhanced expression of inducible nitric oxide synthase (iNOS) in response to Ye antigens in macrophages from mice lacking the tumor necrosis factor receptor p55 (TNFRp55) [31]. These results suggested a role of TNFRp55 and NO in modulating macrophage functions after Ye infection. In addition, we have shown that Ye infection induces local and systemic up-regulation of Galectin-1 (Gal1), an endogenous immunomodulatory lectin, which blunts NO synthesis and limits bacterial clearance [32].

Through binding to β-galactoside-containing glycoconjugates, Gal1 triggers different biological processes including those operating during innate and adaptive immune responses, as well as those involving host-pathogen interactions. Gal1, as well as other members of this lectin family, can cross-link glycosylated receptors, including: the T cell receptor (TCR); pre-B cell receptor (pre-BCR) and CD45, facilitating their cell surface retention and modulating signaling thresholds [33,34]. In this regard, it has been demonstrated that glycan-binding proteins may serve as a bridge that regulates bacterial infection, internalization and immunity [35,36].

Thus, given the emerging roles of Gal1 in infection [33,37–43] and based on our previous results showing that *Y. enterocolitica* induced a YopP dependent positive regulation of Gal1 [32], we hypothesized that Yops could interact with Gal1 and modulate the course of Ye infection. In the present work we studied the interactions between Yops and Gal1, focusing on the role of the Ye virulence factor YopP in shaping the course of early innate immune response upon Ye infection.

2. Materials and Methods

2.1. Bacterial Culture and Purification of Yops

Infection was performed with Ye serotype 0:8 (pYV+, WA-314) (Ye wt) or with Ye WA-314 deficient in YopP (pYV+, WA-C pYVNalrKanr) (Ye $\Delta yopP$) [44], kindly provided by Ingo Autenrieth (Tuebingen, Germany). Bacteria were cultured as previously described [45], diluted 1:20, and incubated at 37 °C with agitation for 2 h (180 rpm). The addition of EGTA (5 mM) for Ca^{2+} chelation, $MgCl_2$ (15 mM), and glucose (0.2%) induced Yops expression and secretion. Bacteria were grown at 37 °C for 2 to 3 h and centrifuged ($10,000 \times g$ for 15 min), and proteins were precipitated from culture supernatants with trichloroacetic acid (TCA) as previously described [46].

2.2. Mice and Infection

C57BL/6 Gal1 knockout ($Lgals1^{-/-}$) mice were kindly provided by F. Poirier (Institute Jacques Monod, Paris, France). C57BL/6 wild-type (WT) mice were purchased from the Animal Facilities of the National University of La Plata, La Plata, Argentina. Breeding colonies were established at the animal facilities of the National University of San Luis (San Luis, Argentina). Mice were housed in a cabinet (Ehret, Emmendingen, Germany) and given ad libitum sterile food and water. Male mice (6–8 wk-old) were used for all the experiments. The Animal Care and Use Committee of the National University of San Luis, Argentina, approved the experimental protocols (Protocol Number: B226/16).

Mice were starved for 2 h before being inoculated orogastrically with 5×10^8 bacteria in 0.2 mL sterile phosphate-buffered saline solution (pH 7.4) using a gastric tube. PBS was given to the control mice. Serial dilutions of the inoculated suspension were plated on Trypticase soy agar (Britania, Buenos Aires, Argentina) to monitor the real number of inoculated bacteria.

The PPs were removed in aseptic conditions and homogenized in PBS. Then, on MacConkey-Igarsan agar, duplicates of 50 μL of serial dilutions of PPs homogenates were plated (Britania, Buenos Aires, Argentina). After 48-h incubation period at 27 °C, colony-forming units (CFU) were counted. The limit of detectable CFU was 25 ($\log_{10} 25 = 1.4$) [47].

2.3. Stimulation of Peritoneal Macrophages

$Lgals1^{-/-}$ and WT resident peritoneal macrophages were isolated from mice of both genotypes using 10 mLof sterile pyrogen-free saline solution, centrifuged twice at $200 \times g$ for 10 min at 4 °C, and re-suspended in DMEM supplemented with 10% heat-inactivated fetal bovine serum (FBS) (Natocor, Córdoba, Argentina), 5 mM L-glutamine, 50 μM 2-ME, 100 IU/mL penicillin, 100 μg/mL streptomycin, and 50 μg/mL gentamicin (Thermo Fisher Scientific, Waltham, Massachusetts, EEUU).This cellular suspension was seeded onto a 24-well culture plate (Costar, Tecnolab, Buenos Aires, Argentina) at 2×10^6 cells per well. After 24 h of incubation at 37 °C in a 5% CO_2 atmosphere, adherent cells were washed three times with saline and incubated for 1 h at 37 °C in a 5% CO_2 atmosphere with or without Ye wt or Ye $\Delta yopP$ (multiplicity of infection, moi: 10:1) in the absence or presence of 5 μM ERK1/2 inhibitor (PD98059) or p38 inhibitor (SB203580) (Calbiochem, San Diego, CA, USA). To eliminate extracellular bacteria, 0.1 g/mL of gentamicin was added. Cells were incubated overnight, and culture supernatants were collected [32].

2.4. NO and Urea Determination

The Griess reaction assay was used to measure nitrite synthesis in macrophage culture supernatants obtained 12 h after Ye infection [32]. In a 96-well flat-bottom plate, 100 μL of culture supernatant was mixed with 100 μL of Griess reagent and incubated for 10 min at room temperature. Absorbance at 550 nm was determined in a plate reader (Bio-Rad, New York, NY, USA). In addition, urea was measured in macrophage culture supernatants using the Urea Color 2R package (Wiener, Rosario, Argentina), according to the manufacturer's instructions.

2.5. Apoptosis Assays

Macrophages isolated from WT or $Lgals1^{-/-}$ mice were infected with Ye wt or with Ye $\Delta yopP$ and incubated in a 5% CO_2 atmosphere. Dimethyl sulfoxide (DMSO) (Sigma, St. Louis, MO, USA) was used as a positive control for apoptosis. Cells (1×10^6) were suspended in binding buffer (10 mM HEPES pH 7.4, 140 mM NaCl, 2.5 mM $CaCl_2$) after being washed twice with PBS. Macrophages were incubated for 15 min at room temperature in the dark with Annexin V-FITC (Sigma, St. Louis, MO, USA). The cells were washed and re-suspended in 500 µL of binding buffer. Finally, macrophages were stained with propidium iodide (PI) (Sigma, St. Louis, MO, USA) and analyzed by flow cytometry using a FACSCalibur cytometer (Becton, Dickinson and Company, Franklin Lakes, NJ, USA).

2.6. Preparation and Purification of RGal1

Recombinant Gal1 (rGal1) was produced and purified as previously described [48]. Briefly, LGALS1 gene was cloned into a pET-3a (+) vector for bacterial expression between the NdeI and BamHI specific recognition sites. The plasmid was first amplified in DH5α *E. coli* and subsequently used for transformation of *E. coli* C41 (DE3) pLysS. The resulting protein was purified by affinity chromatography on a lactosyl-Sepharose resin. Purified Gal1 was dialyzed against PBS (pH 6–9) for 6 h, three times and then subjected to a Polymixin B affinity resin to remove endotoxins from protein solution. Protein was measured by the Pierce BCA Protein Assay Kit (Thermo Fisher, Carlsbad, CA, USA), according to the manufacturer protocol. The recombinant protein was sterilized by passage through a 0.22-µm syringe filter, adjusted to 10 mg/mL in PBS and stored as frozen aliquots until used.

2.7. ELISA Assays

TNF and IL-10 were determined in supernatants of infected WT or $Lgals1^{-/-}$ macrophages using capture ELISA kits (eBioscience, San Diego, CA, USA) according to the manufacturer's instructions. YopP was determined using a modified ELISA protocol described by Chatzipanagiotou et al., 2001 [49]. Briefly, Yops were prepared from Ye wt or Ye $\Delta yopP$. ELISA plates (Corning, Kennebunk, ME, USA) were coated with Yops antigens (10 µg/well) and the binding of rGal1 (10 mg/mL) was detected using rabbit anti-Gal1 antibodies (1/1000). The absorbance was read at 450 nm using a plate reader (Bio-Rad, New York, NY, USA).

2.8. Oxidative Burst Assay

For this assay, we used a protocol described by Schopf et al. (1984) [50] and ROS products were evaluated by the reduction of nitro blue tetrazolium (NBT) (Sigma, St. Louis, MO, USA) to formazan. In all these assays, WT or $Lgals1^{-/-}$ macrophages were infected with Ye wt or with Ye $\Delta yopP$, then, 300 µL of NBT was added and the reaction was stopped with 1N HCl (Tetrahedron, Buenos Aires, Argentina). Dioxane (Dorwill, Buenos Aires, Argentina) was used to obtain formazan, and the absorbance was determined in a microplate reader at 525 nm (Bio-Rad).

2.9. In Vivo and In Vitro Supplementation of RGal1

For in vivo phenotype-rescuing studies, four animal groups were used: groups 1 and 2 were $Lgals1^{-/-}$ mice injected i.p. with rGal1 (3.2 mg/kg) or vehicle control daily for 5 days after Ye wt infection [32]; groups 3 and 4 were $Lgals1^{-/-}$ mice injected i.p. with rGal1 (3.2 mg/kg) or vehicle control daily for 5 days after Ye $\Delta yopP$ infection [32]. Mice were killed five days after infection, and CFU were counted in PPs homogenates as mentioned previously.

For in vitro rescue experiments, peritoneal macrophages from $Lgals1^{-/-}$ mice were pretreated with 4 µg/mL rGal1 for 2 has previously described [51], and then infected with Ye wt or Ye $\Delta yopP$ as outlined above. Then, supernatants were obtained and tested for NO and urea production.

2.10. Analysis of Yops-Gal1 Interactions by SDS-PAGE and Western Blot

Briefly, 25 µL of Yops were added to each well and resolved by 12% SDS-PAGE. Subsequently, bands were transferred onto a PVDF membrane (Bio-Rad, New York, NY, USA), which was then blocked with 3% bovine serum albumin (BSA) and incubated for 18 h at 4 °C in agitation with human rGal1 (6 µg/mL) or with rGal1 plus 30 mM of lactose as a disaccharide competitor. Finally, an in-house generated polyclonal anti-Gal1 antibody [52] was diluted 1:1000 and the reaction was revealed using chemiluminescence detection kit (Amersham Biosciences, London, UK). To evaluate the importance of glycans in Yops-Gal1 interaction, glycan oxidation was achieved by treatment with 10 mM $NaIO_4$ as previously described [53].

2.11. Lectin Blotting

Yops were run in 10% SDS-PAGE, and transferred onto 0.45-µm PVDF membranes (Bio-Rad). Membranes were then blocked with 3% bovine serum albumin (BSA) and strips were probed with the biotinylated lectins listed in Table 1, as previously described [54]. Lectin binding was visualized using horseradish rabbit peroxidase (HRP)-conjugated streptavidin (Sigma) with and C-DiGit® Blot Scanner (LI-COR Biosciences, Lincoln, NE, USA).

Table 1. Lectins used for characterization of carbohydrate structures present in Yops.

Lectin	Ligands Described	Reference
Arachis Hypogaea (Peanut agglutinin) (PNA)	Galβ(1–3)GalNAc Galβ(1–3)GlcNAc Galβ(1–4)GlcNAc Lactose Galactose	[55]
Erythrina crystagalli (ECA)	Galβ(1–4)GlcNAc Lactose > GalNAc > Gal	[56]

2.12. Yops Proteolysis Using Trypsin Digestion

Yops obtained from Ye wt were incubated with or without rGal1 overnight. Samples were subsequently digested with trypsin (200 µg/mL) (Sigma) following the protocol described by Shevchenko et al. (2006) [57]. Digestion products were subjected to separation in denaturing SDS-PAGE and Yops-rGal1 association was evaluated by Western blot using anti-Gal1 antibodies (1:1000). The reaction was revealed by chemiluminescence using the C-DiGit® Blot Scanner (LI-COR Biosciences, EEUU).

2.13. Schiff Staining

Briefly, Yops were added to each well and resolved by SDS-PAGE, and subsequently the gel was immersed in 12.5% trichloroacetic acid overnight and then placed in 1% periodic acid. Finally, the gel was incubated with Schiff's reagent in the dark for 1 h and washed with 0.5% of sodium metabisulphite three times for 10 min followed by distilled water [58].

2.14. Flow Cytometry

Ye (2×10^8 CFU) were fixed in 3% paraformaldehyde for 2 h at room temperature, washed three times in phosphate-buffered saline (PBS), and stored at -80 °C in PBS containing 15% of glycerol. To determine galectin binding, 2×10^7 fixed bacteria were incubated with label free rGal1 as described [59] at a final concentration of 3.3 mM (100 µg/mL) for 1 h at 37 °C. After two washes with PBS/Tween 0.1%, Gal1 binding was detected by incubation with a rabbit anti-human Gal1 antibody for 45 min at 4 °C. Cells were then washed twice in 0.1% PBS-Tween, next, resuspended in 50 µL of PBS with a polyclonal anti-rabbit FITC-conjugate antibody (1/200), and incubated for 30 min on ice. Galectin binding was determined using a flow cytometer (FACSCalibur), and at least 4×10^4 events were recorded. Gal1 binding was evaluated by calculating the Fluorescence Medium Index [% positive gated bacteria multiplied by the geometric mean fluorescence] [60].

2.15. Mass Spectrometry

Proteins were separated through 1D-SDS-PAGE on 10% of polyacrylamide gels. Next, gels were stained with 0.1% of Coomassie R-250. Selected bands were excised from gels and sent to the Center for Chemical and Biological Studies by Mass Spectrometry (CEQUI-BIEM), Faculty of Exact and Natural Sciences, University of Buenos Aires, where protein identification analysis was performed. Briefly, bands were de-stained with 50 mM of ammonium bicarbonate/acetonitrile (50/50% v/v), reduced with DTT, followed by alkylation with iodoacetamide. Trypsin sequencing grade was used for in-gel digestion (Promega, Madison, WI, USA). The tryptic digested peptides were resuspended in 0.1% formic acid in water, injected into Easy nLC 1000 (Thermo Scientific), and analyzed by tandem mass spectrometry using a QExactive Orbitrap mass spectrometer (Thermo Scientific) [61].

2.16. Bioinformatics Analysis

The Mascot Generic Format (mgf) files were extracted from a RAW files using Mascot Distiller program v2.6.2.0 (www.matrixscience.com, original search: 12 December 2019 and corroborated through a most recently search on 11 January 2021), and searched against Yop 20191212 in house customized database (accession WP_010891200.1 from RefSeq, NCBI, 1 sequence, 288 residues) using Mascot server 2.6.2 (www.matrixscience.com, similar date than Distiller) local license. MASCOT server v2.6.2 in MS/MS ion search mode was applied to conduct peptide matches (peptide masses and sequence tags) and protein searches using the database mentioned previously. The following parameters were established for search: carbamidomethyl (C) on cysteine was set as fixed, and variable modifications included asparagines and glutamine deamidation, and methionine oxidation, respectively. Only two missed cleavages were allowed. Monoisotopic masses were counted. The precursor peptide mass tolerance was set at 20 ppm. Fragment mass tolerance was 0.02 Da and the ion score or expected cutoff was set at 5. The MS/MS spectra were searched with MASCOT using a 95% confidence interval (CI%) threshold ($p < 0.05$), while minimum score of 14 was used for peptide identification. Furthermore, the error tolerance mode was set up at MASCOT search to corroborate potential peptides unidentified during the first search.

2.17. Western Blot Analysis of INOS Expression

WT or $Lgals1^{-/-}$ macrophages were infected with Ye wt or Ye $\Delta yopP$. Cell lysates (40 µg of protein/lane) were size fractionated in 12% SDS-polyacrylamide gel electrophoresis and transferred to a PVDF membrane. Membranes were incubated for 90 min in Tris buffered saline (TBS, pH 7.5)-3% milk and then overnight with a 1:200 rabbit antibody against iNOS (Santa Cruz Biotechnology Inc., Santa Cruz, CA, USA). Membranes were washed with TBS-0.05% Tween 20 and incubated with a 1:1000 horseradish peroxidase-conjugated goat anti-rabbit IgG (Sigma, CA, USA). Immunodetection was performed using chemiluminescence, following the protocol provided by the manufacturer. The immunoreactive protein bands were analyzed using the ImageJ software.

2.18. Statistical Analysis

The Mann-Whitney U test or one-way ANOVA with the Dunnett multiple-comparison test were used to determine if the differences between the groups were significant. Results are expressed as the mean ± SEM. All statistical analyses were carried out using Prism version 5.0 (GraphPad, La Jolla, CA, USA). p values < 0.05 were considered statistically significant.

3. Results

3.1. Galectin-1 Binds to YopP in Y. enterocolitica

Previous findings demonstrated the presence of Yops in the membrane fraction of Ye [4]. To explore whether Gal1 can bind to Ye surface proteins, we performed flow cytometry and ELISA using Gal1 as a probe, either with Ye wt or Ye ΔyopP, a genetically modified bacteria devoid of YopP (Figure 1).

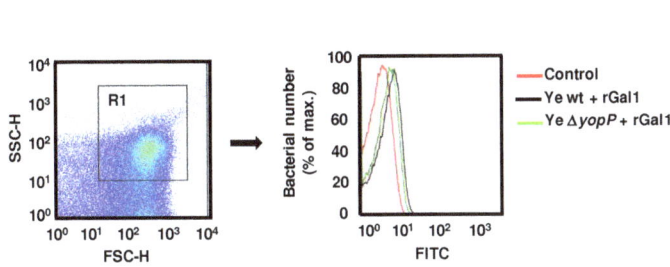

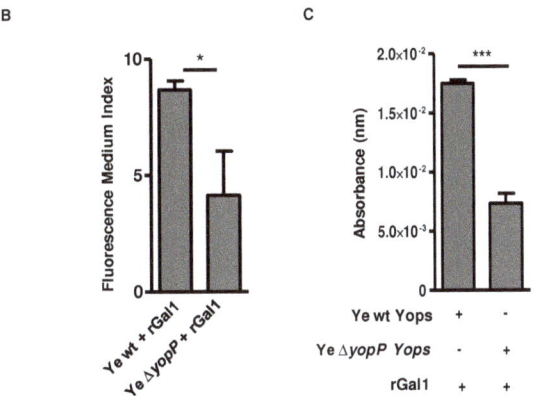

Figure 1. Galectin-1 Binds to Ye YopP. Binding of rGal1 to Ye wt (black) or to Ye ΔyopP (green) is shown. Control (Ye without rGal1) is shown in red. Binding was analyzed by flow cytometry using a FITC-conjugated anti-Gal1 antibody (**A**). Representative flow cytometry analysis of two independent experiments, showing the gate in Region 1 (R1) and the histogram expressing the number of FITC-positive bacteria. (**B**) Binding is expressed as the Fluorescence Medium Index. (**C**) ELISA plates were coated with Yops from Ye wt or Ye ΔyopP (10 µg/well) obtained under the same conditions and incubated with 10 mg/mL of rGal1. Gal1 binding was detected using an anti-Gal1 rabbit polyclonal antibody (1/1000). Data are the mean ± SEM of three independent experiments (**B**,**C**). * $p < 0.05$, *** $p < 0.001$.

We found a significant decrease in Gal1 binding to Ye ΔyopP compared to Ye wt (Figure 1A,B; $p < 0.05$, Figure 1C, $p < 0.001$). Given that secretion of YopP by Ye is significantly lower than secretion of other Yops [62], these results suggest that YopP could mediate Gal1 binding to Ye, although other mediators maybe also contribute to this effect.

3.2. Galectin-1 Recognizes Yops in a Carbohydrate-Dependent Manner

To evaluate potential Gal1 ligands in Ye and given the glycan-binding activity of this protein, we studied whether Gal1-Ye interactions are mediated by specific glycans.

Previous studies demonstrated that Ye spp. presents an alternative bacterial pathway, mediated by a cytoplasmic N-glycosyltransferase, a homolog of *Actinobacillus pleuropneumoniae* HMW1C-Like glycosyltransferase (ApHMW1CLGT). This enzyme uses nucleotide-activated monosaccharides as donors to modify asparagine residues, and transfer glucose and galactose with NX(S/T) as the acceptor sequon [63]. A secondary O-glycosylation activity was described for ApHMW1CLGT transferring a donor sugar to an acceptor sugar, forming di-hexoses on glycoproteins [64]. However, no data are available regarding the O-glycosylation pathway in this bacterium. We found, by means of a classical Schiff staining, that Yops are glycosylated (Figure 2A). In order to evaluate glycan-dependent binding of Gal1 to Yops, these glycoproteins were separated by SDS-PAGE and incubated with rGal1 in the absence or presence of lactose (30 mM) as a competitive carbohydrate inhibitor of Gal1 binding activity (Figure 2A). Gal1-glycan interactions were inhibited by lactose at the level of protein bands corresponding to 14, 25 and 35 kDa, suggesting carbohydrate-dependent binding of this lectin to secreted proteins of Ye. Thus, periodate treatment (which induces glycan oxidation) impaired Gal1 binding, indicating that Gal1-Yop interactions are mediated by specific glycosylated structures (Figure 2A).

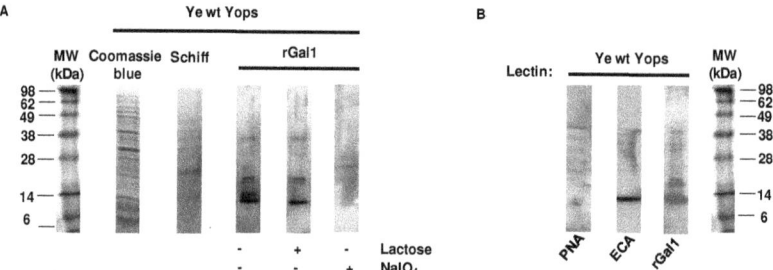

Figure 2. Virulence factors secreted by *Y. enterocolitica* are glycosylated and exhibit Gal1-permissive glycoepitopes. Electrophoresis was performed using an SDS-polyacrylamide gel (25 µL per well of Ye wt Yops; 20–40 µg/well). Subsequently, gels were (**A**) treated with rGal1 (6 µg/mL), rGal1 and lactose (30 mM) or NaIO$_4$ (10 mM) or (**B**) incubated with biotinylated *Peanut agglutinin* (PNA) or *Erythrina cristagalli* lectin (ECA), capable of recognizing disaccharides with lactose-derived structures. Detection was performed by Coomassie blue staining, Schiff staining (**A**) or revealed by chemiluminescence (**B**). Data are representative of two independent experiments.

Based on these findings, we next investigated the presence of Yops glyco-epitopes in electrophoretically-resolved protein bands, using biotinylated plant lectins able to recognize glycan structures permissive for Gal1 binding. Lectin blotting revealed the presence of 14–38 kDa bands that bound to *Erythrina cristagalli* (ECL), a lectin capable of recognizing non-sialylated N-acetyllactosamine (LacNAc, Galβ(1–4)GlcNAc) structures. Notably, *peanut agglutinin* (PNA) reactivity was also (albeit faintly) observed, suggesting that Ye glycoproteins may also display glycans with Galβ(1–3) terminal structures (Figure 2B). These results indicate that β-galactoside residues are exposed in Yops and may act as possible glycoepitopes for binding of host Gal1.

3.3. Mass Spectrometry-Based Proteomics Analysis of Yops

It has been demonstrated that YopP is a critical virulence factor involved in bacterial immune evasion [65] and Gal1 contributes to Ye-driven immunosuppression [32]. Separation of Yops proteins by 1D-SDS-PAGE and identification of bands using nanoLC-MS/MS analysis revealed selected bands corresponding to molecular weights ranging from 30 to 55 kDa (Figure 3A, red square) with nineteen identified peptides matching YopP, including the N- and C-terminus, and representing a 60% of sequence coverage of the protein (Figure 3B and Table 2).

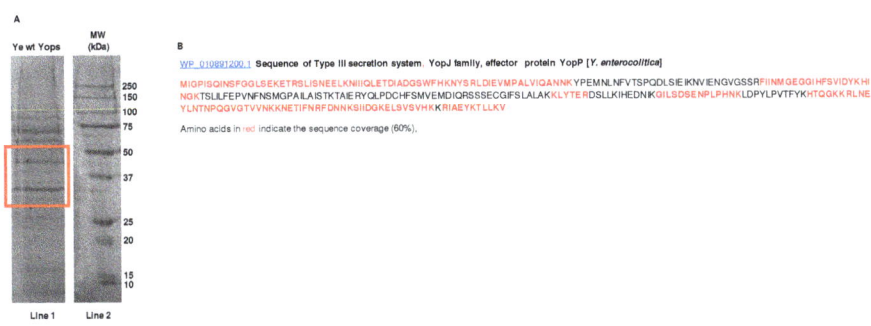

Figure 3. Protein bands derived from Yops selected for MS/MS analysis. Yops were solved in 1D-SDS-PAGE gels and stained with Coomassie Blue G-250 (**A**, Lane 1). Molecular weight markers (MW) are shown in (**A**), Lane 2. The red squares indicate bands subjected to identification through MS/MS analysis. YopP peptides identified (highlighted in red) in the selected bands and protein coverage map of Type III secretion system YopJ family effectors are shown in (**B**).

Table 2. YopP peptides identified through nLC-MS/MS analysis.

Observed [a]	Mass Expt [b]	Mass (Theor) [c]	Delta Error (Da) [d]	Pep_exp_z [e]	Start [f]	End [g]	Peptide Sequence [h]	Modifications [i]
707.816	2120.4261	2120.0208	0.4053	+3	232	250	LNEYLNTNPQG VGTVVNKK	Deamidated (NQ); Lys->CamCys (K)
1123.8687	3368.5843	3367.5	1.0843	+3	100	122	FIINMGEGGIHF SVIDYKHINGK	Deamidated (NQ); Lys->CamCys (K); Hex(1)HexNAc(1) (N); Hex(1)HexNAc(1) (ST); Oxidation (M)
1344.4664	2686.9182	2686.2353	0.6829	+2	48	67	NYSRLDIEVM PALVIQANNK	Deamidated (NQ); Hex(1)HexNAc(1) (N); Lys->CamCys (K)
954.9102	2861.7088	2861.3441	0.3647	+3	258	276	FDNNKSIIDG KELSVSVHK	Deamidated (NQ); 2 Hex(1)HexNAc(1) (ST)
470.2974	938.5802	937.5219	1.0583	+2	284	288	TLLKV	Hex(1)HexNAc(1) (ST)
604.0027	1205.9908	1205.5122	0.4787	+2	182	187	KLYTER	Hex(1)HexNAc(1) (ST); Lys->CamCys (K)
559.2594	1674.7563	1673.6991	1.0572	+3	250	262	KNETIFNR FDNNK	Deamidated (NQ); Lys->CamCys (K)
599.2961	1794.8664	1793.7499	1.1165	+3	21	29	SLISNEELK	Hex(1)HexNAc(1) (N); Hex(1)HexNAc(1) (ST); Lys->CamCys (K)
761.4185	1520.8224	1520.7471	0.0754	+2	200	213	GILSDSEN PLPHNK	Deamidated (NQ)
515.7982	1544.3728	1544.6201	−0.2473	+3	251	262	NETIFNR FDNNK	Deamidated (NQ); Lys->CamCys (K)
1092.9822	3275.9248	3275.4749	0.4499	+3	1	20	MIGPISQINSF GGLSEKETR	Deamidated (NQ); Hex(1)HexNAc(1) (N); 2 Hex(1)HexNAc(1) (ST); Oxidation (M)
525.9216	1574.7429	1574.7069	0.036	+3	263	276	SIIDGKEL SVSVHK	Lys->CamCys (K)
525.2371	2621.1492	2620.2921	0.8571	+5	30	51	NIIQLETDIAD GSWFHKNYSR	Deamidated (NQ)
890.7302	3558.8918	3558.7498	0.142	+4	1	29	MIGPISQINSF GGLSEKETRS LISNEELK	Deamidated (NQ); Hex(1)HexNAc(1) (N); Oxidation (M)
532.2823	1062.55	1062.5193	0.0307	+2	225	230	HTQGKK	Hex(1)HexNAc(1) (ST)
533.9495	1598.8266	1598.8767	−0.05	+1	278	288	RIAEYKTLLK	Hex(1)HexNAc(1) (ST)

Table 2. Cont.

Observed [a]	Mass Expt [b]	Mass (Theor) [c]	Delta Error (Da) [d]	Pep_exp_z [e]	Start [f]	End [g]	Peptide Sequence [h]	Modifications [i]
793.7171	2378.1295	2378.0999	0.0296	+2	2	17	IGPISQINS FGGLSEK	Deamidated (NQ); 2 Hex(1)HexNAc(1) (ST)
642.8429	1283.6713	1282.4918	1.1795	+3	225	230	HTQGKKR	Hex(1)HexNAc(1) (ST); 2 Lys->CamCys (K)
793.4021	2377.1845	2376.9625	0.2219	+3	258	268	FDNNK SIIDGK	Hex(1)HexNAc(1) (N); Hex(1)HexNAc(1) (ST); Lys->CamCys (K)

[a] The m/z (mass to charge state ratio) observed at the mass spectrometer, [b] The experimental mass of the peptide measured at the mass spectrometer, [c] The theoretical mass of the peptide obtained from the data base after in silico digestion, [d] The error of the peptide mass in Da calculated from theoretical mass minus experimental mass obtained from the mass spectrometry analysis, [e] Peptide charge state after nanoESI ionization, [f] Position of the first residue of the peptide identified in the whole protein sequence, [g] Position of the last residue of the peptide identified in the whole protein sequence, [h] Peptide sequence retrieved from the protein database after the bioinformatics analysis, [i] Potential modifications observed at the peptide identified.

3.4. Gal1 Protects Yops from Protease Degradation

It has been well established that certain members of the galectin family, such as galectin-4 (Gal4), protect the brush border enzymes in the small intestine of the action of proteinases and lipases through binding to these enzymes [66]. To investigate whether Ye can take advantage of Gal1-glycan interactions and protect Yops from degradation, Yops were incubated with rGal1 and then treated with trypsin (200 µg/mL), separated by SDS-PAGE, and incubated with polyclonal anti-Gal1 antibodies. The results demonstrate the binding of Gal1 to two particular protein bands running in 14 and 35 kDa (Figure 4), suggesting that this lectin might protect these glycoproteins from protease digestion. Since purified Yops could also contain other proteins. Future studies should be conducted to analyze their identity.

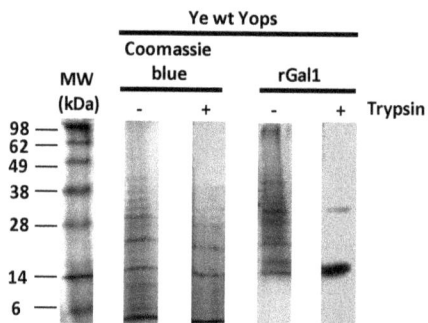

Figure 4. Gal1 protects virulence factors secreted by Ye from protease degradation. Electrophoresis was performed in an SDS-polyacrylamide gel (25 µL per well of Ye wt Yops; 20–40 µg/well). Subsequently, gels were treated with trypsin (200 µg/mL). Detection was performed by Coomassie blue staining or revealed by chemiluminescence. Data are representative of two independent experiments.

3.5. Gal1 and YopP Control Y. enterocolitica Infection by Decreasing NO Production

To further address the functional relevance of Gal1-YopP interactions during Ye infection, we first evaluated the impact this endogenous lectin in oxidative burst and inflammatory response. Given that superoxide (O_2^-) and tumor necrosis factor (TNF) contribute to innate responses of resident macrophages [67], we evaluated the O_2^- and TNF production by Ye-infected resident macrophages in the presence or absence of Gal1. We found that O_2^- and TNF were not significantly different in $Lgals1^{-/-}$ macrophages infected with Ye wt or Ye $\Delta yopP$ compared with macrophages isolated from WT infected mice (Figure 5A,B). Interestingly, in spite of the ability of YopP to inhibit TNF through MAPKs [14], we found no significant difference in TNF production by Ye ΔyopP-infected WT macrophages compared to WT macrophages infected with Ye wt (Figure 5A). In this sense, the production of IL-10 was evaluated, given its well-established role in attenuating TNF synthesis [30]. We studied IL-10 production byn WT and $Lgals1^{-/-}$ macrophages infected with Ye wt or Ye ΔyopP, and found no significant changes in its synthesis (Figure 5C) Then, we analyzed NO production by WT or $Lgals1^{-/-}$ peritoneal macrophages after in vitro Ye infection. Remarkably, Gal1 and YopP induced a substantial regulation of NO and urea production (Figure 5D,E; $p < 0.05$). However, no significant differences in apoptosis were detected between WT and $Lgals1^{-/-}$ macrophages infected in vitro with Ye ΔyopP or Ye wt (Figure 5F).

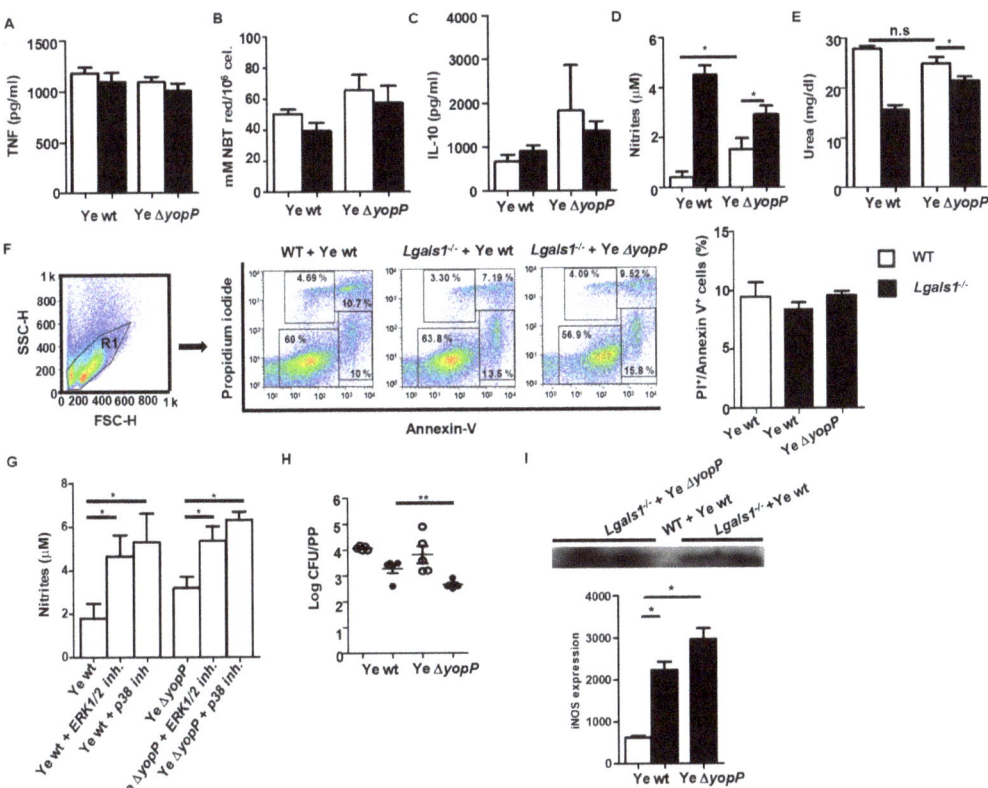

Figure 5. Lack of Gal1 and YopP induces NO Production and Confers Protection Against Ye. WT and $Lgals1^{-/-}$ macrophages were infected with Ye wt or Ye ΔyopP for 1 h. TNF and IL-10 were determined in culture supernatants by ELISA (**A,C**). Superoxide anion was determined as described in *Materials and methods* (**B**). NO production was measured in WT and $Lgals1^{-/-}$ macrophages after Ye Wt or Ye ΔyopP infection for 1 h (**D**). Urea was determined in culture supernatant as an indirect assessment of arginase activity (**E**). Macrophages were isolated from $Lgals1^{-/-}$ or WT mice, infected in vitro with Ye wt or Ye ΔyopP for 2 h, stained with annexin-V and propidium iodide and analyzed by flow cytometry. In the gated Region 1 (R1), the percentage of annexin-V$^+$ propidium iodide$^+$ cells are shown (right panel) (**F**). NO production in macrophages infected in vitro with Ye wt or Ye ΔyopP in the absence or presence of ERK1/2 or p38 inhibitors (**G**). CFU were evaluated in PPs of mice infected with Ye wt or with Ye ΔyopP after 5 days. Limit of detectable CFU was 25 ($\log_{10}25 = 1.4$) (**H**). After infection, macrophage lysates were analyzed by Western blot using specific antibodies against iNOS (**I**). Data are the mean ± SEM of three independent experiments (**A–F** right panel and **I** bottom panel), representative of three independent experiments (**F** left panel and **I** upper panel) or representative of two independent experiments (**H**, $n = 5$ mice per group). * $p < 0.05$, ** $p < 0.01$.

To investigate possible mechanisms underlying this immunomodulatory effect, we inhibited ERK1/2 or p38 signaling pathways and then infected macrophages with Ye wt or Ye $\Delta yopP$. Ye-driven suppression of NO synthesis was significantly prevented when both signaling pathways were interrupted (Figure 5G, $p < 0.05$). To evaluate whether the lack of Gal1 and/or YopP influences the clearance of Ye, we assessed bacterial load in PPs of WT or $Lgals1^{-/-}$ mice after 5 days of infection with Ye wt or with Ye $\Delta yopP$. Significantly lower numbers of CFU were detected in PPs of Ye $\Delta yopP$ infected $Lgals1^{-/-}$ mice (Figure 5H, $p < 0.01$).

To confirm the effect of Ye $\Delta yopP$ infection and Gal1 on NO, iNOS expression was evaluated by Western blot. We observed inhibition of iNOS expression when WT macrophages were infected with Ye wt. On the contrary, an increased expression of iNOS was detected when $Lgals1^{-/-}$ macrophages were infected with Ye wt and Ye $\Delta yopP$ (Figure 5I, $p < 0.05$).

3.6. Exogenous Supplementation of rGal1 Does Not Influence the Protective Anti-Y. enterocolitica Response Observed in the Absence of YopP

To evaluate the role of Gal1 and YopP in hindering anti-Y. enterocolitica immunity, we explored whether exogenous rGal1 could override the protective effect observed in $Lgals1^{-/-}$ hosts infected with Ye $\Delta yopP$. In this regard, we have previously demonstrated that administration of exogenous rGal1 in Ye wt-infected $Lgals1^{-/-}$ mice abolished protection compared with untreated control $Lgals1^{-/-}$ mice [32]. However, the administration of rGal1 to $Lgals1^{-/-}$ mice infected with Ye $\Delta yopP$ showed a similar CFU number in PPs (Figure 6A), and no significant differences were observed in both NO and urea production when compared with the control group (Figure 6B,C), suggesting that the exogenous lectin does not restore the phenotype generated by Gal1 and/or YopP deficiency.

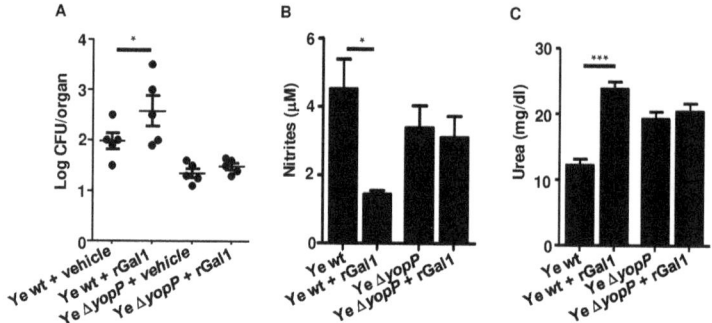

Figure 6. Exogenous Supplementation of rGal1 does not Revert the Protective Anti-Ye Response Observed in the Absence of YopP. $Lgals1^{-/-}$ mice were treated i.p. with rGal1 or vehicle control daily for 5 d, starting on the day of infection. Mice were euthanized at day 5, and CFU were determined in homogenates of PPs. Limit of detectable CFU was 25 ($\log_{10}25 = 1.4$) (**A**). $Lgals1^{-/-}$ macrophages were infected with Ye wt or Ye $\Delta yopP$ for 1 h in the presence or absence of rGal1. NO production was determined in supernatants using Griess assay (**B**). Urea was determined in culture supernatants as an indirect evaluation of arginase activity (**C**). Data are representative of two independent experiments ($n = 5$ mice per group) (**A**). Data are mean ± SEM of two independent experiments (**B,C**). * $p < 0.05$, *** $p < 0.001$.

4. Discussion

Ye are Gram-negative bacteria that invade the intestine and use the type III protein secretion machinery to deliver bacterial effector proteins to host cells [2,4]. Similar to other microbes, the mechanisms underlying infection and immune evasion processes may involve bacterial glycoproteins recognized by host lectins, [36,43]. Several innate and adaptive immune cells, including macrophages, dendritic cells (DCs), and activated B and T cells, are an important source of Gal1 secretion [52,68–70]. In turn, this endogenous lectin controls the magnitude and nature of immune responses through diverse mechanisms including modulation of M1-M2 macrophage polarization, DC immunogenicity, regulatory T (Treg) cell expansion, T helper cell differentiation, and apoptosis [51,70–76]. Interestingly, Gal1 and its glycosylated ligands could be potentially used by pathogens as a glyco-checkpoint to subvert innate and adaptive immune programs [77]. In this sense, bacterial proteins such as Chlamydial membrane proteins MOMP and OmcB showed a permissive glycosylation pattern for Gal1 binding [43], and Gal-1 expressed by human cervical epithelial cells binds to the virulence factor lipophosphoglycan of *Trichomonas vaginalis* in a carbohydrate-dependent manner [78]. Additionally, Nita-Lazar et al. showed that upon influenza infection, *Streptococcus pneumoniae* adhesion to the airway epithelial surface is enhanced via the coordinated action of host galectins and viral and pneumococcal neuraminidases [79]. In this study, we provide evidence that Yops-secreted proteins from Ye- may bind Gal1 through carbohydrate-dependent mechanisms. Even though data on Ye glycosylation are still scarce, our findings showed the presence of permissive glycoepitopes for Gal1 binding in Yops, and particularly the relevant role of YopP in Gal1 binding to Ye, as demonstrated in binding experiments with Ye $\Delta yopP$. Although alternative proteins, other than Yops, could be secreted from Ye [80], optimal culture conditions are offered for Yops secretion, among them, the addition of EGTA (5 mM) for Ca^{2+} chelation, $MgCl_2$ (15 mM), and glucose (0.2%) [46]. Under these conditions, Yops represent a major component of the Ye secretome [81]. In this sense, a mass spectrometry-based identification of YopP showed several potentially glycosylated peptides; however, the poor number of b and y ion series during fragmentation, as well as low signal-to-noise ratio, hampered their full characterization. Further structural studies using a pre-enrichment technique and other strategies to improve the detection and analysis of glycopeptides would be relevant for a more complete understanding of Ye glycosylation pathways. Additionally, immunoprecipitation would also be useful to specifically verify these interactions.

Interestingly, we previously demonstrated that YopP up-regulates Gal1 expression in mouse splenocytes [32]. Here we found that Gal1 binds to the Ye surface and that lack of the critical effector protein YopP disrupts this association, highlighting specific interactions between YopP and Gal1. Although the functional role of Gal1-YopP interactions is unknown, previous studies showed a scaffold role for this lectin in other cell systems [82,83]. In this sense, we found that Gal1 prevents trypsin degradation of Yops. This finding is consistent with previous results demonstrating the biological relevance of this lectin in resistance to trypsin [84] and elastase [65] digestion. In this regard, using the software PeptideCutter (ExPASy), we identified several cleavage sites for trypsin on YopP and other Yops sequence (data not shown). Thus, Gal1 binding to Yops could represent an evolutionarily conserved mechanism to render bacterial virulence factors resistant to proteases implicated in infection.

It has been well established that macrophages confer early protection during the course of Ye infection [85]. NO synthesized by inducible iNOS is a major effector pathway of inflammatory macrophages; this inflammatory mediator plays essential roles in anti-microbial responses and host defense. Arginase catalyzes the alternative arginine metabolic pathway, which converts arginine to ornithine and urea [86]. Gal1 regulates L-arginine metabolism in peritoneal macrophages and microglia in this fashion by shifting the balance from classically-activated M1-type toward alternatively-activated M2-type macrophages and microglia [51,76]. On the other hand, previous studies suggested that Yops corresponding to pathogenic *Yersinia* spp. inhibit LPS-mediated production of NO by macrophages [30]. Likewise, in the present study we found that YopP inhibited NO production and increased urea levels in a coordinate fashion with Gal1. In agreement with our findings, Silva Monnazzi et al. and Tansini et al. demonstrated that NO production in murine macrophages is suppressed by *Y. pseudotuberculosis* and that the YopP counterpart, YopJ, could be, at least in part, responsible of such effect [87,88]. In addition, we observed increased NO production by WT macrophages after inhibition of p38 and ERK1/2 signaling pathways and subsequent infection with Ye wt or Ye $\Delta yopP$. These results are in agreement with our previous results showing that Gal1 production is regulated, at least in part, through p38 and ERK1/2 signaling pathways [32]. Although high levels of nitrogen species can damage basic cellular components and trigger cell death in macrophages [89], no significant differences in apoptosis were detected in $Lgals1^{-/-}$ peritoneal macrophages infected with Ye $\Delta yopP$ compared with the WT counterparts. In this regard, NO is a multifaceted molecule with dichotomous regulatory functions. Whereas it promotes apoptosis in several cell types, it prevents execution of cell death programs in other settings through specific inhibition of caspases [90].

Interestingly, Boland et al. (1998) showed that YopP inhibits TNF release by infected macrophages. Moreover, Giordano et al. (2011) [91] reported increased expression of pro-inflammatory cytokines, including TNF, in iNOS deficient innate immune cells. These results indicate that NO can inhibit the production of pro-inflammatory cytokines, which are usually produced by M1 macrophages [92] These data are in agreement with the unaltered TNF levels in absence of YopP, which could be due to the increased amounts of NO production.

A remarkable feature of macrophages is their plasticity. The classically-activated proinflammatory M1-type macrophages constitute one end of the spectrum, while alternatively activated anti-inflammatory M2 macrophages are on the other. Pro-inflammatory cytokines and mediators such as TNF and ROS are synthesized by M1 macrophages while anti-inflammatory factors such as IL-10, TGF-β, and arginase are considerably expressed in M2 macrophages [93]. To determine if a pro- or anti-inflammatory condition prevails in the presence of Gal1, we evaluate iNOS expression and IL-10 production in macrophages. We observed a key role of Gal1 in the negative modulation of iNOS expression, this finding is in agreement with those obtained by Starossom et al. (2012) who demonstrated that expression of iNOS mRNA was significantly decreased by Gal1 in M1 mice microglia [76]. In future studies, it would be useful to evaluated the polarization and inflammatory state of macrophages by monitoring their gene expression profile. On the other hand, it is well-known that IL-10 inhibits macrophage function and controls inflammation [92]. Moreover, several results showed that apoptotic macrophages trigger production of IL-10 [20,22]. We demonstrated that rGal1 supplementation restored NF-kB activation. TNF synthesis, and IL-6 production in PPs from $Lgals\text{-}1^{-/-}$ mice to levels comparable to those attained in WT hosts [32]. Here, we observed that administration of exogenous rGal1 to $Lgals\text{-}1^{-/-}$ macrophages infected with Ye $\Delta yopP$ was not sufficient to restore decreased NO production and increased urea levels. Moreover, we observed that administration of exogenous Gal1 did not thwart the antibacterial protective effect unleashed in the absence of endogenous Gal1 and YopP. These results suggest that Gal1 and the bacterial virulence factor, YopP, might be crucial to regulate Ye pathogenesis using a coordinated mechanism, as has been reported for YopJ and IKKβ, MKK1, MKK2, MKK3, MKK4, MKK5, MKK6 [16,94] and IP6 [19]. Thus, in response to Ye infection, Gal1 and the virulence factor YopP may limit anti-bacterial responses. Conversely, deficiency in YopP or Gal1 controls the clearance of Ye and increases NO production. Additionally, our results suggest that ERK1/2 and p38 pathways mediate inhibition of NO production driven by Ye through mechanisms that could potentially involve regulation of Gal1 expression [95].

Thus, host derived Gal1 and glycosylated ligands may contribute to Ye infection by associating with YopP. In this regard, our studies identify glycosylation-dependent interactions between endogenous Gal1 and Yops that may play an important role during Ye infection through modulation of NO production. These findings may have critical implications in the design of tailored therapies aimed at controlling anti-bacterial responses during Ye infection. However, in spite of considerable progress, the clinical implications of our findings as well as the molecular mechanisms underlying YopP-Gal-1 interactions remain to be further investigated.

Author Contributions: Conceptualization, G.A.R., M.S.D.G. and R.C.D.; methodology, J.M.P.S. and K.V.M.; validation, B.L.J. and R.C.D.; formal analysis, B.L.J., R.J.E., J.E.S., M.D.P., E.C., K.V.M., M.S.D.G. and R.C.D.; investigation, B.L.J., R.J.E., M.S.D.G. and R.C.D.; resources, J.M.P.S., M.D.P.,

E.C., G.A.R., K.V.M. and M.S.D.G.; writing—original draft preparation, K.V.M., M.S.D.G. and R.C.D.; writing—review and editing, B.L.J., R.J.E., J.E.S., M.D.P., E.C., G.A.R., K.V.M., M.S.D.G. and R.C.D.; visualization, B.L.J., R.J.E., J.E.S., E.C., G.A.R., M.S.D.G. and R.C.D.; supervision, G.A.R. and M.S.D.G.; funding acquisition, G.A.R., M.S.D.G. and R.C.D. All authors have read and agreed to the published version of the manuscript.

Funding: This research was funded by grants from Consejo Nacional de Investigaciones Científicas y Técnicas (CONICET; PIP 00815), Universidad Nacional de San Luis, PROICO 2–1218, Fundación Sales and Fundación Bunge y Born.

Institutional Review Board Statement: The study was conducted according to the guidelines of the Declaration of Helsinki and approved by the The Animal Care and Use Committee of the National University of San Luis, Argentina (Protocol Number: B226/16).

Informed Consent Statement: Not applicable.

Acknowledgments: We are grateful to Ingo Autenrieth for providing Ye wt and Ye yopP strains. We also thank Gabinete de Asesoramiento en Escritura Científica en Inglés (GAECI) staff for writing assistance and Messrs. Ryan Johnson and Bill Conn from USD-IT Research Computing for their help in server operation and maintenance.

Conflicts of Interest: The authors declare no conflict of interest.

References

1. Wren, B.W. The *Yersiniae*—A model Genus to Study the Rapid Evolution of Bacterial Pathogens. *Nat. Rev. Microbiol.* **2003**, *1*, 55–64. [CrossRef] [PubMed]
2. Autenrieth, I.B.; Firsching, R. Penetration of M Cells and Destruction of Peyer's Patches by *Yersinia Enterocolitica*: An ultrastructural and Histological Study. *J. Med. Microbiol.* **1996**, *44*, 285–294. [CrossRef] [PubMed]
3. Bottone, E.J. Yersinia Enterocolitica: Overview and Epidemiologic Correlates. *Microbes Infect.* **1999**, *1*, 323–333. [CrossRef]
4. Cornelis, G.R.; Boland, A.; Boyd, A.P.; Geuijen, C.; Iriarte, M.; Neyt, C.; Sory, M.-P.; Stainier, I. The Virulence Plasmid of Yersinia, an Antihost Genome. *Microbiol. Mol. Biol. Rev.* **1998**, *62*, 1315–1352. [CrossRef] [PubMed]
5. Alonso, A.; Bottini, N.; Bruckner, S.; Rahmouni, S.; Williams, S.; Schoenberger, S.P.; Mustelin, T. Lck Dephosphorylation AT Tyr-394 and Inhibition of T Cell Antigen Receptor Signaling by Yersinia phosphatase yopH. *J. Biol. Chem.* **2004**, *279*, 4922–4928. [CrossRef] [PubMed]
6. Grosdent, N.; Maridonneau-Parini, I.; Sory, M.P.; Cornelis, G.R. Role of Yops and adhesins in resistance of *Yersinia enterocolitica* to phagocytosis. *Infect. Immunity* **2002**, *70*, 4165–4176. [CrossRef]
7. Viboud, G.I.; Bliska, J.B. *Yersinia* outer Proteins: Role in Modulation of Host Cell Signaling Responses and Pathogenesis. *Annu. Rev. Microbiol.* **2005**, *59*, 69–89. [CrossRef] [PubMed]
8. Grabowski, B.; Schmidt, M.A.; Rüter, C. Immunomodulatory *Yersinia* Outer Proteins (Yops)—Useful Tools for Bacteria and Humans Alike. *Virulence* **2017**, *8*, 1124–1147. [CrossRef]
9. Ruckdeschel, K.; Mannel, O.; Richter, K.; Jacobi, C.A.; Trülzsch, K.; Rouot, B.; Heesemann, J. Yersinia Outer Protein P of *Yersinia Enterocolitica* Simultaneously Blocks the nuclear factor-κb pathway and exploits lipopolysaccharide Signaling to Trigger Apoptosis in Macrophages. *J. Immunol.* **2001**, *166*, 1823–1831. [CrossRef] [PubMed]
10. Erfurth, S.E.; Gröbner, S.; Kramer, U.; Gunst, D.S.; Soldanova, I.; Schaller, M.; Autenrieth, I.B.; Borgmann, S. *Yersinia enterocolitica* Induces Apoptosis and Inhibits Surface Molecule Expression and Cytokine Production in Murine Dendritic Cells. *Infect. Immun.* **2004**, *72*, 7045–7054. [CrossRef] [PubMed]
11. Thiefes, A.; Wolf, A.; Doerrie, A.; Grassl, G.A.; Matsumoto, K.; Autenrieth, I.; Bohn, E.; Sakurai, H.; Niedenthal, R.; Resch, K.; et al. The Yersinia Enterocolitica Effector YopP Inhibits Host Cell Signalling by Inactivating the Protein Kinase TAK1 in the IL-1 Signalling Pathway. *EMBO Rep.* **2006**, *7*, 838–844. [CrossRef]
12. Mittal, R.; Peak-Chew, S.-Y.; McMahon, H.T. Acetylation of MEK2 and I B kinase (IKK)Activation Loop Residues by YopJ Inhibits Signaling. *Proc. Natl. Acad. Sci. USA* **2006**, *103*, 18574–18579. [CrossRef]
13. Autenrieth, S.E.; Soldanova, I.; Rösemann, R.; Gunst, D.; Zahir, N.; Kracht, M.; Ruckdeschel, K.; Wagner, H.; Borgmann, S.; Autenrieth, I.B. *Yersinia Enterocolitica* Yopp Inhibits Map Kinase-Mediated Antigen Uptake in Dendritic Cells. *Cell. Microbiol.* **2007**, *9*, 425–437. [CrossRef]
14. Boland, A.; Cornelis, G.R. Role of Yopp in Suppression of Tumor Necrosis Factor Alpha Release by Macrophages during *Yersinia* Infection. *Infect. Immun.* **1998**, *66*, 1878–1884. [CrossRef] [PubMed]
15. Ruckdeschel, K.; Harb, S.; Roggenkamp, A.; Hornef, M.; Zumbihl, R.; Köhler, S.; Heesemann, J.; Rouot, B. Yersinia Enterocolitica Impairs Activation of Transcription Factor Nf-Kb: Involvement in the Induction of Programmed Cell Death and in the Suppression of the Macrophage Tumor Necrosis Factor α Production. *J. Exp. Med.* **1998**, *187*, 1069–1079. [CrossRef] [PubMed]
16. Orth, K. Inhibition of the Mitogen-Activated Protein Kinase Kinase Superfamily by a *Yersinia* Effector. *Science* **1999**, *285*, 1920–1923. [CrossRef]

17. Ma, K.W.; Ma, W. YopJ Family Effectors Promote Bacterial Infection through a Unique Acetyltransferase Activity. *Microbiol. Mol. Biol. Rev.* **2016**, *80*, 1011–1027. [CrossRef] [PubMed]
18. Shaw, A.S.; Filbert, E.L. Scaffold proteins and immune-cell signaling. *Nat. Rev. Immunol.* **2009**, *9*, 47–56. [CrossRef] [PubMed]
19. Mittal, R.; Peak-Chew, S.Y.; Sade, R.S.; Vallis, Y.; McMahon, H.T. The Acetyltransferase Activity of the Bacterial Toxin YopJ of *Yersinia* Is Activated by Eukaryotic Host Cell Inositol Hexakisphosphate. *J. Biol. Chem.* **2010**, *285*, 19927–19934. [CrossRef]
20. Fink, S.L.; Cookson, B.T. Apoptosis, Pyroptosis, and Necrosis: Mechanistic Description of Dead and Dying Eukaryotic Cells. *Infect. Hnmun.* **2005**, *73*, 1907–1916. [CrossRef]
21. Bergsbaken, T.; Cookson, B.T. Innate immune response during Yersinia infection: Critical modulation of cell death mechanisms through phagocyte activation. *Leukoc. Biol.* **2009**, *86*, 1153–1158. [CrossRef] [PubMed]
22. Philip, N.H.; Brodsky, I.E. Cell death programs in *Yersinia* immunity and pathogenesis. *Front. Cell. Infect. Microbiol.* **2012**, *2*, 149. [CrossRef]
23. Monack, D.M.; Mecsas, J.; Bouley, D.; Falkow, S. *Yersinia*-induced apoptosis in vivo aids in the establishment of a systemic infection of mice. *J. Expr. Med.* **1998**, *188*, 2127–2137. [CrossRef] [PubMed]
24. Bergsbaken, T.; Cookson, B.T. Macrophage Activation Redirects *Yersinia*-Infected Host Cell Death from Apoptosis to Caspase-1-Dependent Pyroptosis. *PLoS Pathog.* **2007**, *3*, e161. [CrossRef]
25. Hanski, C.; Naumann, M.; Grützkau, A.; Pluschke, G.; Friedrich, B.; Hahn, H.; Riecken, E.O. Humoral and Cellular Defense against Intestinal Murine Infection with Yersinia Enterocolitica. *Infect. Immun.* **1991**, *59*, 1106–1111. [CrossRef]
26. Autenrieth, I.B.; Hantschmann, P.; Heymer, B.; Heesemann, J. Immunohistological Characterization of the Cellular Immune Response against *Yersinia Enterocolitica* in Mice: Evidence for the Involvement of t Lymphocytes. *Immunobiology* **1993**, *187*, 1–16. [CrossRef]
27. Olson, R.M.; Dhariwala, M.O.; Mitchell, W.J.; Anderson, D.M. *Yersinia Pestis* Exploits Early Activation of myd88 for Growth in the Lungs during Pneumonic Plague. *Infect. Immun.* **2019**, *87*. [CrossRef] [PubMed]
28. Mantovani, A.; Sica, A.; Sozzani, S.; Allavena, P.; Vecchi, A.; Locati, M. The Chemokine System in Diverse Forms of Macrophage Activation and Polarization. *Trends Immunol.* **2004**, *25*, 677–686. [CrossRef] [PubMed]
29. Tufano, M.A.; Rossano, F.; Catalanotti, P.; Liguori, G.; Marinelli, A.; Baroni, A.; Marinelli, P. Properties of *Yersinia Enterocolitica* Porins: Interference with Biological Functions of Phagocytes, Nitric Oxide Production and Selective Cytokine Release. *Res. Microbiol.* **1994**, *145*, 297–307. [CrossRef]
30. Carlos, I.Z.; Silva Monnazzi, L.G.; Falcão, D.P.; de Medeiros, B.M.M. TNF-α, H_2O_2 and No Response of Peritoneal Macrophages to *Yersinia Enterocolitica* O:3 Derivatives. *Microbes Infect.* **2004**, *6*, 207–212. [CrossRef] [PubMed]
31. Eliçabe, R.J.; Arias, J.L.; Rabinovich, G.A.; Di Genaro, M.S. TNFRp55 Modulates Il-6 and Nitric oxide Responses Following *Yersinia* Lipopolysaccharide Stimulation in Peritoneal Macrophages. *Immunobiology* **2011**, *216*, 1322–1330. [CrossRef]
32. Davicino, R.C.; Méndez-Huergo, S.P.; Eliçabe, R.J.; Stupirski, J.C.; Autenrieth, I.; Di Genaro, M.S.; Rabinovich, G.A. Galectin-1–Driven Tolerogenic Programs Aggravate *Yersinia enterocolitica* infection by Repressing antibacterial immunity. *J. Immunol.* **2017**, *199*, 1382–1392. [CrossRef]
33. Baum, L.G.; Garner, O.B.; Schaefer, K.; Lee, B. Microbe-Host Interactions Are Positively and Negatively Regulated BY Galectin-Glycan Interactions. *Front. Immunol.* **2014**, *5*. [CrossRef] [PubMed]
34. Cerliani, J.P.; Blidner, A.G.; Toscano, M.A.; Croci, D.O.; Rabinovich, G.A. Translating the 'Sugar Code' into Immune and Vascular Signaling Programs. *Trends Biochem. Sci.* **2017**, *42*, 255–273. [CrossRef] [PubMed]
35. Raman, R.; Tharakaraman, K.; Sasisekharan, V.; Sasisekharan, R. Glycan–Protein Interactions in viral pathogenesis. *Curr. Opin. Struct. Biol.* **2016**, *40*, 153–162. [CrossRef]
36. Davicino, R.C.; Eliçabe, R.J.; Di Genaro, M.S.; Rabinovich, G.A. Coupling Pathogen Recognition to Innate Immunity through Glycan-Dependent Mechanisms. *Int. Immunopharmacol.* **2011**, *11*, 1457–1463. [CrossRef]
37. Rabinovich, G.A.; Gruppi, A. Galectins as immunoregulators during infectious Processes: From Microbial Invasion to the Resolution of the Disease. *Parasite Immunol.* **2005**, *27*, 103–114. [CrossRef]
38. Vasta, G.R. Roles of Galectins in Infection. *Nat. Rev. Microbiol.* **2009**, *7*, 424–438. [CrossRef] [PubMed]
39. Liu, F.-T.; Rabinovich, G.A. Galectins: Regulators of Acute and Chronic Inflammation. *Ann. N. Y. Acad. Sci.* **2010**, *1183*, 158–182. [CrossRef]
40. Salatino, M.; Rabinovich, G.A. Fine-Tuning Antitumor Responses through the Control of Galectin–Glycan Interactions: An Overview. *Methods Mol. Biol.* **2010**, 355–374. [CrossRef]
41. Chen, H.-Y.; Weng, I.-C.; Hong, M.-H.; Liu, F.-T. Galectins as Bacterial Sensors in the HOST Innate Response. *Curr. Opin. Microbiol.* **2014**, *17*, 75–81. [CrossRef] [PubMed]
42. Méndez-Huergo, S.P.; Blidner, A.G.; Rabinovich, G.A. Galectins: Emerging Regulatory Checkpoints Linking tumor Immunity and angiogenesis. *Curr. Opin. Immunol.* **2017**, *45*, 8–15. [CrossRef] [PubMed]
43. Lujan, A.L.; Croci, D.O.; Tudela, J.A.G.; Losinno, A.D.; Cagnoni, A.J.; Mariño, K.V.; Damiani, M.T.; Rabinovich, G.A. Glycosylation-Dependent Galectin–Receptor Interactions Promotechlamydia Trachomatisinfection. *Proc. Natl. Acad. Sci. USA* **2018**, *115*. [CrossRef] [PubMed]
44. Adkins, I.; Köberle, M.; Gröbner, S.; Bohn, E.; Autenrieth, I.B.; Borgmann, S. *Yersinia* Outer Proteins E, H, P, and T Differentially Target the Cytoskeleton and Inhibit Phagocytic Capacity of Dendritic Cells. *Int. J. Med. Microbiol.* **2007**, *297*, 235–244. [CrossRef] [PubMed]

45. Eliçabe, R.J.; Cargnelutti, E.; Serer, M.I.; Stege, P.W.; Valdez, S.R.; Toscano, M.A.; Rabinovich, G.A.; Di Genaro, M.S. Lack of Tnfr P55 Results in Heightened Expression of Ifn-γ and IL-17 during the Development of Reactive Arthritis. *J. Immunol.* **2010**, *185*, 4485–4495. [CrossRef] [PubMed]
46. Trülzsch, K.; Sporleder, T.; Igwe, E.I.; Rüssmann, H.; Heesemann, J. Contribution of the Major Secreted Yops of Yersinia enterocolitica O:8 to Pathogenicity in the mouse Infection Model. *Infect. Immun.* **2004**, *72*, 5227–5234. [CrossRef] [PubMed]
47. Autenrieth, I.B.; Beer, M.; Bohn, E.; Kaufmann, S.H.; Heesemann, J. Immune Responses to *Yersinia enterocolitica* in Susceptible BALB/c and Resistant C57BL/6 Mice: An Essential Role for gamma interferon. *Infect. Immun.* **1994**, *62*, 2590–2599. [CrossRef] [PubMed]
48. Sáez, J.M.P.; Hockl, P.F.; Cagnoni, A.J.; Huergo, S.P.M.; García, P.A.; Gatto, S.G.; Cerliani, J.P.; Croci, D.O.; Rabinovich, G.A. Characterization of a neutralizing anti-human galectin-1 monoclonal antibody with angioregulatory and immunomodulatory activities. *Angiogenesis* **2021**, *24*, 1–5. [CrossRef] [PubMed]
49. Chatzipanagiotou, S.; Legakis, J.N.; Boufidou, F.; Petroyianni, V.; Nicolaou, C. Prevalence of Yersinia Plasmid-Encoded Outer protein (Yop) Class-Specific Antibodies in Patients with Hashimoto's Thyroiditis. *Clin. Microbiol. Infect.* **2001**, *7*, 138–143. [CrossRef]
50. Schopf, R.E.; Mattar, J.; Meyenburg, W.; Scheiner, O.; Hammann, K.P.; Lemmel, E.-M. Measurement of the Respiratory Burst in Human Monocytes and Polymorphonuclear Leukocytes by Nitro Blue Tetrazolium Reduction and Chemiluminescence. *J. Immunol. Methods* **1984**, *67*, 109–117. [CrossRef]
51. Correa, S.G. Opposite Effects of Galectin-1 on Alternative Metabolic Pathways of l-Arginine in Resident, Inflammatory, and Activated Macrophages. *Glycobiology* **2002**, *13*, 119–128. [CrossRef] [PubMed]
52. Rabinovich, G.A.; Iglesias, M.M.; Modesti, N.M.; Castagna, L.F.; Wolfenstein-Todel, C.; Riera, C.M.; Sotomayor, C.E. Activated rat macrophages produce a galectin-1-like protein that induces apoptosis of T cells: Biochemical and functional characterization. *J. Immunol.* **1998**, *160*, 4831–4840.
53. Reinhold, B.B.; Hauer, C.R.; Plummer, T.H.; Reinhold, V.N. Detailed Structural Analysis of a novel, Specific O-Linked glycan from the prokaryote Flavobacterium Meningosepticum. *J. Biol. Chem.* **1995**, *270*, 13197–13203. [CrossRef]
54. Croci, D.O.; Cerliani, J.P.; Dalotto-Moreno, T.; Méndez-Huergo, S.P.; Mascanfroni, I.D.; Dergan-Dylon, S.; Toscano, M.A.; Caramelo, J.J.; García-Vallejo, J.J.; Ouyang, J.; et al. Glycosylation-Dependent Lectin-Receptor Interactions preserve Angiogenesis IN Anti-VEGF Refractory Tumors. *Cell* **2014**, *156*, 744–758. [CrossRef] [PubMed]
55. Sharma, V.; Srinivas, V.R.; Adhikari, P.; Vijayan, M.; Surolia, A. Molecular Basis of Recognition by Gal/Galnac Specific Legume Lectins: Influence of Glu 129 on the Specificity of *Peanut agglutinin* (pna) towards c2-substituents of Galactose. *Glycobiology* **1998**, *8*, 1007–1012. [CrossRef] [PubMed]
56. Debray, H.; Montreuil, J.; Lis, H.; Sharon, N. Affinity of Four Immobilized *Erythrina* lectins toward Various N-linked Glycopeptides and Related Oligosaccharides. *Carbohydr. Res.* **1986**, *151*, 359–370. [CrossRef]
57. Shevchenko, A.; Tomas, H.; Havli, J.; Olsen, J.V.; Mann, M. In-Gel Digestion for Mass Spectrometric Characterization of Proteins and Proteomes. *Nat. Protoc.* **2006**, *1*, 2856–2860. [CrossRef] [PubMed]
58. Nakamura, H.; Kiyoshi, M.; Anraku, M.; Hashii, N.; Oda-Ueda, N.; Ueda, T.; Ohkuri, T. Glycosylation Decreases Aggregation and Immunogenicity of Adalimumab Fab Secreted from *Pichia Pastoris*. *J. Biochem.* **2020**, *169*, 435–443. [CrossRef] [PubMed]
59. Andre, S.; Pei, Z.; Siebert, H.; Ramstrom, O.; Gabius, H. Glycosyldisulfides from Dynamic Combinatorial Libraries as o-Glycoside Mimetics for Plant and Endogenous Lectins: Their Reactivities in Solid-Phase and Cell Assays and Conformational Analysis by Molecular Dynamics Simulations. *Bioorganic Med. Chem.* **2006**, *14*, 6314–6326. [CrossRef] [PubMed]
60. Findlow, J.; Taylor, S.; Aase, A.; Horton, R.; Heyderman, R.; Southern, J.; Andrews, N.; Barchha, R.; Harrison, E.; Lowe, A.; et al. Comparison and Correlation of Neisseria meningitidis serogroup b immunologic assay results and human antibody responses following three Doses of the Norwegian Meningococcal Outer Membrane Vesicle Vaccine MenBvac. *Infect. Immun.* **2006**, *74*, 4557–4565. [CrossRef]
61. Pizarro-Guajardo, M.; Ravanal, M.C.; Paez, M.D.; Callegari, E.; Paredes-Sabja, D. Identification of *Clostridium Difficile* Immunoreactive Spore Proteins of the Epidemic Strain r20291. *Proteom. Clin. Appl.* **2018**, *12*, 1700182. [CrossRef] [PubMed]
62. Wilharm, G.; Lehmann, V.; Krauss, K.; Lehnert, B.; Richter, S.; Ruckdeschel, K.; Heesemann, J.; Trülzsch, K. *Yersinia Enterocolitica* Type III Secretion Depends on the Proton Motive Force but Not on the flagellar motor Components MotA and MotB. *Infect. Immun.* **2004**, *72*, 4004–4009. [CrossRef] [PubMed]
63. Schwarz, F.; Fan, Y.-Y.; Schubert, M.; Aebi, M. Cytoplasmic n-Glycosyltransferase of *Actinobacillus pleuropneumoniae* Is an Inverting Enzyme and Recognizes the NX(S/T) Consensus Sequence. *J. Biol. Chem.* **2011**, *286*, 35267–35274. [CrossRef] [PubMed]
64. Choi, K.-J.; Grass, S.; Paek, S.; St. Geme, J.W.; Yeo, H.-J. The *Actinobacillus Pleuropneumoniae* HMW1C-Like Glycosyltransferase Mediates N-linked Glycosylation of the *Haemophilus Influenzae* Hmw1 Adhesin. *PLoS ONE* **2010**, *5*. [CrossRef] [PubMed]
65. Sing, A.; Roggenkamp, A.; Geiger, A.M.; Heesemann, J. *Yersinia enterocolitica* evasion of the Host Innate Immune Response by v Antigen-Induced Il-10 Production of Macrophages Is Abrogated in Il-10-Deficient Mice. *J. Immunol.* **2002**, *168*, 1315–1321. [CrossRef] [PubMed]
66. Cao, Z.Q.; Guo, X.L. The role of galectin-4 in physiology and diseases. *Protein Cell* **2016**, *7*, 314–324. [CrossRef] [PubMed]
67. Keeney, J.T.R.; Miriyala, S.; Noel, T.; Moscow, J.A.; St. Clair, D.K.; Butterfield, D.A. Superoxide Induces Protein Oxidation in Plasma and Tnf-α Elevation in Macrophage Culture: Insights into Mechanisms of Neurotoxicity Following Doxorubicin Chemotherapy. *Cancer Lett.* **2015**, *367*, 157–161. [CrossRef] [PubMed]

68. Toscano, M.A.; Campagna, L.; Molinero, L.L.; Cerliani, J.P.; Croci, D.O.; Ilarregui, J.M.; Fuertes, M.B.; Nojek, I.M.; Fededa, J.P.; Zwirner, N.W.; et al. Nuclear Factor (NF)-KB controls expression of the immunoregulatory glycan-binding protein Galectin-1. *Mol. Immunol.* **2011**, *48*, 1940–1949. [CrossRef]
69. Zuñiga, E.; Rabinovich, G.A.; Iglesias, M.M.; Gruppi, A. Regulated expression of galectin-1 during B-cell activation and implications for T-cell apoptosis. *J. Leukoc. Biol.* **2001**, *70*, 73–79.
70. Ilaregui, J.M.; Croci, D.O.; Bianco, G.A.; Toscano, M.A.; Salatino, M.; Vermeulen, M.E.; Geffner, J.R.; Rabinovich, G.A. Tolerogenic Signals Delivered by Dendritic Cells to T Cells through a Galectin-1-driven immunoregulatory Circuit Involving Interleukin 27 and Interleukin 10. *Nat. Immunol.* **2009**, *10*, 981–991. [CrossRef] [PubMed]
71. Toscano, M.A.; Bianco, G.A.; Ilarregui, J.M.; Croci, D.O.; Correale, J.; Hernandez, J.D.; Zwirner, N.W.; Poirier, F.; Riley, E.M.; Baum, L.G.; et al. Differential Glycosylation OF Th1, Th2 and Th-17 Effector Cells selectively Regulates Susceptibility to Cell Death. *Nat. Immunol.* **2007**, *8*, 825–834. [CrossRef]
72. Toscano, M.A.; Commodaro, A.G.; Ilarregui, J.M.; Bianco, G.A.; Liberman, A.; Serra, H.M.; Hirabayashi, J.; Rizzo, L.V.; Rabinovich, G.A. Galectin-1 Suppresses Autoimmune retinal disease by promoting Concomitant th2- and T Regulatory-Mediated Anti-Inflammatory Responses. *J. Immunol.* **2006**, *176*, 6323–6332. [CrossRef] [PubMed]
73. Poncini, C.V.; Ilarregui, J.M.; Batalla, E.I.; Engels, S.; Cerliani, J.P.; Cucher, M.A.; van Kooyk, Y.; González-Cappa, S.M.; Rabinovich, G.A. *Trypanosoma Cruzi* infection Imparts a Regulatory Program in Dendritic Cells and T Cells via Galectin-1–Dependent Mechanisms. *J. Immunol.* **2015**, *195*, 3311–3324. [CrossRef] [PubMed]
74. Cedeno-Laurent, F.; Opperman, M.; Barthel, S.R.; Kuchroo, V.K.; Dimitroff, C.J. Galectin-1 Triggers an Immunoregulatory Signature in Th Cells Functionally Defined BY IL-10 Expression. *J. Immunol.* **2012**, *188*, 3127–3137. [CrossRef] [PubMed]
75. Dalotto-Moreno, T.; Croci, D.O.; Cerliani, J.P.; Martinez-Allo, V.C.; Dergan-Dylon, S.; Méndez-Huergo, S.P.; Stupirski, J.C.; Mazal, D.; Osinaga, E.; Toscano, M.A.; et al. Targeting Galectin-1 Overcomes Breast Cancer-Associated immunosuppression and prevents Metastatic Disease. *Cancer Res.* **2012**, *73*, 1107–1117. [CrossRef] [PubMed]
76. Starossom, S.C.; Mascanfroni, I.D.; Imitola, J.; Cao, L.; Raddassi, K.; Hernandez, S.F.; Bassil, R.; Croci, D.O.; Cerliani, J.P.; Delacour, D.; et al. Galectin-1 Deactivates Classically Activated Microglia and Protects from Inflammation-Induced Neurodegeneration. *Immunity* **2012**, *37*, 249–263. [CrossRef] [PubMed]
77. Keir, M.E.; Butte, M.J.; Freeman, G.J.; Sharpe, A.H. PD-1 and Its Ligands in Tolerance and Immunity. *Annu. Rev. Immunol.* **2008**, *26*, 677–704. [CrossRef] [PubMed]
78. Okumura, C.Y.; Baum, L.G.; Johnson, P.J. Galectin-1 on Cervical Epithelial Cells Is a Receptor for the Sexually Transmitted Human parasite *Trichomonas vaginalis*. *Cell. Microbiol.* **2008**, *10*, 2078–2090. [CrossRef]
79. Nita-Lazar, M.; Banerjee, A.; Feng, C.; Amin, M.N.; Frieman, M.B.; Chen, W.H.; Cross, A.S.; Wang, L.-X.; Vasta, G.R. Desialylation of Airway Epithelial Cells during Influenza Virus Infection Enhances Pneumococcal Adhesion via Galectin Binding. *Mol. Immunol.* **2015**, *65*, 1–16. [CrossRef]
80. Mahdavi, A.; Szychowski, J.; Ngo, J.T.; Sweredoski, M.J.; Graham, R.L.; Hess, S.; Schneewind, O.; Mazmanian, S.K.; Tirrell, D.A. Identification of Secreted Bacterial Proteins by Noncanonical Amino Acid Tagging. *Proc. Natl. Acad. Sci. USA* **2013**, *111*, 433–438. [CrossRef] [PubMed]
81. Heesemann, J.; Gross, U.; Schmidt, N.; Laufs, R. Immunochemical Analysis of Plasmid-Encoded Proteins Released by Enteropathogenic *Yersinia* Sp. Grown in Calcium-Deficient Media. *Infect. Immun.* **1986**, *54*, 561–567. [CrossRef] [PubMed]
82. Plowman, S.J.; Hancock, J.F. Ras Signaling from Plasma Membrane AND Endomembrane Microdomains. *Biochim. Biophys. Acta (BBA)—Mol. Cell Res.* **2005**, *1746*, 274–283. [CrossRef]
83. Belanis, L.; Plowman, S.J.; Rotblat, B.; Hancock, J.F.; Kloog, Y. Galectin-1 Is a Novel Structural Component and a Major Regulator of H-Ras Nanoclusters. *Mol. Biol. Cell* **2008**, *19*, 1404–1414. [CrossRef] [PubMed]
84. Wei, W.; Behloul, N.; Baha, S.; Liu, Z.; Aslam, M.S.; Meng, J. Dimerization: A Structural Feature for the Protection of Hepatitis e Virus Capsid Protein against Trypsinization. *Sci. Rep.* **2018**, *8*. [CrossRef]
85. Tumitan, A.R.; Monnazzi, L.G.; Ghiraldi, F.R.; Cilli, E.M.; de Medeiros, B.M. Pattern of Macrophage Activation Inyersinia-Resistant Andyersinia-Susceptible Strains of Mice. *Microbiol. Immunol.* **2007**, *51*, 1021–1028. [CrossRef] [PubMed]
86. Liew, F.Y.; Li, Y.; Moss, D.; Parkinson, C.; Rogers, M.V.; Moncada, S. Resistance Toleishmania Major Infection Correlates with the Induction of Nitric Oxide Synthase in Murine Macrophages. *Eur. J. Immunol.* **1991**, *21*, 3009–3014. [CrossRef] [PubMed]
87. Monnazzi, L.G.; Carlos, I.Z.; deMedeiros, B.M. Influence of *Yersinia pseudotuberculosis* Outer Proteins (YOPS) on Interleukin-12, Tumor Necrosis Factor Alpha and Nitric Oxide Production BY Peritoneal Macrophages. *Immunol. Lett.* **2004**, *94*, 91–98. [CrossRef]
88. Tansini, A.; de Medeiros, B.M. Susceptibility to *Yersinia Pseudotuberculosis* infection Is Linked to the Pattern of Macrophage Activation. *Scand. J. Immunol.* **2009**, *69*, 310–318. [CrossRef]
89. Virág, L.; Jaén, R.I.; Regdon, Z.; Boscá, L.; Prieto, P. Self-Defense of Macrophages against Oxidative Injury: Fighting for Their Own Survival. *Redox Biol.* **2019**, *26*, 101261. [CrossRef]
90. Kim, P.K.M.; Zamora, R.; Petrosko, P.; Billiar, T.R. The Regulatory Role of Nitric Oxide in Apoptosis. *Int. Immunopharmacol.* **2001**, *1*, 1421–1441. [CrossRef]
91. Giordano, D.; Li, C.; Suthar, M.S.; Draves, K.E.; Ma, D.Y.; Gale, M., Jr.; Clark, E.A. Nitric oxide controls an inflammatory-like Ly6C(hi)PDCA1+DC subset that regulates Th1 immune responses. *J. Leukoc. Biol.* **2011**, *89*, 443–455. [CrossRef]
92. Xue, Q.; Yan, Y.; Zhang, R.; Xiong, H. Regulation of iNOS on Immune Cells and Its Role in Diseases. *Int. J. Mol. Sci.* **2018**, *19*, 3805. [CrossRef]

93. Hu, G.; Su, Y.; Kang, B.H.; Fan, Z.; Dong, T.; Brown, D.R.; Cheah, J.; Wittrup, K.D.; Chen, J. High-throughput phenotypic screen and transcriptional analysis identify new compounds and targets for macrophage reprogramming. *Nat. Commun.* **2021**, *12*, 773. [CrossRef] [PubMed]
94. Zheng, Y.; Lilo, S.; Brodsky, I.E.; Zhang, Y.; Medzhitov, R.; Marcu, K.B.; Bliska, J.B. A Yersinia effector with Enhanced Inhibitory Activity on the Nf-Kb PATHWAY Activates the nlrp3/Asc/Caspase-1 Inflammasome in Macrophages. *PLoS Pathog.* **2011**, *7*. [CrossRef] [PubMed]
95. Fuertes, M.B.; Molinero, L.L.; Toscano, M.A.; Ilarregui, J.M.; Rubinstein, N.; Fainboim, L.; Zwirner, N.W.; Rabinovich, G.A. Regulated Expression of Galectin-1 during T-Cell Activation Involves lck and Fyn Kinases and Signaling through mek1/Erk, P38 MAP Kinase And p70S6kinase. *Mol. Cell. Biochem.* **2004**, *267*, 177–185. [CrossRef] [PubMed]

Article

Expression, Regulation, and Functions of the Galectin-16 Gene in Human Cells and Tissues

Jennifer D. Kaminker and Alexander V. Timoshenko *

Department of Biology, The University of Western Ontario, London, ON N6A 5B7, Canada; jkamink@uwo.ca
* Correspondence: atimoshe@uwo.ca; Tel.: +1-519-661-2111 (ext. 88900)

Abstract: Galectins comprise a family of soluble β-galactoside-binding proteins, which regulate a variety of key biological processes including cell growth, differentiation, survival, and death. This paper aims to address the current knowledge on the unique properties, regulation, and expression of the galectin-16 gene (*LGALS16*) in human cells and tissues. To date, there are limited studies on this galectin, with most focusing on its tissue specificity to the placenta. Here, we report the expression and 8-Br-cAMP-induced upregulation of *LGALS16* in two placental cell lines (BeWo and JEG-3) in the context of trophoblastic differentiation. In addition, we provide the results of a bioinformatics search for *LGALS16* using datasets available at GEO, Human Protein Atlas, and prediction tools for relevant transcription factors and miRNAs. Our findings indicate that *LGALS16* is detected by microarrays in diverse human cells/tissues and alters expression in association with cancer, diabetes, and brain diseases. Molecular mechanisms of the transcriptional and post-transcriptional regulation of *LGALS16* are also discussed based on the available bioinformatics resources.

Keywords: galectin; *LGALS16*; placenta; brain tissues; cell differentiation; transcription factor; miRNA

Citation: Kaminker, J.D.; Timoshenko, A.V. Expression, Regulation, and Functions of the Galectin-16 Gene in Human Cells and Tissues. *Biomolecules* 2021, *11*, 1909. https://doi.org/10.3390/biom11121909

Academic Editor: Lu-Gang Yu

Received: 9 December 2021
Accepted: 16 December 2021
Published: 20 December 2021

Publisher's Note: MDPI stays neutral with regard to jurisdictional claims in published maps and institutional affiliations.

Copyright: © 2021 by the authors. Licensee MDPI, Basel, Switzerland. This article is an open access article distributed under the terms and conditions of the Creative Commons Attribution (CC BY) license (https://creativecommons.org/licenses/by/4.0/).

1. Introduction

Galectins comprise a family of soluble β-galactoside binding proteins, which regulate key biological processes including cell growth, differentiation, apoptosis, and immune responses [1–4]. Sixteen galectin genes have been identified in animal kingdoms, 12 of which are expressed in humans. Galectins share a conserved carbohydrate recognition domain (CRD) and they are subcategorized into prototype, tandem-repeat, or chimeric type according to their number of CRDs and structural features. Prototype galectins contain one CRD and include galectins -1, -2, -5, -7, -10, -11, -13, -14, -15, and -16. Tandem-repeat galectins contain two homologous CRDs connected by a linker of ~70 amino acids and include galectins-4, -6, -8, -9, and -12. The only chimera-type galectin is galectin-3, which contains one CRD linked to a non-lectin N-terminal proline/glycine-rich domain. Galectins form a network of proteins to perform glycan-dependent and glycan-independent functions both intra- and extracellularly [3–5]. Intracellularly, galectins have multiple binding partners and primarily function via glycan-independent mechanisms to regulate processes such as cell growth, apoptosis, and pre-mRNA splicing among others [4–6]. Extracellular galectins are secreted from cells through unconventional mechanisms [3,7] and can bind to glycoligands on the cell surface or glycoproteins in the extracellular matrix to promote cell adhesion and migration [8] or bind to specific cell surface receptors to facilitate their cross-linking and transmembrane signaling [3,8–10].

Galectin expression profiles vary significantly between different cells and tissues. Some galectins are commonly expressed with low tissue specificity, e.g., galectin-1 and galectin-3, while others are highly-tissue specific [3]. *LGALS16* was characterized in placental tissue by Than and co-authors [11] and together with two other galectins (*LGALS13* and *LGALS14*) was found to be upregulated in differentiated trophoblast cells to confer immunotolerance at the maternal–fetal interface [12]. These three galectin genes are located in a cluster of four human protein-coding galectin genes on chromosome 19 and they are proposed to have evolutionarily emerged to sustain hemochorial placentation in anthropoids [11]. The correct expression of placenta-specific galectins is an important part of proper reprogramming of the transcriptional activity of the trophoblast [12]. This involves the differentiation and fusion of villous cytotrophoblasts into a multinucleated syncytium that is in direct contact with maternal blood and is responsible for facilitating gas, nutrient, and waste exchange between the mother and fetus, mediating hormonal regulation, and forming an immunological barrier during pregnancy [12]. Differentiated extravillous trophoblasts proliferate, invade, and remodel the maternal spiral arteries to provide blood flow and nutrients to the fetus [13]. Dysregulation of this placenta-specific gene cluster containing LGALS16 is associated with disorders such as preeclampsia, which can be highly fatal for both the mother and fetus [12–14].

Currently, experimental studies on LGALS16 are limited, although multiple microarray datasets and bioinformatics resources contain relevant information. Here, we use experimental and bioinformatics approaches for examining expression, regulation, and functions of LGALS16 to position this galectin within the complex galectin network in cells and to identify directions for future studies.

2. Materials and Methods

2.1. Bioinformatics Data and Tools

Microarray and RNA-sequencing data were extracted from the Gene Expression Omnibus (GEO) Profiles, which contained 287 datasets for LGALS16 (accessed on 2 November 2021), considering the following criteria: (1) inclusion of only controls and untreated cell/tissue samples, (2) inclusion of only cases with positive gene expression values for matched ACTB (a housekeeping gene), LGALS1 (a low tissue specific galectin), and LGALS16 genes, and (3) deletion of few datasets, which report enormous deviations (>100-folds) from average expression levels of ACTB and LGALS1 genes. In silico prediction of transcription factor binding sites in LGALS16 gene DNA sequence was performed with PROMO version 3.0.2 software, which utilized version 8.3 of TRANSFAC database [15,16]. The dissimilarity index for the transcription factor search was set at 0% to limit the number of non-specific matches. Ensembl Release 104 was used to extract the sequence of the 2 kb promoter region of the gene (accessed on 24 August 2021). Four different online platforms were used and compared to predict putative miRNA targets for LGALS16 including Diana Tools [17], miRabel [18], miRDB [19], and TargetScan [20]. The Human Protein Atlas (HPA) [21], GenBank [22], and Protein Data Bank (PDB) [23] were exploited for searching the relevant structures, sequences, and expression patterns of LGALS16 based on the gene symbol.

2.2. Cell Cultures

Placenta choriocarcinoma BeWo and JEG-3 cell lines (kindly provided by Dr. Renaud, Department of Anatomy and Cell Biology, Western University, London, ON, Canada) were cultured in Dulbecco's Modified Eagle Medium/Ham's F12 medium and RPMI-1640 medium, respectively, supplemented with 10% or 8% fetal bovine serum, 100 IU/mL penicillin, and 100 µg/mL streptomycin. Cell cultures were maintained in a CO_2-incubator at 37 °C and 5% CO_2. To induce trophoblastic differentiation, cells were grown in 6-well plates and treated with 250 µM of 8-Br-cAMP (cat. # B7880, Sigma-Aldrich, Oakville, ON, Canada) for 48 h (BeWo cells) or 36 h (JEG-3 cells). Over the time of these treatments, cell culture media was replaced one time for BeWo cells after 24 h of growth and two times (every 12 h) for JEG-3 cells to avoid accumulation of acidic metabolites.

2.3. Gene Expression Analysis

The total RNA pools were isolated from cell monolayers using TRIzol® reagent (cat. # 15596018, Ambion, Carlsbad, CA, USA) and 1 µg was used for cDNA synthesis with the Advanced cDNA Synthesis Kit (cat. # 801-100, Wisent, Montreal, QC, Canada). The conventional and quantitative polymerase chain reaction (PCR) analyses were used to assess the mRNA expression levels for following genes: *ACTB*, *CGB3/5* (biomarkers of trophoblastic differentiation), and *LGALS16*. The oligonucleotide PCR primers for *LGALS16* (forward 5'-ATTTGCGAGTGCACTTAGGC-3' and reverse 5'-GACACACGTAGATGCGCAAG-3', PCR amplicon length of 132 bp) targeting exon 3 (Figure 1) were designed using Primer-BLAST tool at NCBI [24]. Oligonucleotide primers for *ACTB* (forward 5'-TCAGCAAGCAGGAGTA TGACGAG-3' and reverse 5'-ACATTGTGAACTTTGGGGGATG-3', PCR amplicon length of 265 bp) and *CGB3/5* (forward 5'-CCTGGCCTTGTCTACCTCTT-3' and reverse 5'-GGCTT TATACCTCGGGGTTG-3', PCR amplicon length of 109 bp) were available elsewhere [25,26]. To run conventional PCR, reaction mixes (10 µL 2X Taq FroggaMix (cat. # FBTAQM, FroggaBio, Toronto, ON, Canada), 2 µL forward and reverse primer mixture from 10 µM stock, 7 µL nuclease free water, and 1 µL template cDNA) were loaded into a T100 Thermal Cycler (Bio-Rad Laboratories, Mississauga, ON, Canada) and amplified using the following PCR regime: 26 cycles of 94 °C for 3 min, 94 °C for 30 s, 56 °C seconds, 72 °C for 60 s, 72 °C for 10 min, and held at 4 °C. The PCR products were separated on a 2% agarose gel as described earlier [27] and the gel was imaged using the Molecular Imager® Gel Doc™ XR+ (Bio-Rad) to confirm the expected size of PCR amplicons. The quantitative PCR was performed in the CFX Connect™ Thermocycler and quantified as described previously [28] using the SsoAdvanced Universal SYBR® Supermix kit (cat. # 1725274, Bio-Rad Laboratories, Mississauga, ON, Canada). To assess the expression of 84 genes encoding human transcription factors, the RT^2 Profiler™ PCR Array Kit (cat. # PAHS-075ZD-2, Qiagen, Toronto, ON, Canada) was used following the protocols provided by the manufacturer.

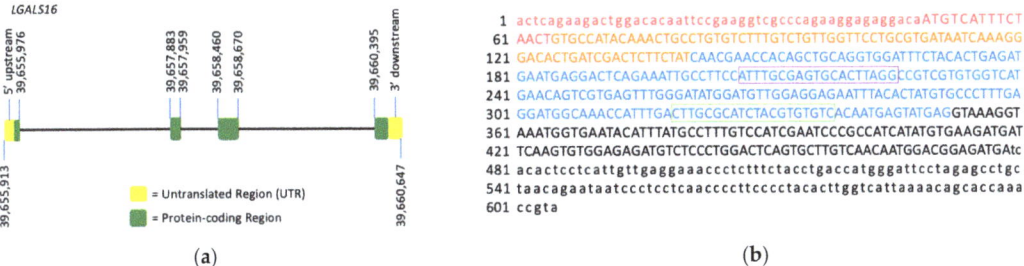

Figure 1. *LGALS16* gene structure and the mRNA sequence. (**a**) *LGALS16* (4735 bp) is located on chromosomal band 19q13.2 and contains 4 exons (ENSG00000249861). (**b**) NCBI reference sequence of *LGASL16* mRNA (NM_001190441.3). Each exon is highlighted with red, orange, blue, and black representing exons 1, 2, 3, and 4, respectively. The protein coding sequence (CDS) is indicated in capitals while UTRs in small characters. The oligonucleotide sequences for PCR amplification are boxed.

2.4. Statistical Analysis

Statistical analysis was performed using GraphPad Prism 9 for Windows, version 9.1.2 (GraphPad Software, San Diego, CA, USA) and the data were presented as mean ± SD. One-way analysis of variance (ANOVA) was used to determine statistical significance across treatments followed by Tukey's honestly significant difference test to detect which means were statistically significant at a value of $p < 0.05$.

3. Results and Discussion

3.1. Molecular Characteristics of Galectin-16 Gene and Recombinant Protein

The *LGALS16* gene structure and molecular details were described by Than and coauthors [11]. *LGALS16* (4735 bp) is located on chromosomal band 19q13.2, spans from bases 39,655,913 to 39,660,647, and contains 4 exons (Figure 1a,b). *LGALS16* is found only in primates and is part of the chromosome 19 gene cluster containing four protein-coding genes (*LGALS10*, *LGALS13*, *LGALS14*, *LGALS16*) [11,12,14,29]. The diversification and evolutionary origin of this cluster, including *LGALS16*, is thought to be related to placenta development and mediated by transposable long interspersed nuclear elements

(LINEs), which are commonly found at the boundaries of large inversions and gene duplication units [11,30,31]. The relevant rearrangements and subsequent gains and losses of duplicated genes and pseudogenes are proposed to have enabled anthropoids to sustain highly invasive placentation and placental phenotypes, such as longer gestation for larger offspring and an increased body to brain size ratio [11].

To the best of our knowledge, no studies are available on native galectin-16 at the protein level whereas recombinant protein has been produced and tested. The crystal structure of recombinant galectin-16 and its mutants was solved by Si and co-authors [32]. Recombinant galectin-16 is a monomeric protein, which is composed of 142 amino acids and has a typical galectin structure of the CRD β-sandwich with two sheets formed by six β-strands on the concave side (S1–S6) and five β-strands on the convex side (F1–F5) (Figure 2). This group also showed that galectin-16 lacks lactose-binding ability unless arginine (Arg55) is replaced with asparagine in S4 β-strand. In comparison, an earlier report showed that recombinant galectin-16 and two other human galectins (galectin-13 and galectin-14) can bind lactose–agarose beads and are efficiently and competitively eluted by lactose [11]. More insights into this discrepancy are required considering multiple interfering factors, mutations/replacements of amino acids within the CRD, and different study designs. Regardless, both glycan-dependent and glycan-independent interactions might be essential for galectin-16 similar to other galectins [3].

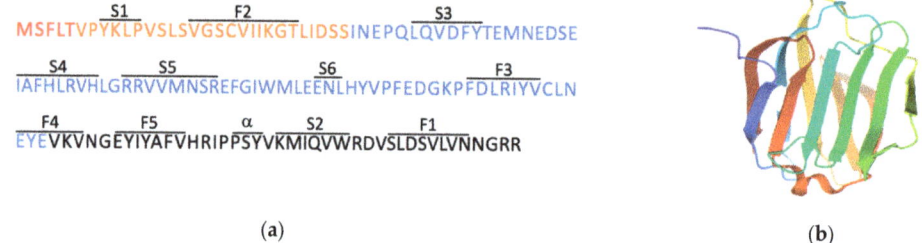

(a)   (b)

Figure 2. Protein sequence and structure of recombinant galectin-16. (**a**) The 142 amino acid sequence is 16.6 kDa for the galectin-16 protein. Each color corresponds to the exon from which the amino acids were encoded with red, orange, blue, and black representing exons 1, 2, 3, and 4, respectively. The anti-parallel β-sheets of F-face (F1–F5) and S-face (S1–S6) strands as well as a short α-helix are showed. (**b**) The crystal structure was extracted from Protein Data Bank (available online: rcsb.org, accessed on 6 September 2021), PDB ID: 6LJP.

3.2. Expression Patterns and Functions of LGALS16 in Cells and Tissues

Experimental studies focusing on *LGALS16* are limited and an essential source of relevant information about this gene is Gene Expression Omnibus (GEO), a data repository for microarray and RNA-sequencing data [33]. Overall, 287 datasets are available on GEO (November 2021 search) reporting *LGALS16* expression in 52 types of tissues and various cell lines based on the following platforms: Affymetrix Human Genome ($n = 27$), Affymetrix Human Gene ($n = 151$), Agilent ($n = 31$), Human Unigene ($n = 1$), Illumina Human ($n = 82$), MCI Human ($n = 1$), NuGO ($n = 1$), and Sentrix Human ($n = 15$). Quantification of differences in *LGALS16* expression between different platforms is challenging. However, evaluation of gene expression values within the same GEO datasets demonstrates that *LGALS16* can be classified as a gene with relatively low expression in comparison with *LGALS1* (a widely expressed galectin with a low tissue specificity) and *ACTB* (a common housekeeping gene) (Table 1). Indeed, regardless of the platform, average GEO percentile rank of expression for *LGALS16* measured with different arrays ranged 4–32% on a scale of 1–100% while the range was 63–100% for *LGALS1* and 94–100% for *ACTB*. Available GEO profiles do not contain relevant datasets with *LGALS16* for placenta for comparison, however, the Human Protein Atlas (HPA) reports tissue-specific overexpression of *LGALS16* in placenta followed by brain tissues and retina (Figure 3a). The biological meaning and reasons of overexpression of *LGALS16* in these diverse tissues is unknown and requires further investigations in the context of developmental biology. For instance, the complex mechanisms of the placenta–brain axis of cell development [34] could be addressed in terms of the unique association of *LGALS16* with these tissues.

Table 1. Comparative expression of *LGALS16* in human tissues and cells from the Gene Expression Omnibus database.

Names of Cells or Tissues	GEO Accession Number	ACTB	LGALS1	LGALS16	Sample Size
Acetabular labrum cells	GDS5427 [a]	12.682 ± 0.150	12.057 ± 0.107	2.949 ± 0.0093	3
Acute lymphoblastic leukemia cell line RS4;11	GDS4043 [b]	13.861 ± 0.017	11.487 ± 0.033	0.4023 ± 0.607	2
Acute myeloblastic leukemia cell line Kasumi-1	GDS5600 [a]	11.965 ± 0.025	6.455 ± 0.299	2.918 ± 0.036	3
Acute promyelocytic leukemia cell line NB4	GDS4180 [a]	13.130 ± 0.035	10.823 ± 0.031	3.650 ± 0.108	3
Adipocyte progenitor cells (subcutaneous)	GDS5171 [a]	13.523 ± 0.038	13.397 ± 0.112	4.597 ± 0.251	6
Adipocyte progenitors from deep neck	GDS5171 [a]	13.469 ± 0.057	13.208 ± 0.177	4.505 ± 0.094	6
Bone marrow CD34+ cells (chronic myeloid leukemia)	GDS4756 [a]	13.524	11.137	3.050	1
Bone marrow plasma cells	GDS4968 [a]	11.990 ± 0.226	8.714 ± 0.515	3.052 ± 0.257	5
Brain frontal cortex	GDS4758 [a]	13.402 ± 0.125	96.333 ± 0.840	4.632 ± 0.249	18
Brain hippocampus	GDS4758 [a]	13.477 ± 0.130	11.133 ± 0.375	4.659 ± 0.300	10
Brain hippocampus	GDS4879 [a]	12.113 ± 0.409	9.076 ± 0.232	3.177 ± 0.177	19
Brain temporal cortex	GDS4758 [a]	13.560 ± 0.131	11.189 ± 0.280	4.749 ± 0.193	19
Breast cancer cell line MCF-7	GDS2759 [b]	15.884 ± 0.030	13.752 ± 0.153	6.053 ± 0.237	2
Breast cancer cell line MCF-7	GDS4972 [a]	13.029 ± 0.038	12.439 ± 0.083	3.892 ± 0.066	3
Breast cancer cell line MCF-7	GDS4090 [a]	13.087 ± 0.019	9.566 ± 0.100	2.827 ± 0.405	3
Breast cancer cell line MDA-MB-231	GDS4800 [a]	13.875 ± 0.007	13.565 ± 0.042	5.189 ± 0.085	3
Bronchial smooth muscle primary cells	GDS4803 [a]	11.629 ± 0.175	11.533 ± 0.041	3.181 ± 0.095	3
Bronchopulmonary neuroendocrine cell line NCI-H727	GDS4330 [a]	11.978	5.715	3.808	1
Burkitt lymphoma cell line Namalwa	GDS4978 [a]	13.468 ± 0.187	8.005 ± 0.073	3.916 ± 0.297	3
Burkitt lymphoma cell line Raji	GDS4978 [a]	13.367 ± 0.093	8.052 ± 0.141	3.962 ± 0.019	3
Colorectal adenocarcinoma cell line SW620	GDS5416 [e]	16.400 ± 0.362	17.280 ± 0.043	2.766 ± 0.554	2
Embryonic kidney cell line HEK-293	GDS4233 [a]	10.330 ± 0.050	7.109 ± 0.098	3.757 ± 0.328	4
Endothelial progenitor cells	GDS3656 [c]	15.397 ± 0.174	13.845 ± 0.457	8.018 ± 0.103	11
Esophagus biopsies	GDS4350 [a]	12.617 ± 0.230	8.062 ± 0.507	3.255 ± 0.208	8
Gastrointestinal neuroendocrine cell line KRJ-1	GDS4330 [a]	12.135	9.592	2.859	1
Germinal center B cells	GDS4977 [a]	9.793 ± 0.373	8.438 ± 0.225	6.723 ± 0.538	5
Gingival fibroblasts	GDS5811 [a]	13.628 ± 0.101	13.770 ± 0.174	3.674 ± 0.140	2
Heart (left ventricle)	GDS4772 [a]	11.293 ± 0.361	10.672 ± 0.377	2.941 ± 0.030	5
Heart (left ventricle)	GDS4314 [a]	12.142 ± 0.365	11.052 ± 0.223	3.344 ± 0.154	5
Heart (right ventricular)	GDS5610 [a]	11.930 ± 0.255	10.934 ± 0.044	3.637 ± 0.181	2
Hepatocellular carcinoma cell line HepG2	GDS5340 [a]	13.259 ± 0.039	11.256 ± 0.054	4.281 ± 0.327	3
Microglia cell line HMO6	GDS4151 [a]	13.545	12.231	2.979	1
Keratinocytes	GDS4426 [a]	12.679 ± 0.056	11.147 ± 0.236	3.804 ± 0.138	6
Lung carcinoma cell line A549	GDS4997 [a]	10.970 ± 0.044	12.187 ± 0.049	2.418 ± 0.072	3
Lung carcinoma cell line H460	GDS5247 [a]	12.504 ± 0.043	11.111 ± 0.063	3.439 ± 0.117	3
Lung microvascular endothelial cell line CC-2527	GDS2987 [b]	32,061 ± 7366	15,158 ± 2227	8.100 ± 9.051	2
Lymphoblastoid cell line TK6	GDS4915 [a]	13.365 ± 0.061	11.161 ± 0.323	4.005 ± 0.327	2
Lymphoblastoid cell line TK6	GDS4916 [a]	13.940 ± 0.058	12.023 ± 0.130	4.061 ± 0.357	2
Medulloblastoma tumor tissue	GDS4469 [a]	13.099 ± 0.302	9.490 ± 0.801	4.005 ± 0.839	15
Melanoma cell line A-375	GDS5085 [a]	13.888 ± 0.011	13.474 ± 0.101	4.618 ± 0.045	3
Melanoma cell line FEMX-I	GDS3489 [d]	16.04 ± 0.354	16.04 ± 0.354	0.550 ± 1.061	2
Melanoma cell line Hs294T	GDS5670 [a]	11.353 ± 0.245	10.349 ± 0.097	2.149 ± 0.585	2
Microglia cell line HMO6	GDS4151 [a]	13.545	12.231	2.979	1
Myotubes from musculus obliquus internus	GDS5378 [a]	13.224 ± 0.099	12.925 ± 0.114	2.840 ± 0.057	4
Pancreatic neuroendocrine cell line QGP-1	GDS4330 [a]	12.057	5.749	3.031	1
Peripheral blood CD34+ cells (chronic myeloid leukemia)	GDS4756 [a]	13.414 ± 0.049	11.144 ± 0.578	2.974 ± 0.140	2
Peripheral blood CD4+ T cells	GDS5544 [a]	13.598 ± 0.053	9.707 ± 0.247	4.584 ± 0.126	4
Peripheral blood cells	GDS4240 [a]	11.825 ± 0.084	7.307 ± 0.154	1.506 ± 0.112	7
Renal adenocarcinoma cell line 786-O	GDS5810 [a]	12.902 ± 0.030	12.809 ± 0.015	5.753 ± 0.031	2
Retinal pigment epithelia primary cells	GDS4224 [a]	13.407 ± 0.110	11.842 ± 0.449	3.468 ± 0.367	4
Retinal pigmented epithelium cell line ARPE-19	GDS4224 [a]	13.288	11.946	3.646	1
Skeletal muscle (vastus lateralis) primary cells	GDS4920 [a]	13.649 ± 0.084	13.385 ± 0.114	4.609 ± 0.136	12

Table 1. Cont.

Names of Cells or Tissues	GEO Accession Number	ACTB	LGALS1	LGALS16	Sample Size
Skeletal muscle tissue	GDS4841 [a]	9.400 ± 0.190	11.486 ± 0.247	2.786 ± 0.355	5
Skin cancer cell line RT3Sb	GDS5381 [a]	13.409 ± 0.062	8.775 ± 0.114	3.539 ± 0.252	4
Skin epidermis	GDS3806 [c]	15.139 ± 0.141	9.534 ± 0.370	7.909 ± 0.469	7
Visceral adipose tissue (omentum)	GDS4857 [a]	11.875 ± 0.352	11.488 ± 0.416	4.666 ± 0.754	8

Notes: The means ± SD of available gene expression values are shown. The GEO datasets originated from different platforms: [a] Affymetrix Human Gene 1.0 ST Array, [b] Sentrix Human-6 Expression BeadChip, [c] Sentrix HumanRef-8 Expression BeadChip, [d] MCI Human HEEBOChip 42k oligo array, and [e] Agilent-014850 Whole Human Genome Microarray 4 × 44K G4112F.

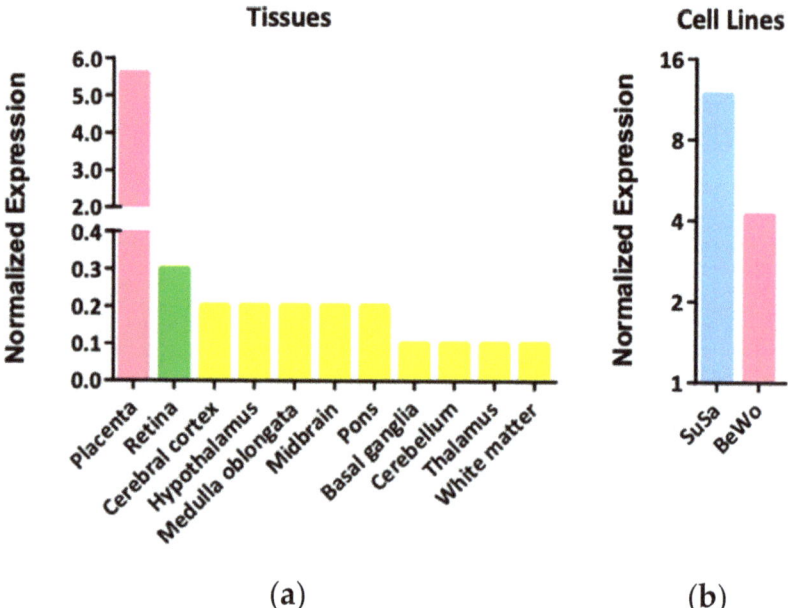

Figure 3. The normalized expression of *LGALS16* mRNA in human tissues and cells from HPA datasets. (**a**) *LGALS16*-positive cases out of 55 tissue types; (**b**) *LGALS16*-positive cases out of 69 cell lines. The data were retrieved on 28 November 2021.

In comparison with tissues, HPA reports the expression of *LGALS16* mRNA only in two human cell lines including placental choriocarcinoma cell line BeWo and testicular teratoma cell line SuSa, which probably can be used as appropriate systems to explore the biological role of *LGALS16* gene (Figure 3b). Human syncytiotrophoblasts, which are terminally differentiated placental cells, can also serve as a strong positive control for *LGALS16* overexpression [10–12].

To develop experimental models for studying *LGALS16* functions and regulation, we examined the gene expression in BeWo cells and an additional placental cell line JEG-3 in the context of trophoblastic differentiation. The expression of *LGALS16* mRNA was significantly increased in both cell lines after 36 h (JEG-3 cells) and 48 h (BeWo cells) treatment with a potent cell-permeable and metabolically stable activator of cAMP-dependent protein kinase 8-Br-cAMP (250 µM), which coincided with upregulation of *CGB3/5*, genes encoding chorionic gonadotropin subunits 3 and 5 (Figure 4). As chorionic gonadotropin is one of the biomarkers of placenta and trophoblastic differentiation, our results suggest classifying *LGALS16* to the same category of biological molecules. Other studies also reported significant upregulation of *LGALS16* in association with processes of cellular differentiation, even if the basal levels were relatively low. Thus, treatment of BeWo cells with forskolin, an inducer of cyclic adenosine 3′,5′-monophosphate (cAMP), stimulated trophoblastic differentiation and simultaneous *LGALS16* overexpression [12]. An interesting example of *LGALS16* upregulation was reported in a model of intestinal differentiation of Caco-2 cells induced by a combined treatment with dexamethasone and p44/42 MAPK inhibitor PD98059 [35]. Therefore, *LGALS16* may deserve further attention as a factor associated with processes of cellular differentiation and tissue development.

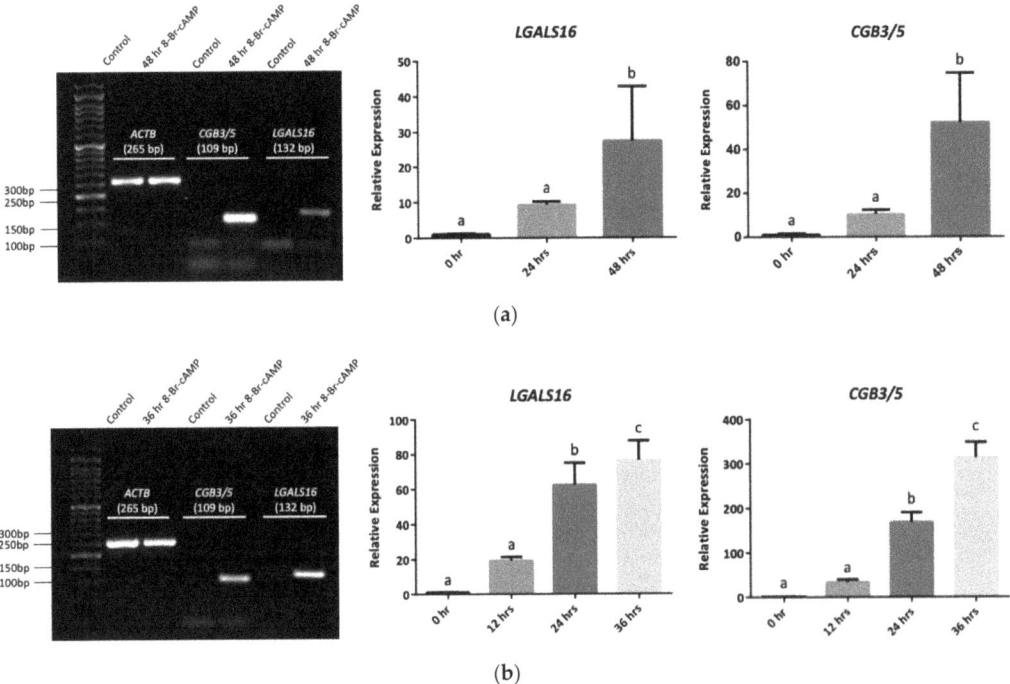

Figure 4. *LGALS16* expression in human placental choriocarcinoma cell lines, BeWo and JEG-3. Cells were treated with 8-Br-cAMP (250 µM) for different periods of time to induce syncytiotrophoblast differentiation. (**a**) BeWo cells ($n = 4$); (**b**) JEG-3 cells ($n = 3$). Agarose gels on the left confirm the expected size of PCR amplicons. Bar graphs show the fold changes in the expression of *LGALS16* and *CGB3/5* genes obtained by qPCR, which were quantified by the Livak method ($2^{-\Delta\Delta CT}$) using *ACTB* as a reference gene. Data are presented as means ± SD; means with the same letter are not significantly different from each other (Tukey's post hoc HSD test, $p > 0.05$).

An important function of galectin-16 as well as placental galectin-13 and galectin-14 is the ability to induce apoptosis of CD3+ T cells, which was detected by flow cytometry of cells double-stained with annexin-V-FITC and propidium iodide [11]. Considering the high expression of galectin-16 in differentiated trophoblasts, the apoptotic mechanism might contribute to the immune tolerance at the maternal-fetal interface reducing the danger of maternal immune attacks on the fetus and enabling anthropoid primates to evolve long gestation periods while retaining highly invasive placentation. The details of this regulation are obscure since there are no studies addressing the secretion of galectin-16 from trophoblasts. However, intracellular EGFP-tagged recombinant galectin-16 was readily localized in the nucleus and cytoplasm of transfected cells including, HeLa, 293T, HCT-116, SMMC-7721 and Jurkat cells [32]. In fact, the nuclear staining was much stronger than in the cytoplasm suggesting that the transport of galectin-16 into the nucleus might play a role in regulating intranuclear processes. These authors showed that the binding partner of galectin-16 is c-Rel, a member of the NF-κB family of transcription factors (TFs), which is involved in the regulation of multiple processes such as apoptosis, inflammation, immune responses, tumorigenesis, cell growth and differentiation [32,36]. All NF-κB family members, including c-Rel, have a conserved N-terminal DNA-binding/dimerization domain, known as the Rel homology domain (RHD) [37]. Recombinant galectin-16 strongly binds to the RHD which might inhibit c-Rel and prevent activation of anti-apoptotic genes, such as Bcl-2 and Bcl-xL, promoting T-cell apoptosis during pregnancy [32]. An additional aspect of *LGALS16* functions may contribute to the rescue of glucose restriction-induced cell death in a model of a whole genome gain-of-function CRISPR activation using human mitochondrial disease complex I mutant cells [38].

3.3. Transcriptional and Post-Transcriptional Regulation of LGALS16

3.3.1. Transcription Factors

Multiple TFs can be involved in the regulation of *LGALS16* expression based on the presence of specific response elements in the promoter regions of the gene. Original analysis of retrotransposons within the 10 kb 5′ UTR by Than and co-authors demonstrated that the *LGALS16* promoter has binding sites for GATA2, TEF5, and ESRRG, which are also involved in the regulation of important trophoblast-specific genes such as *ERVWE1* (marker of cell fusion), *CGA*, and *CGB3* (markers of chorionic gonadotropin production, a hormone released by differentiated trophoblasts to maintain pregnancy) [12]. The contribution of these TFs in regulating *LGALS16* expression was claimed to vary, especially with decreased regulation from GATA2, due to the specific layout and properties of transposable elements (L1PA6 and L1PREC2) within the 5′UTR of this gene as compared to two other placental genes, *LGALS13* and *LGALS14*. Additional shared TFs for the placental galectin gene cluster include TFAP2A and GCM1, which have binding sites within ALU transposable elements next to L1PREC2. Experimental evidence of this regulation was confirmed in a model of forskolin-induced differentiation of primary trophoblasts, which revealed time-dependent upregulation of *LGALS16* in parallel with the expression of *TEAD3*, *ESRRG*, *GCM1*, and *ERVWE1* [12]. It is interesting to note that this study did not reveal the effect of 5-azacytidin on *LGALS16* expression in BeWo trophoblast cells as compared to other upregulated placental galectins, which suggested a minor role of DNA methylation in the context of *LGALS16* regulation.

To enrich this analysis, we used human choriocarcinoma cell line JEG-3 and Qiagen RT^2 Profiler™ PCR Array to test changes in the mRNA transcript levels of 84 TFs during trophoblastic differentiation induced by 8-Br-cAMP. Overall, 60 TFs were upregulated in this assay including three top genes encoding Jun B proto-oncogene (*JUNB*), SMAD family member 9 (*SMAD9*), and activating transcription factor 3 (*ATF3*) (Figure 5). Since all of these three genes are expressed in placenta and brain tissues [21,39–43], which are *LGALS16*-positive, this observation provides a new insight into possible transcriptional regulation of this gene. JUNB and ATF3 belong to a family of TFs with a basic leucine zipper DNA binding domain, with JUNB preferentially binding to the 12-O-tetradecanoylphorbol-13-acetate response element sequence and ATF3 binding to the cAMP response element in promoters with the consensus sequence, TGACGTCA [44]. They are subunits of activating protein 1 (AP-1) TFs, which function as homodimers or heterodimers in association with other members of JUN, FOS, ATF, and MAF protein families [44]. JUNB was reported to be directly involved in processes of trophoblastic cell syncytialization [45,46], while upregulation of ATF3 was associated with cellular stress responses [41,47,48], decidualization [47], and preeclampsia [49]. In comparison, SMAD9 is activated by bone morphogenic proteins (BMPs), a subfamily of the transforming growth factor-β (TGF-β) family [40,50]. Although some BMPs such as BMP-4 can be regulated in a downstream manner from the cAMP pathway [51], the connection between SMAD9 and cAMP is still unclear. GATA2 was found to be slightly upregulated, which may suggest that the enrichment of L1PREC2 in the 5′UTR still plays a role in regulating *LGALS16* despite the insertion of L1PA6 [12]. Interestingly, CREB1, a major regulator downstream of the cAMP pathway was not upregulated in this RT-qPCR array analysis suggesting that post-translational modification and transcriptional activation might be essential for this TF.

Figure 5. Changes in the expression of genes encoding TFs in JEG-3 cells. Cells were treated 8−Br−cAMP (250 μM) for 36 h to induce trophoblastic differentiation and Qiagen RT² Profiler™ PCR Array kit was used to assess fold changes in gene expression between differentiated and control cells presented as a heatmap.

We further analyzed the 2 kb region upstream from the transcription start site of *LGALS16* by extracting the sequence from Ensembl (Release 104) and performing in silico analysis of TF-binding sites using PROMO virtual laboratory with a 0% dissimilarity index. Putative binding sites for seventeen TFs were identified, which may represent specific response elements, enhancers, or silencers (Figure 6). To reveal common patterns in the expression of the predicted TFs, we watched for their protein levels in two *LGALS16*-positive tissues, the cerebellum and placenta, using the expression scores (high, medium, low) available at HPA. Within this set of data, two TFs (CEBPβ and TFII-I) were characterized by high protein expression levels, seven TFs had variable levels (GR, NFAT1, p53, STAT4, TCF-4E, TFIID, and YY1), four TFs (ERα, FOXP3, Pax5, and PR A) showed low expression, and no HPA data were available for GRα, GRβ, PR B, and XBP-1 in these tissues (Figure 7). Thus, the role of the predicted TFs in tissue-specific transcriptional regulation of *LGALS16* can be different and remains to be studied.

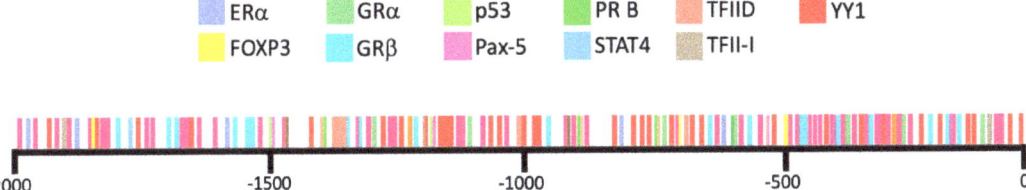

Figure 6. In silico screening of putative transcription factor binding sites for the *LGALS16* gene. There are multiple binding sites for 17 transcription factors within the 2 kb promoter region upstream the transcription start site of *LGALS16* gene as detected by PROMO.

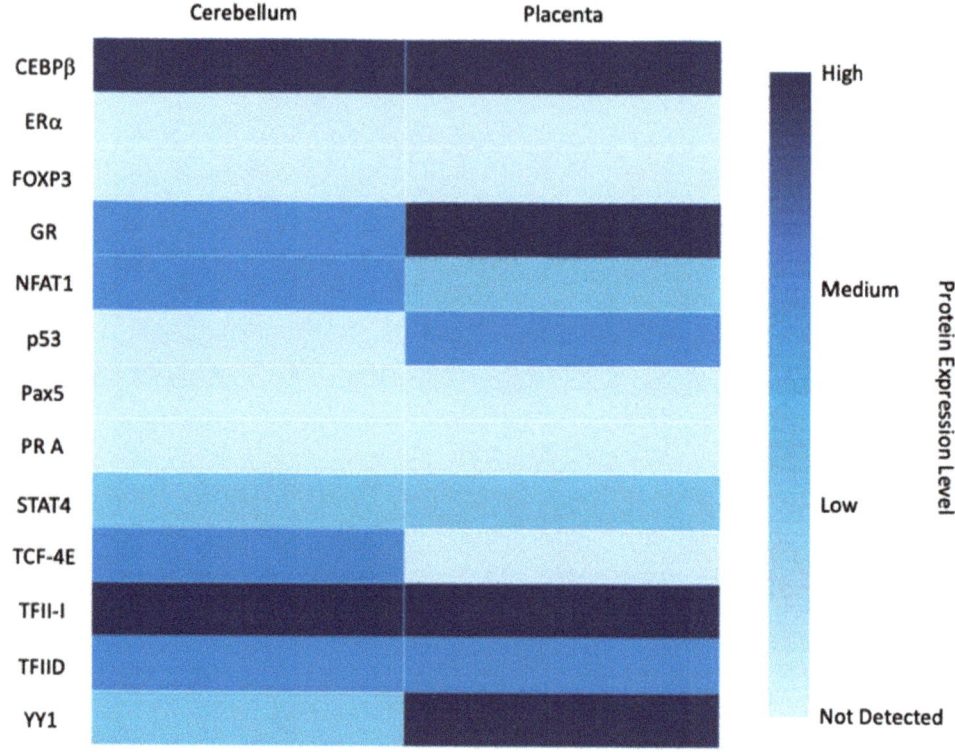

Figure 7. Protein expression patterns of predicted transcription factors for *LGALS16* regulation in the cerebellum and placenta.

3.3.2. miRNAs

Post-transcriptional control of mRNA availability for protein synthesis depends on miRNAs which can hybridize to complementary sequences in protein-coding mRNAs at the 3′ untranslated region and either block protein translation or induce mRNA degradation [52]. Multiple miRNAs were predicted to target the *LGALS16* transcript by bioinformatics tools, such as Diana Tools [17], miRabel [18], miRDB [19], and TargetScan [20], which use different algorithms and methods. A robust application of these tools using default options shows that five miRNAs (hsa-miR-3155a, hsa-miR-3155b, hsa-miR-4689, hsa-miR-4778-5p, hsa-miR-6783-5p) are predicted by all four of these online platforms (Figure 8). These miRNAs among others can be considered as perspective candidates for regulating the stability and/or translational potential of the *LGALS16* transcripts, especially in relevant tissues such as placenta and brain. Indeed, a significant decrease in hsa-miR-4778-5p expression during gestation in exosomes from maternal blood was associated with preterm birth pregnancies [53]. Expression of hsa-miR-3155a was significantly upregulated in the anterior cingulate cortex of deceased patients with major depressive disorder [54]. In comparison, the expression of hsa-miR-4689 was downregulated in exosomes isolated from the plasma of patients with mesial temporal lobe epilepsy with hippocampal sclerosis compared to controls [55]. Differential expression of exosomal hsa-miR-4689 and hsa-miR-6783-5p was reported in patients with intracranial aneurysms [56]. Human miRNA tissue atlas confirms expression of hsa-miR-3155a, hsa-miR-3155b, hsa-miR-4689, and hsa-miR-4778-5p in brain among other tissues at variable levels [57]. Unraveling possible mechanisms of miRNA-mediated regulation of galectin-16 in these tissues awaits future research.

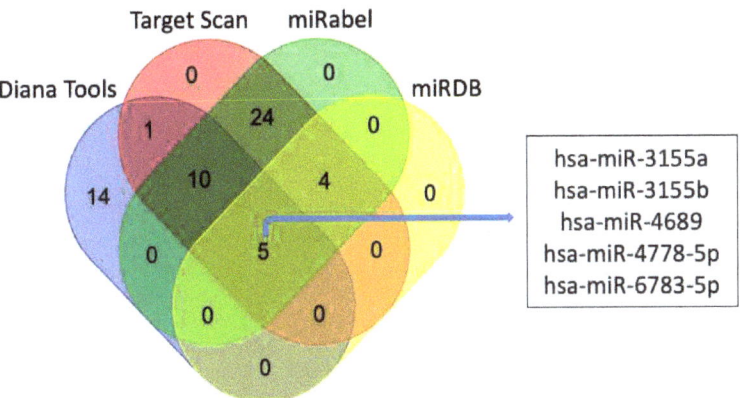

Figure 8. Putative miRNAs targeting *LGALS16* mRNA transcript. Five miRNAs are unanimously predicted by four different online platforms such as Diana Tools, miRabel, miRDB, and TargetScan.

3.4. LGALS16 and Human Diseases

Dysregulation of the placenta-specific gene cluster containing *LGALS16* is associated with a pregnancy complication known as preeclampsia, which can be highly fatal for both the mother and fetus. As such, *LGALS16* together with *LGALS13* and *LGALS14* were confirmed to satisfy the criteria of placenta enriched genes in a comprehensive study of RNA-Seq datasets from 302 placental biopsies [58]. However, although increasing expression of *LGALS13*, *LGALS14*, and *LGALS16* was observed during forskolin-induced syncytialization and differentiation of primary trophoblasts and BeWo cells in culture, only *LGALS13* and *LGALS14* were downregulated in preeclampsia with no significant changes of *LGALS16* [12]. Remarkably, *LGALS16* does not show sex-biased expression depending on the chromosomal sex of the fetus while *LGALS13* and *LGALS14* are notably elevated in fetal male placentas based on the chorionic villus transcriptome [59]. These aspects of galectin network regulation remain unclear in the context of placental disorders and development.

Alterations in the expression or mutations of *LGALS16* have been also reported for several other diseases and based mostly on microarray and RNA-Seq analysis, although the application of this gene as a biomarker is still unknown. Gene expression profiling with RNA sequencing data revealed that *LGALS16* was detected as an upregulated gene in fusiform gyrus tissue sections of 219 autopsy-confirmed Alzheimer's cases versus 70 neurologically normal age-matched controls [60]. *LGALS16* was also recognized as a brain tissue-specific gene within genome-wide associations with several neuroimaging psychiatric traits [61]. Further, *LGALS16* was expressed two-fold higher in chronic myeloid leukemia granulocytes compared to controls [62]. Copy number variations were identified in chromosome 19 for multiple genes including *LGALS16* in association with clinical features, such as histological type, ethnicity, disease stage, and familial history, of breast cancer using tumor samples from a Brazilian cohort [63]. In addition, *LGALS16* was determined to be a moderate impact variant associated with autism spectrum disorder, consisting of a missense single nucleotide variant (SNV), which was reported as detrimental by bioinformatic tools SIFT and PolyPhen-2 [64]. *LGALS16* also had greater SNVs within the 3′ flank region with one or more mutations in patients with diffuse large B-cell lymphoma [65]. Moreover, a *LGALS16* SNP was revealed to be associated with insulin secretion in a cohort of African Americans [66]. This study also showed that interactions between this *LGALS16* SNP and others, such as an intergenic SNP upstream of the LYPLAL1 gene, have also been associated with type 2 diabetes risk. The *LGALS16* transcript was one of the top 50 down-regulated mRNA present in the exosomes isolated from the cerebrospinal fluid in patients with meningeal carcinomatosis in comparison with healthy controls [67].

4. Conclusions

Although the galectin-16 gene was described more than 10 years ago [11], the regulation, functions, and clinical aspects of this tissue-specific molecule are largely unexplored. Primary association of LGALS16 with placental tissue has been challenged by its detection in brain tissues and several cancer cell lines as followed from available microarray and RNA-seq databases. There are bioinformatics indications that the expression of LGALS16 changes in association with Alzheimer's disease, chronic myeloid leukemia, breast cancer, B-cell lymphoma, and type 2 diabetes. Although LGALS16 was not significantly impacted at the gene level in preeclampsia, there remain questions regarding regulation at the protein level, which cannot be properly addressed at this time due to the absence of commercially available specific galectin-16 antibodies. The results obtained with recombinant galectin-16 are promising, but there is still a gap in our understanding of why the expression of endogenous galectin-16 protein has not been reported. Nevertheless, among the possible functions of galectin-16 in these and other tissues, its contribution to the regulation of cellular differentiation and programmed cell death (apoptosis) warrants special attention. Lastly, the use of proper cell culture models and the examination of multiple factors (transcription regulators and miRNA) is evidently the first line of study to position galectin-16 within a complex galectin network in cells. The generation of galectin-16-specific antibody and LGALS16 knockout cell lines using CRISPR/Cas9 technology might be required steps to unravel the role and significance of this molecule in the context of cell biology.

Author Contributions: Conceptualization, A.V.T.; methodology, A.V.T. and J.D.K.; formal analysis, A.V.T. and J.D.K.; investigation, J.D.K.; writing—original draft preparation, J.D.K.; writing—review and editing, A.V.T. and J.D.K.; supervision, A.V.T.; funding acquisition, A.V.T. All authors have read and agreed to the published version of the manuscript.

Funding: This research was supported in part by a Discovery Grant from The Natural Sciences and Engineering Research Council of Canada (RGPIN-2019-06628) to A.V.T., J.D.K. is a recipient of the Canada Graduate Scholarship—Master's Program from The Natural Sciences and Engineering Research Council of Canada.

Data Availability Statement: The data presented in this study are available on request from the corresponding author.

Acknowledgments: We would like to thank Ahmad Butt and Haya Tawfik for their assistance in the collecting records for Table 1.

Conflicts of Interest: The authors declare no conflict of interest.

References

1. Timoshenko, A.V. Towards molecular mechanisms regulating the expression of galectins in cancer cells. *Cell. Mol. Life Sci.* **2015**, *72*, 4327–4340. [CrossRef] [PubMed]
2. Allo, V.C.M.; Toscano, M.A.; Pinto, N.; Rabinovich, G.A. Galectins: Key players at the frontiers of innate and adaptive immunity. *Trends Glycosci. Glycotechnol.* **2018**, *30*, SE97–SE107. [CrossRef]
3. Johannes, L.; Jacob, R.; Leffler, H. Galectins at a glance. *J. Cell Sci.* **2018**, *131*, jcs208884. [CrossRef] [PubMed]
4. Tazhitdinova, R.; Timoshenko, A.V. The emerging role of galectins and O-GlcNAc homeostasis in processes of cellular differentiation. *Cells* **2020**, *9*, 8. [CrossRef] [PubMed]
5. Vladoiu, M.C.; Labrie, M.; St-Pierre, Y. Intracellular galectins in cancer cells: Potential new targets for therapy (Review). *Int. J. Oncol.* **2014**, *44*, 1001–1014. [CrossRef]
6. Patterson, R.J.; Haudek, K.C.; Voss, P.G.; Wang, J.L. Examination of the role of galectins in pre-mRNA splicing. *Methods Mol. Biol.* **2015**, *1207*, 431–449.
7. Popa, S.J.; Stewart, S.E.; Moreau, K. Unconventional secretion of annexins and galectins. *Semin. Cell Dev. Biol.* **2018**, *83*, 42–50. [CrossRef] [PubMed]
8. He, J.; Baum, L.G. Galectin interactions with extracellular matrix and effects on cellular function. *Methods Enzymol.* **2006**, *417*, 247–256.
9. Nabi, I.R.; Shankar, J.; Dennis, J.W. The galectin lattice at a glance. *J. Cell Sci.* **2015**, *128*, 2213–2219. [CrossRef]
10. Than, N.G.; Romero, R.; Kim, C.J.; McGowen, M.R.; Papp, Z.; Wildman, D.E. Galectins: Guardians of eutherian pregnancy at the maternal-fetal interface. *Trends Endocrinol. Metab.* **2012**, *23*, 23–31. [CrossRef] [PubMed]

11. Than, N.G.; Romero, R.; Goodman, M.; Weckle, A.; Xing, J.; Dong, Z.; Xu, Y.; Tarquini, F.; Szilagyi, A.; Gal, P.; et al. A primate subfamily of galectins expressed at the maternal-fetal interface that promote immune cell death. *Proc. Natl. Acad. Sci. USA* **2009**, *106*, 9731–9736. [CrossRef]
12. Than, N.G.; Romero, R.; Xu, Y.; Erez, O.; Xu, Z.; Bhatti, G.; Leavitt, R.; Chung, T.H.; El-Azzamy, H.; LaJeunesse, C.; et al. Evolutionary origins of the placental expression of chromosome 19 cluster galectins and their complex dysregulation in preeclampsia. *Placenta* **2014**, *35*, 855–865. [CrossRef]
13. Pollheimer, J.; Vondra, S.; Baltayeva, J.; Beristain, A.G.; Knöfler, M. Regulation of placental extravillous trophoblasts by the maternal uterine environment. *Front. Immunol.* **2018**, *9*, 2597. [CrossRef]
14. Blois, S.M.; Dveksler, G.; Vasta, G.R.; Freitag, N.; Blanchard, V.; Barrientos, G. Pregnancy galectinology: Insights into a complex network of glycan binding proteins. *Front. Immunol.* **2019**, *10*, 1166. [CrossRef] [PubMed]
15. Messeguer, X.; Escudero, R.; Farré, D.; Núñez, O.; Martínez, J.; Albà, M.M. PROMO: Detection of known transcription regulatory elements using species-tailored searches. *Bioinformatics* **2002**, *18*, 333–334. [CrossRef] [PubMed]
16. Farré, D.; Roset, R.; Huerta, M.; Adsuara, J.E.; Roselló, L.; Albà, M.M.; Messeguer, X. Identification of patterns in biological sequences at the ALGGEN server: PROMO and MALGEN. *Nucleic Acids Res.* **2003**, *31*, 3651–3653. [CrossRef] [PubMed]
17. Paraskevopoulou, M.D.; Georgakilas, G.; Kostoulas, N.; Vlachos, I.S.; Vergoulis, T.; Reczko, M.; Filippidis, C.; Dalamagas, T.; Hatzigeorgiou, A.G. DIANA-microT web server v5.0: Service integration into miRNA functional analysis workflows. *Nucleic Acids Res.* **2013**, *41*, W169–W173. [CrossRef] [PubMed]
18. Quillet, A.; Saad, C.; Ferry, G.; Anouar, Y.; Vergne, N.; Lecroq, T.; Dubessy, C. Improving bioinformatics prediction of microRNA targets by ranks aggregation. *Front. Genet.* **2020**, *10*, 1330. [CrossRef] [PubMed]
19. Chen, Y.; Wang, X. miRDB: An online database for prediction of functional microRNA targets. *Nucleic Acids Res.* **2020**, *48*, D127–D131. [CrossRef] [PubMed]
20. Agarwal, V.; Bell, G.W.; Nam, J.W.; Bartel, D.P. Predicting effective microRNA target sites in mammalian mRNAs. *eLife* **2015**, *4*, e05005. [CrossRef]
21. Uhlén, M.; Fagerberg, L.; Hallström, B.M.; Lindskog, C.; Oksvold, P.; Mardinoglu, A.; Sivertsson, A.; Kampf, C.; Sjöstedt, E.; Asplund, A. Tissue-based map of the human proteome. *Science* **2015**, *347*, 1260419. [CrossRef]
22. Clark, K.; Karsch-Mizrachi, I.; Lipman, D.J.; Ostell, J.; Sayers, E.W. GenBank. *Nucleic Acids Res.* **2016**, *44*, D67–D72. [CrossRef] [PubMed]
23. Berman, H.M.; Westbrook, J.; Feng, Z.; Gilliland, G.; Bhat, T.N.; Weissig, H.; Shindyalov, I.N.; Bourne, P.E. The protein data bank. *Nucleic Acids Res.* **2000**, *28*, 235–242. [CrossRef] [PubMed]
24. Ye, J.; Coulouris, G.; Zaretskaya, I.; Cutcutache, I.; Rozen, S.; Madden, T.L. Primer-BLAST: A tool to design target-specific primers for polymerase chain reaction. *BMC Bioinform.* **2012**, *13*, 134. [CrossRef] [PubMed]
25. Timoshenko, A.V. Chitin hydrolysate stimulates VEGF-C synthesis by MDA-MB-231 breast cancer cells. *Cell Biol. Int.* **2011**, *35*, 281–286. [CrossRef]
26. Renaud, S.J.; Chakraborty, D.; Mason, C.W.; Rumi, M.A.; Vivian, J.L.; Soares, M.J. OVO-like 1 regulates progenitor cell fate in human trophoblast development. *Proc. Natl. Acad. Sci. USA* **2015**, *112*, E6175–E6184. [CrossRef] [PubMed]
27. Timoshenko, A.V.; Lanteigne, J.; Kozak, K. Extracellular stress stimuli alter galectin expression profiles and adhesion characteristics of HL-60 cells. *Mol. Cell Biochem.* **2016**, *413*, 137–143. [CrossRef]
28. Sherazi, A.A.; Jariwala, K.A.; Cybulski, A.N.; Lewis, J.W.; Karagiannis, J.; Cumming, R.C.; Timoshenko, A.V. Effects of global O-GlcNAcylation on galectin gene-expression profiles in human cancer cell lines. *Anticancer Res.* **2018**, *38*, 6691–6697. [CrossRef] [PubMed]
29. Ely, A.Z.; Moon, J.M.; Sliwoski, G.R.; Sangha, A.K.; Shen, X.-X.; Labella, A.L.; Meiler, J.; Capra, J.A.; Rokas, A. The impact of natural selection on the evolution and function of placentally expressed galectins. *Genome Biol. Evol.* **2019**, *11*, 2574–2592. [CrossRef]
30. Singer, M.F. SINEs and LINEs: Highly repeated short and long interspersed sequences in mammalian genomes. *Cell* **1982**, *28*, 433–434. [CrossRef]
31. Weckselblatt, B.; Rudd, M.K. Human structural variation: Mechanisms of chromosome rearrangements. *Trends Genet.* **2015**, *31*, 587–599. [CrossRef] [PubMed]
32. Si, Y.; Yao, Y.; Ayala, G.J.; Li, X.; Han, Q.; Zhang, W.; Xu, X.; Tai, G.; Mayo, K.H.; Zhou, Y.; et al. Human galectin-16 has a pseudo ligand binding site and plays a role in regulating c-Rel mediated lymphocyte activity. *Biochim. Biophys. Acta Gen. Subj.* **2021**, *1865*, 129755. [CrossRef] [PubMed]
33. Barrett, T.; Wilhite, S.E.; Ledoux, P.; Evangelista, C.; Kim, I.F.; Tomashevsky, M.; Marshall, K.A.; Phillippy, K.H.; Sherman, P.M.; Holko, M.; et al. NCBI GEO: Archive for functional genomics data sets—Update. *Nucleic Acids Res.* **2013**, *41*, D991–D995. [CrossRef]
34. Rosenfeld, C.S. The placenta-brain-axis. *J. Neurosci. Res.* **2021**, *99*, 271–283. [CrossRef]
35. Inamochi, Y.; Mochizuki, K.; Goda, T. Histone code of genes induced by co-treatment with a glucocorticoid hormone agonist and a p44/42 MAPK inhibitor in human small intestinal Caco-2 cells. *Biochim. Biophys. Acta* **2014**, *1840*, 693–700. [CrossRef] [PubMed]
36. Park, M.H.; Hong, J.T. Roles of NF-κB in cancer and inflammatory diseases and their therapeutic approaches. *Cells* **2016**, *5*, 15. [CrossRef]
37. Hayden, M.S.; Ghosh, S. NF-κB in immunobiology. *Cell Res.* **2011**, *21*, 223–244. [CrossRef]

38. Balsa, E.; Perry, E.A.; Bennett, C.F.; Jedrychowski, M.; Gygi, S.P.; Doench, J.G.; Puigserver, P. Defective NADPH production in mitochondrial disease complex I causes inflammation and cell death. *Nat. Commun.* **2020**, *11*, 2714. [CrossRef]
39. Nuzzo, A.M.; Giuffrida, D.; Zenerino, C.; Piazzese, A.; Olearo, E.; Todros, T.; Rolfo, A. JunB/Cyclin-D1 imbalance in placental mesenchymal stromal cells derived from preeclamptic pregnancies with fetal-placental compromise. *Placenta* **2014**, *35*, 483–490. [CrossRef]
40. Han, Y.M.; Romero, R.; Kim, J.S.; Tarca, A.L.; Kim, S.K.; Draghici, S.; Kusanovic, J.P.; Gotsch, F.; Mittal, P.; Hassan, S.S.; et al. Region-specific gene expression profiling: Novel evidence for biological heterogeneity of the human amnion. *Biol. Reprod.* **2008**, *79*, 954–961. [CrossRef]
41. Knyazev, E.N.; Zakharova, G.S.; Astakhova, L.A.; Tsypina, I.M.; Tonevitsky, A.G.; Sukhikh, G.T. Metabolic reprogramming of trophoblast cells in response to hypoxia. *Bull. Exp. Biol. Med.* **2019**, *166*, 321–325. [CrossRef] [PubMed]
42. Walcott, B.P.; Winkler, E.A.; Zhou, S.; Birk, H.; Guo, D.; Koch, M.J.; Stapleton, C.J.; Spiegelman, D.; Dionne-Laporte, A.; Dion, P.A.; et al. Identification of a rare BMP pathway mutation in a non-syndromic human brain arteriovenous malformation via exome sequencing. *Hum. Genome Var.* **2018**, *5*, 18001. [CrossRef] [PubMed]
43. Ma, S.; Pang, C.; Song, L.; Guo, F.; Sun, H. Activating transcription factor 3 is overexpressed in human glioma and its knockdown in glioblastoma cells causes growth inhibition both in vitro and in vivo. *Int. J. Mol. Med.* **2015**, *35*, 1561–1573. [CrossRef]
44. Garces de Los Favos Alonso, I.; Liang, H.C.; Turner, S.D.; Lagger, S.; Merkel, O.; Kenner, L. The role of activator protein-1 (AP-1) family members in CD30-positive lymphomas. *Cancers* **2018**, *10*, 93. [CrossRef]
45. Shankar, K.; Kang, P.; Zhong, Y.; Borengasser, S.J.; Wingfield, C.; Saben, J.; Gomez-Acevedo, H.; Thakali, K.M. Transcriptomic and epigenomic landscapes during cell fusion in BeWo trophoblast cells. *Placenta* **2015**, *36*, 1342–1351. [CrossRef]
46. Cheng, Y.-H.; Richardson, B.D.; Hubert, M.A.; Handwerger, S. Isolation and characterization of the human syncytin gene promoter. *Biol. Reprod.* **2004**, *70*, 694–701. [CrossRef] [PubMed]
47. Wang, Z.; Liu, Y.; Liu, J.; Kong, N.; Jiang, Y.; Jiang, R.; Zhen, X.; Zhou, J.; Li, C.; Sun, H.; et al. ATF3 deficiency impairs the proliferative-secretory phase transition and decidualization in RIF patients. *Cell Death Dis.* **2021**, *12*, 387. [CrossRef]
48. Jadhav, K.; Zhang, Y. Activating transcription factor 3 in immune response and metabolic regulation. *Liver Res.* **2017**, *1*, 96–102. [CrossRef]
49. Moslehi, R.; Mills, J.L.; Signore, C.; Kumar, A.; Ambroggio, X.; Dzutsev, A. Integrative transcriptome analysis reveals dysregulation of canonical cancer molecular pathways in placenta leading to preeclampsia. *Sci. Rep.* **2013**, *3*, 2407. [CrossRef]
50. Tsukamoto, S.; Mizuta, T.; Fujimoto, M.; Ohte, S.; Osawa, K.; Miyamoto, A.; Yoneyama, K.; Murata, E.; Machiya, A.; Jimi, E.; et al. Smad9 is a new type of transcriptional regulator in bone morphogenetic protein signaling. *Sci. Rep.* **2014**, *4*, 7596. [CrossRef]
51. Heo, K.S.; Fujiwara, K.; Abe, J. Disturbed-flow-mediated vascular reactive oxygen species induce endothelial dysfunction. *Circ. J.* **2011**, *75*, 2722–2730. [CrossRef] [PubMed]
52. Fabian, M.R.; Sonenberg, N.; Filipowicz, W. Regulation of mRNA translation and stability by microRNAs. *Annu. Rev. Biochem.* **2010**, *79*, 351–379. [CrossRef] [PubMed]
53. Menon, R.; Debnath, C.; Lai, A.; Guanzon, D.; Bhatnagar, S.; Kshetrapal, P.K.; Sheller-Miller, S.; Salomon, C.; Garbhini Study Team. Circulating exosomal miRNA profile during term and preterm birth pregnancies: A longitudinal study. *Endocrinology* **2019**, *160*, 249–275. [CrossRef]
54. Yoshino, Y.; Roy, B.; Dwivedi, Y. Altered miRNA landscape of the anterior cingulate cortex is associated with potential loss of key neuronal functions in depressed brain. *Eur. Neuropsychopharmacol.* **2020**, *40*, 70–84. [CrossRef]
55. Yan, S.; Zhang, H.; Xie, W.; Meng, F.; Zhang, K.; Jiang, Y.; Zhang, X.; Zhang, J. Altered microRNA profiles in plasma exosomes from mesial temporal lobe epilepsy with hippocampal sclerosis. *Oncotarget* **2017**, *8*, 4136–4146. [CrossRef] [PubMed]
56. Liao, B.; Zhou, M.X.; Zhou, F.K.; Luo, X.M.; Zhong, S.X.; Zhou, Y.F.; Qin, Y.S.; Li, P.P.; Qin, C. Exosome-derived miRNAs as biomarkers of the development and progression of intracranial aneurysms. *J. Atheroscler. Thromb.* **2020**, *27*, 545–610. [CrossRef]
57. Ludwig, N.; Leidinger, P.; Becker, K.; Backes, C.; Fehlmann, T.; Pallasch, C.; Rheinheimer, S.; Meder, B.; Stähler, C.; Meese, E.; et al. Distribution of miRNA expression across human tissues. *Nucleic Acids Res.* **2016**, *44*, 3865–3877. [CrossRef]
58. Gong, S.; Gaccioli, F.; Dopierala, J.; Sovio, U.; Cook, E.; Volders, P.J.; Martens, L.; Kirk, P.D.W.; Richardson, S.; Smith, G.C.S.; et al. The RNA landscape of the human placenta in health and disease. *Nat. Commun.* **2021**, *12*, 2639. [CrossRef] [PubMed]
59. Braun, A.E.; Muench, K.L.; Robinson, B.G.; Wang, A.; Palmer, T.D.; Winn, V.D. Examining Sex Differences in the Human Placental Transcriptome During the First Fetal Androgen Peak. *Reprod Sci.* **2021**, *28*, 801–818. [CrossRef]
60. Vastrad, B.; Vastrad, C. Bioinformatics analyses of significant genes, related pathways and candidate prognostic biomarkers in Alzheimer's disease. *BioRxiv* **2021**. [CrossRef]
61. Zhao, B.; Shan, Y.; Yang, Y.; Zhaolong, Y.; Li, T.; Wang, X.; Luo, T.; Zhu, Z.; Sullivan, P.; Zhao, H.; et al. Transcriptome-wide association analysis of brain structures yields insights into pleiotropy with complex neuropsychiatric traits. *Nat. Commun.* **2021**, *12*, 2878. [CrossRef] [PubMed]
62. Čokić, V.P.; Mojsilović, S.; Jauković, A.; Kraguljac-Kurtović, N.; Mojsilović, S.; Šefer, D.; Mitrović Ajtić, O.; Milošević, V.; Bogdanović, A.; Đikić, D.; et al. Gene expression profile of circulating CD34(+) cells and granulocytes in chronic myeloid leukemia. *Blood Cells Mol. Dis.* **2015**, *55*, 373–381. [CrossRef] [PubMed]
63. Rodrigues-Peres, R.M.; de Carvalho, B.S.; Anurag, M.; Lei, J.T.; Conz, L.; Gonçalves, R.; Cardoso Filho, C.; Ramalho, S.; de Paiva, G.R.; Derchain, S.; et al. Copy number alterations associated with clinical features in an underrepresented population with breast cancer. *Mol. Genet. Genomic. Med.* **2019**, *7*, e00750. [CrossRef] [PubMed]

64. Santos, J.X.; Rasga, C.; Marques, A.R.; Martiniano, H.F.M.C.; Asif, M.; Vilela, J.; Oliveira, G.; Vicente, A.M. A role for gene-environment interactions in Autism Spectrum Disorder is suggested by variants in genes regulating exposure to environmental factors. *BioRxiv* **2019**. [CrossRef]
65. Arthur, S.E.; Jiang, A.; Grande, B.M.; Alcaide, M.; Cojocaru, R.; Rushton, C.K.; Mottok, A.; Hilton, L.K.; Kumar Lat, P.; Zhao, E.Y. Genome-wide discovery of somatic regulatory variants in diffuse large B-cell lymphoma. *Nat. Commun.* **2018**, *9*, 4001. [CrossRef] [PubMed]
66. Keaton, J.M.; Hellwege, J.N.; Ng, M.C.; Palmer, N.D.; Pankow, J.S.; Fornage, M.; Wilson, J.G.; Correa, A.; Rasmussen-Torvik, L.J.; Rotter, J.I.; et al. Genome-wide interaction with insulin secretion loci reveals novel loci for type 2 diabetes in African Americans. *PLoS ONE* **2016**, *11*, e0159977. [CrossRef]
67. Cheng, P.; Feng, F.; Yang, H.; Jin, S.; Lai, C.; Wang, Y.; Bi, J. Detection and significance of exosomal mRNA expression profiles in the cerebrospinal fluid of patients with meningeal carcinomatosis. *J. Mol. Neurosci.* **2021**, *71*, 790–803. [CrossRef]

Article

Structural Characterization of Rat Galectin-5, an N-Tailed Monomeric Proto-Type-like Galectin

Federico M. Ruiz [1], Francisco J. Medrano [1], Anna-Kristin Ludwig [2], Herbert Kaltner [2], Nadezhda V. Shilova [3], Nicolai V. Bovin [3], Hans-Joachim Gabius [2] and Antonio Romero [1,*]

[1] Department of Structural and Chemical Biology, CIB Margarita Salas, CSIC, Ramiro de Maeztu 9, 28040 Madrid, Spain; fruiz@cib.csic.es (F.M.R.); fjmedrano@cib.csic.es (F.J.M.)
[2] Physiological Chemistry, Department of Veterinary Sciences, Ludwig-Maximilians-University Munich, Lena-Christ-Str. 48, 82152 Planegg-Martinsried, Germany; Anna-Kristin.Ludwig@tiph.vetmed.uni-muenchen.de (A.-K.L.); kaltner@tiph.vetmed.uni-muenchen.de (H.K.); gabius@tiph.vetmed.uni-muenchen.de (H.-J.G.)
[3] Shemyakin & Ovchinnikov Institute of Bioorganic Chemistry, Russian Academy of Sciences, 16/10 Miklukho-Maklaya str., 117437 Moscow, Russia; pumatnv@gmail.com (N.V.S.); professorbovin@yandex.ru (N.V.B.)
* Correspondence: romero@cib.csic.es

Abstract: Galectins are multi-purpose effectors acting via interactions with distinct counterreceptors based on protein-glycan/protein recognition. These processes are emerging to involve several regions on the protein so that the availability of a detailed structural characterization of a full-length galectin is essential. We report here the first crystallographic information on the N-terminal extension of the carbohydrate recognition domain of rat galectin-5, which is precisely described as an N-tailed proto-type-like galectin. In the ligand-free protein, the three amino-acid stretch from Ser2 to Ser5 is revealed to form an extra β-strand (F0), and the residues from Thr6 to Asn12 are part of a loop protruding from strands S1 and F0. In the ligand-bound structure, amino acids Ser2–Tyr10 switch position and are aligned to the edge of the β-sandwich. Interestingly, the signal profile in our glycan array screening shows the sugar-binding site to preferentially accommodate the histo-blood-group B (type 2) tetrasaccharide and N-acetyllactosamine-based di- and oligomers. The crystal structures revealed the characteristically preformed structural organization around the central Trp77 of the CRD with involvement of the sequence signature's amino acids in binding. Ligand binding was also characterized calorimetrically. The presented data shows that the N-terminal extension can adopt an ordered structure and shapes the hypothesis that a ligand-induced shift in the equilibrium between flexible and ordered conformers potentially acts as a molecular switch, enabling new contacts in this region.

Keywords: β-hairpin; β-sandwich; blood group B; lectin; sugar code

1. Introduction

Storage of biological information involves more than nucleic acids and proteins. The ubiquity of occurrence, the enormous diversity already at the level of oligomers and the fine-tuned spatiotemporal regulation of the appearance of distinct structures are solid arguments for a fundamental functional meaning of the glycan part of cellular glycoconjugates [1–6]. Indeed, by molecular complementarity of oligosaccharides with a contact region in the carbohydrate recognition domains (CRDs) of sugar-binding proteins (lectins), glycan-encoded messages are 'read' and 'translated' into cellular effects [6–8]. Toward this end, triggering specific bioeffects, not only the selection of the binding partner(s), appears to matter. Furthermore, the lectin's design, modularity and quaternary structure are first revealed in the case of the tetrameric leguminous lectin concanavalin A by a lower extent of crosslinking of certain cell surface receptors [9,10]. Fittingly, the context of presentation of the CRD shows a wide range of variability within lectin families. When considering the emerging multifunctionality of lectins, regions not involved in glycan binding can also affect their mode of action. Variability in design and the potential of regions beyond the glycan-binding site to be a physiologically relevant call for a detailed structural analysis within the lectin families in all their naturally occurring forms.

Focusing on the adhesion/growth-regulatory ga(lactose-binding) lectins, their common CRD is presented in three types of protein architecture, i.e., as (non)covalently associated homo/heterodimers (proto or tandem-repeat types) or as the chimera-type galectin-3 (Gal-3) with its N-terminal stalk attached to the CRD [11–14]. This highly dynamic, over 100-amino-acid-long sequence is composed of sections of known functionality; i.e., non-triple helical collagen-like repeats (for self-association) and an N-terminal peptide with two sites for serine phosphorylation (for intracellular compartmentalization) [12,15–17]. Two members of this lectin family are peculiar: galectin-related protein (GRP) and rat galectin-5 (rGal-5) present a short N-terminal extension (of up to 37 amino acids) of the canonical CRD of unknown function, which is clearly a challenge to study. Their special status as N-tailed proto-type-like proteins thus prompted to accomplish structural characterization of the full-length protein. In particular, it is of interest to define the structural features of the N-terminal extension, if adopted. Since respective attempts had so far been unsuccessful in the cases of human and chicken GRP, which had been crystallized as a truncated version [18,19], rGal-5 is the remaining target protein to try a full characterization of an N-tailed proto-type-like galectin.

This lectin was first purified from rat lung (denoted as RL-18) [20]. Sequencing of the cDNA from a rat reticulocyte library identified a strong homology (over 80%) to the C-terminal CRD of the tandem-repeat-type galectin-9 (Gal-9C), and monitoring among mammalian genomes disclosed its status as being uniquely present in rats [21–24]. The exon profile assumed that the rGal-5 gene originated from a species-specific gene duplication event followed by partial deletion to maintain the first exon coding for 13 amino acids and then the three exons of Gal-9C [12,24] (Figure 1). Notably, duplications and copy number variability of the galectin genes between species are not uncommon among mammals [25], rGal-5 being a specially processed species-specific form.

In solution, the current status of analysis describes rGal-5 as a monomer with a weak haemagglutinin activity [20,22,26]. Non-sialylated glycan termini are binding partners, especially when clustered, as is the case for N-acetyllactosamine (LacNAc) of the three complex-type N-glycans of the nonavalent pan-galectin-binding glycoprotein asialofetuin, and rGal-5 binding distinguishes late-stage apoptotic from secondary necrotic peripheral blood lymphocytes [27–30]. Selectivity in glycan binding is also implied in rGal-5's involvement in the sorting processes during reticulocyte membrane remodeling by exosomal release [31].

In our study, we first report the binding specificity of rGal-5 through a glycan array, followed by further analysis of the ligand binding by isothermal titration calorimetry (ITC). The top places of a blood-group tetrasaccharide in the array test are due to its high affinity relative to LacNAc and is envisioned by evidence-based docking. Having succeeded in obtaining crystals of full-length rGal-5, structural information is then presented on the lectin in the absence and in the presence of lactose (Lac). The detected difference in the arrangement of the N-terminal tail in ligand-free and -loaded protein might suggest a molecular switch controlling contact formation in this area.

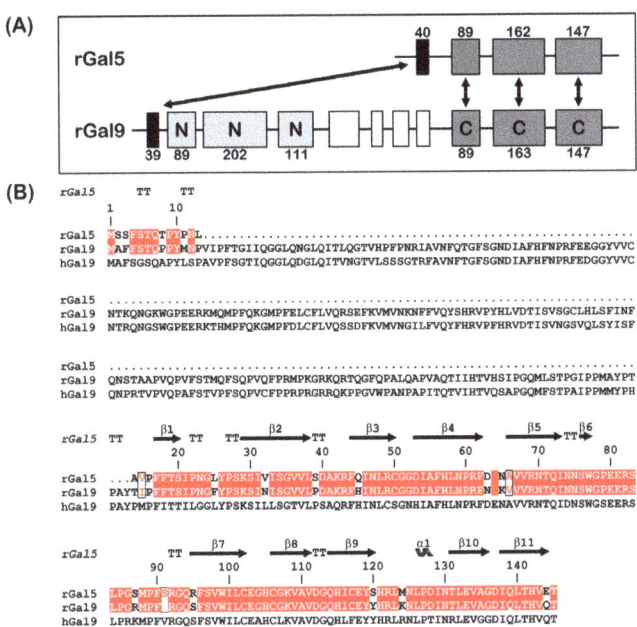

Figure 1. (**A**) Gene structures for rGal-5 and -9. The size of the exons (boxes) is indicated, and introns are drawn as lines (not drawn to scale). The N- and C-terminal CRDs of rGal-9 are labelled and given in different grey levels. Homologous exons are indicated (double arrows). (**B**) Sequence alignment of rGal-5 with rat and human Gal-9. Strictly conserved residues (red) and similar residues (boxed red letters) between rat proteins are shown. The upper lane represents the secondary structure elements of rGal-5 (α represents α-helices, β represents β-sheets and TT represents β-turns).

2. Materials and Methods

2.1. Protein Production and Purification

Recombinant production and purification by affinity chromatography followed by controls to ascertain purity were performed as previously described [26,32]). Afterwards, the rGal-5 and GRIFIN protein samples were extensively dialyzed at 277 K against 5 mM sodium phosphate buffer (pH 7.2), 0.2 M NaCl and 4 mM β-mercaptoethanol (PBS$_\beta$). Finally, the galectin-containing solutions were concentrated using Amicon Ultra 10,000 MCWO centrifugal filter units (Millipore, Darmstadt, Germany), and then loaded into a Hi-Prep 16/60 Sephacryl S100 column (GE Healthcare, Freiburg, Germany) equilibrated with 20 mM Na-K phosphate buffer (pH 7.0), 150 mM NaCl and 4 mM β-mercaptoethanol. To obtain the cGRIFIN–tetrasaccharide complex, the purified protein sample was then incubated with the sugar at a 1:2 molar ratio in the same buffer for 10 min in ice. In the case of the rGal-5–lactose complex, the sample was purified by affinity chromatography using home-made lactosylated Sepharose 4B [33] immediately after the first dialysis. The protein bound to the resin was eluted with PBS$_\beta$ and 200 mM lactose, concentrated and loaded into a Hi-Prep 16/60 Sephacryl S100 column (GE Healthcare, Freiburg, Germany) equilibrated with the previously described buffer supplemented with 5 mM lactose. All solutions of sugar-free and -loaded proteins were concentrated to a final concentration of 15 mg/mL for the screening of crystallization conditions.

2.2. Glycan Array Measurements

The array consisted of 609 compounds covering glycans and polysaccharides printed onto commercial NHS-activated Slide H (Schott Nexterion, Jena, Germany), tested with 100 μg/mL biotinylated rGal-5 and involving Alexa555-labeled streptavidin (Thermo

Fisher Scientific, Eugene, OR, USA) as second-step reagent for Innoscan 1100AL scanner-based (Innopsys, Carbonne, France) signal qualification, expressed in medium relative fluorescence units (RFU) and medium absolute deviation (MAD), as described previously [34,35]. When the fluorescence intensity exceeded the background value by a factor of five, the respective signal was considered to be significant.

2.3. Crystallization, Data Collection and Processing

Protein crystals were grown at 295 K by the vapor diffusion method. Specifically, rGal-5 and rGal-5–lactose crystals grew in 10% PEG 8000, 100 mM Tris pH 7.0 and 200 mM MgCl$_2$. Crystals were soaked in this solution supplemented with 30% v/v glycerol as cryoprotectant. GRIFIN–tetrasaccharide crystals were grown in 15% w/v PEG 400 and 100 mM MES pH 6.5. The same solution supplemented with 25% w/v PEG 400 was used as cryoprotectant. All crystals were flash-cooled by immersion in liquid nitrogen. Diffraction data were collected at the beamlines BM14 of the ESRF Synchrotron (Grenoble, France) and BL13-XALOC of the ALBA Synchrotron (Cerdanyola del Valles, Barcelona, Spain). Crystallographic data were processed using XDS [36] and Aimless [37]. Details of the diffraction data are presented in Table 1.

Table 1. Data collection and refinement statistics.

	rGal-5	rGal-5 + Lactose	cGRIFIN + Tetrasaccharide
Data collection			
Space group	P 2$_1$	P 22$_1$2$_1$	P 22$_1$2$_1$
Cell dimensions			
a, b, c (Å)	66.1, 68.1, 95.4	39.0, 65.8, 112.7	39.1, 70.6, 87.7
A, β, γ (°)	β = 91.8		
Resolution(Å) [a]	33.03–1.70	39.02–1.90	43.85–1.13
	(1.76–1.70)	(1.97–1.90)	(1.17–1.13)
Total reflections	402258 (17733)	46992 (4574)	181470 (17698
Unique reflections	92950 (9247)	23584 (2322)	90857 (8884)
R$_{merge}$	0.057 (0.457)	0.060 (0.252)	0.015 (0.305)
R$_{meas}$	0.077 (0.623)	0.085(0.947)	0.021 (0.431)
CC 1/2	0.99 (0.84)	0.99 (0.84)	1.00 (0.90)
Completeness (%)	99.75 (99.21)	99.78 (99.74)	99.13 (97.73)
<I/σ(I)>	14.01 (2.80)	7.60 (2.74)	22.0 (2.0)
Wilson B-factor	13.96	15.69	9.99
Multiplicity	4.1 (3.7)	2.0 (2.0)	2.0 (2.0)
Refinement			
R$_{work}$	0.17 (0.23)	0.17 (0.22)	0.15 (0.42)
R$_{free}$	0.21 (0.28)	0.23 (0.29)	0.17 (0.43)
N° atoms (non-hydrogens)	7547	2533	2926
Protein	6659	2212	2350
Ligands	42	47	184
Water	846	274	482
Protein residues	834	279	276
Average B factor (Å^2)	18.06	16.67	14.52
Protein atoms	16.60	15.61	12.39
Ligands	31.46	15.19	14.45
Water	28.88	25.47	24.93
R.m.s. deviations			
Bond lengths (Å)	0.006	0.007	0.007
Bond angles (°)	0.81	0.88	1.05
Ramachandran statistics			
Favoured (%)	98	98	95.59
Outliers (%)	0.5	1.2	0.37
Clashscore	5.7	1.79	3.56
PDB code	5JP5	5JPG	7P8H

[a] Values in parentheses are for the highest-resolution shell.

2.4. Structure Determination and Refinement

The Molecular Replacement Method was used to solve the structures. A poly-Ala model based on the structure of Gal-9C (PDB entry 3NV1 [38]) was used to determine the structure of rGal-5. The PDB entry 5NMJ [32] was used as the search model to solve the cGRIFIN–tetrasaccharide structure. Structural refinements were carried out using Phenix [39]. Manual building, addition of water molecules and placement of ligands were done using Coot [40]. Details of the model refinements are given in Table 1. Protein–protein interactions, in particular those engaging the rGal-5 N-terminal residues, were analyzed using the PISA web server [41]. Figures for structural representation were drawn with the Pymol program [42].

2.5. Analytical Ultracentrifugation

Galectin-containing samples were diluted to a final protein concentration of 0.45 mg/mL in buffers for size-exclusion chromatography and pre-cleared by a centrifugation step at 16,000× g. Sedimentation velocity experiments were run at 293 K in an Optima KL-I analytical ultracentrifuge (Beckman Coulter, Krefeld, Germany) with an An50-Ti rotor and standard double-sector Epon-charcoal center pieces (1.2 cm optical path length). Measurements were performed at 48,000 rpm, registering successive entries every minute at 280 nm. Rayleigh interferometric detection was used to monitor the profile of the concentration gradient as a function of time and radial position, and the data were analyzed using SedFit software (Version 14.7).

2.6. SAXS Experiments

SAXS data were collected at the beamline BM29 (ESRF Synchrotron, Grenoble, France) using the BioSAXS robot and a Pilatus 1M detector (Dectris, Switzerland) with synchrotron radiation at a wavelength of λ = 1.000 Å and a sample-detector distance of 2.867 m [43]. Each measurement consisted of 10 frames of 1 s exposure each for a 100 µL sample flowing through a 1-mm-diameter capillary during X-ray exposure. Buffer scattering was measured immediately before each measurement of the corresponding protein sample at 277 K. The obtained scattering images were spherically averaged and the buffer scattering intensities subtracted using in-house software. Protein samples were prepared at concentrations of 4 mg/mL and 6 mg/mL in 20 mM Na-K phosphate buffer at pH 7.0 containing 150 mM NaCl and 5 mM lactose. Particle envelopes were generated ab initio using the program DAMMIF [44]. Multiple runs were performed to generate 20 independent model shapes that were combined and filtered to produce an averaged model using the DAMAVER software package [45].

2.7. ITC Measurements

The spacered B (type 2) tetrasaccharide was synthesized as described previously [46]. Titrations were monitored in a PEAQ-ITC calorimeter (Malvern, Westborough, MA, USA), using a galectin-containing solution of 250 µL in PBS (10 mM Na_2HPO_4, 2 mM KH_2PO_4, 137 mM NaCl and 3 mM KCl at pH 6.8) containing 10 mM β-mercaptoethanol and injections at 150 s intervals of 2 µL ligand-containing solution (up to adding 36.4 µL) at 25 °C (750 rpm), as described [33,47]. The protein concentration was 125 µM, and the ligand concentration in the syringe was 2.5 mM. In each titration, a fitted offset parameter was applied to account for potential background. Data processing was performed using the MiroCal PEAQ-ITC Analysis software.

3. Results

3.1. Glycan Array Data

rGal-5 is first tested to determine its binding profile to chip-presented substances, mostly glycans up to the molecular mass of bacterial polysaccharides. By using an array platform with 609 compounds, the spacered histo-blood-group B (type 2) tetrasaccharide, LacNAc-based dimers and the xenoantigen with α1,3-linked galactose added to a LacNAc core were found to be frontrunners in terms of signal intensity, together with several bacterial polysaccharides (Figure 2; for a complete listing of compounds and signal intensities, please see Supplementary Material, Table S1). rGal-5, in contrast to GRP, which has lost the ability to bind β-galactosides [19], thus presents a profile with typical selectivity among this class of glycans. To report the contact pattern between rGal-5 and the selected carbohydrate ligands, we then carried out systematic screening to find conditions for crystallization. In these experimental series, we used the full-length protein to obtain structural information on the N-terminal tail.

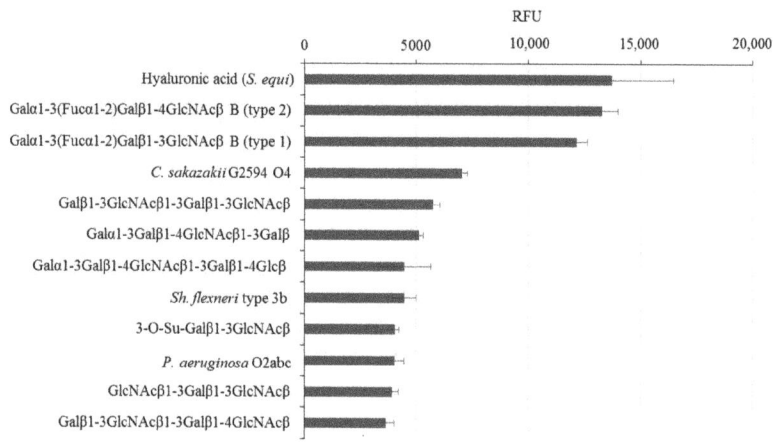

Figure 2. Top-12 glycans in the composition of the glycan array that exhibit binding with rGal-5.

3.2. Overall Crystallographic Structure of Full-Length rGal-5

Ligand-free full-length rGal-5 crystallizes in the monoclinic P2$_1$ space group and diffracts to a resolution of 1.7 Å. An estimation of the crystal solvent content suggested the presence of six galectin molecules in the asymmetric unit (Figure 3A). The rGal-5–lactose complex crystals belong to space group P22$_1$2$_1$, with only two molecules present in the asymmetric unit (Figure 4A), and diffract to a 1.9 Å resolution.

The overall fold in ligand-free (Figure 3B) and -loaded rGal-5 (Figure 4A) is composed of two antiparallel β-sheets (F1 to F5 and S1 to S6 strands) that form the characteristic β-sandwich structure. A short 3$_{10}$ helix is placed between strands F5 and S2. Beyond analyzing the architecture of the contact site for glycans (see below), these crystals offered the opportunity to examine whether the N-terminal extension presents well-ordered elements or high flexibility.

3.3. Structure of Ligand-Free rGal-5

In the ligand-free structure, six rGal-5 molecules are arranged in the asymmetric unit, as shown in Figure 3A. The core of the different protein units can readily be superimposed onto each other, as revealed by the low average root mean square deviation (RMSD) value among them of 0.26 Å for all Cα atoms. Differences are attributed mainly to the N-terminus, and the six protein monomers can be divided into two groups (Figure 3A), based on the experimental electron density. In the first group (chains A–C), the 11 residues at the N-terminus are not visible in the electron density map. Their likely extended and flexible structure in solution can indeed be derived from our SAXS data: the ab initio model of rGal-5 calculated on this basis exhibits, expectably, a globular shape. Most interestingly, though, a cylindrical extension on its top was seen in the model. When placing the CRD within the spherical region, the extended N-terminal section matches the geometry of the cylindrical part of the SAXS model (see below).

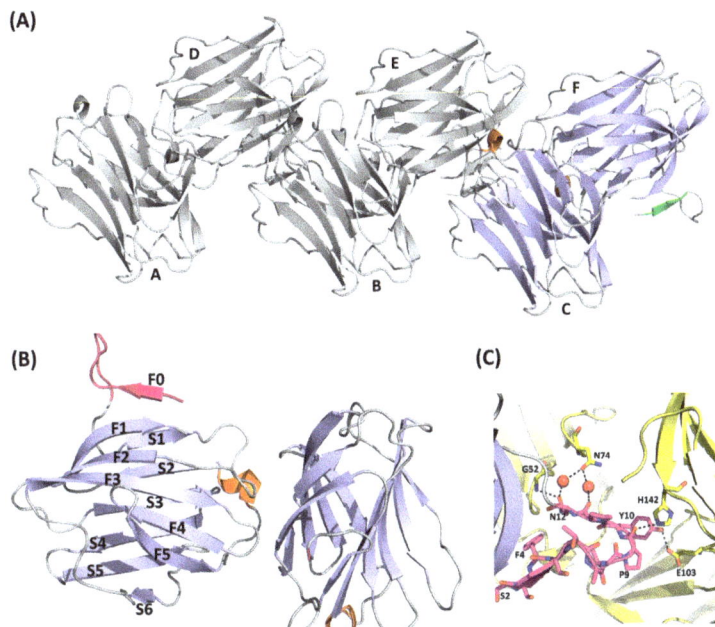

Figure 3. Structure of ligand-free rGal-5. (**A**) Ribbon diagram of the asymmetric unit of the crystals containing six molecules(A–F) of rGal-5 (helix in orange, β-strands in blue and the N-terminal extension in green). (**B**) Strands (labelled F0 to F5 on one side and S1 to S6 on the other) forming the characteristic β-sheets are labelled (helix in orange, β-strands in green and the N-terminal extension in magenta). An extra strand is found, named F0, placed immediately in front of a loop protruding from the CRD. (**C**) Intermolecular interactions stabilize the extended loop conformation. These interactions involve residues Gly52 (G52), Pro9 (P9), His142 (H142), Tyr10 (Y10), Glu103 (E103), Asn12 (N12), Gly52 (G52), Asn74 (N74). Symmetry-related molecules are shown in yellow.

When inspecting the second group (chains D–F), the electron density for this peptide stretch was clearly observed: the three amino acids from Ser2 to Ser5 form an extra β-strand, named F0, running in antiparallel direction to the C-terminal F1 strand. Residues Thr6 to Asn12 are in a loop, placed in parallel to the axis of the β-sandwich and protruding more than 10 Å from the S1 and F0 strands (Figure 3B).

The conformation within this loop is stabilized by interactions with symmetry-related molecules: hydrogen bonds between Pro9–His142, Tyr10–Glu103 and Asn12–Gly52 as well as a water bridge between Asn12 and Asn74 (Figure 3C).

3.4. Structure of Ligand-Loaded rGal-5

The two CRDs present in the asymmetric unit, which have bound lactose (Lac), exhibit very similar features to ligand-free rGal-5 (Figure 4A), with an RMSD value of only 0.3 Å for all Cα atoms. One of the two monomers in the asymmetric unit exhibits strong electron density for the first 10 residues so that their structure could be modelled. Intriguingly, these residues run parallel to the edge of the β-sandwich (Figure 4A) instead of forming the F0 strand and the protuberant loop observed in the ligand-free state. Intramolecular (hydrogen bonds between Ser2 and Ser3 with Ser39) and intermolecular contacts with symmetry-related molecules (hydrogen bond between Thr6 and Asp40) stabilize this special spatial arrangement (Figure 4B).

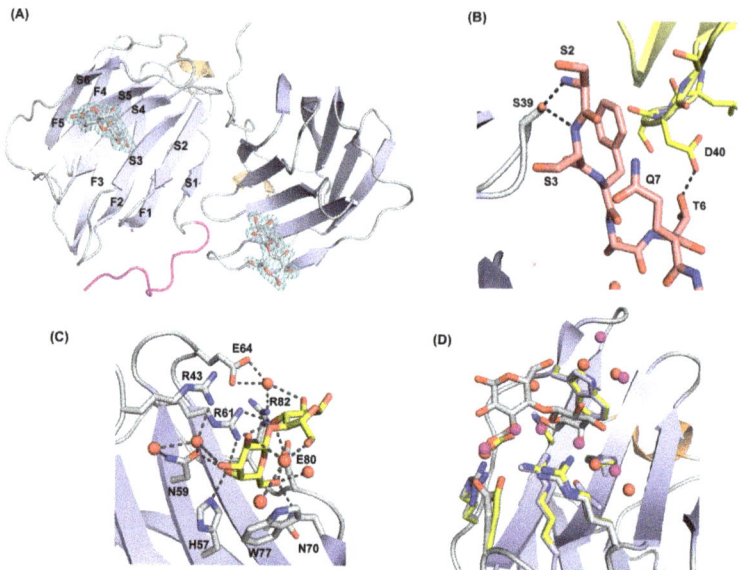

Figure 4. Structure of the rGal-5–lactose complex. (**A**) Overall architecture of the asymmetric unit of the rGal-5–lactose complex, with two CRDs; in one of them, the N-terminal residues (in magenta) adopt an extended geometry interacting with the edge of the β-sandwich. Strands that form the characteristic β-sheets are labelled (S1 to S6 on one strand, F1 to F5 on the other), and lactose molecules represented in sticks showing the 1.9 Å resolution $2F_o$-F_c electron density map (in blue) contoured at 1.0 σ. (**B**) Inter- and intra-molecular interactions that stabilize the extended conformation of the N-terminal residues. These interactions involve residues Ser2 (S2), Ser3 (S3), Ser39 (S39), Thr6 (T6) and Asp40 (D40). Symmetry-related molecules are represented in different colors. (**C**) Close-up view of the ligand-binding site of rGal-5 to show interactions between the protein and lactose. Key protein residues [His57 (H57), Asn59 (N59), Arg61 (R61), Asn70 (N70), Glu80 (E80) Arg82 (R82), Arg43 (R43), Glu64 (E64) and Trp77 (W77)] and lactose are represented in stick mode and water molecules as red spheres. (**D**) Superposition of the ligand-binding sites of rGal-5 (yellow) and the rGal-5–lactose complex (grey). Side-chain positions of residues at this site are not affected by ligand binding, indicating a preformed geometry. In the ligand-free structure, the position of the residues is kept by interactions with water or glycerol molecules. Water molecules from this last structure are shown in purple for clarity.

The carbohydrate-binding site in the concave face of the β-sheet is constituted by β-strands S4 to S6. The amino acids of the signature sequence, i.e., His57, Asn59, Arg61, Asn70, Glu80 and Arg82, directly interact with lactose through hydrogen bonding interactions. Additionally, Arg43, Gln45 and Glu64 form water-mediated hydrogen bonds with the ligand. As commonly found in galectin–lactose complexes, the indole ring of Trp77 stacks to the β-face of the pyranose ring of galactose (Figure 4C). In the absence of lactose, these residues form contacts with water (or glycerol molecules under conditions used for crystallization) molecules. Only minor rearrangements are observed for residues Arg43 and Glu64, which interact through water molecules with lactose (Figure 4D). Moving beyond defining the contact pattern, the thermodynamics of the ligand binding was analyzed by ITC.

3.5. ITC Measurements

rGal-5 interact with Lac with a dissociation constant of 136 ± 16 µM, which is lowered in the case of LacNAc to 30.5 ± 1.9 µM and 5.5 ± 0.6 µM for the blood-group B tetrasaccharide (Table 2 and Supplementary Material, Figure S2). This stepwise affinity enhancement can be explained by the increased number of contacts that these ligands make with additional amino acids. In this case, we used its complex with an avian galectin cGRIFIN (see below) shown in Figure 5 and the respective model building to obtain the relevant information for the new additional contacts, as described below, between the tetrasaccharide and rGal5.

Table 2. ITC data for ligand binding to recombinant rGal-5 (at 25 °C).

Ligand	K_d (µM)	Stoichiometry	ΔG^0_{obs} (kcal/mol)	ΔH^0_{obs} (kcal/mol)	$-T\Delta S^0_{obs}$ (kcal/mol)
Lactose	121 ± 5	0.95 ± 0.05	−5.35	−4.99 ± 0.08	−0.36
	151 ± 5	0.99 ± 0.06	−5.22	−5.03 ± 0.41	−0.18
LacNAc	28.6 ± 2.0	0.92 ± 0.01	−6.1	−10.0 ± 0.2	3.87
	32.3 ± 7.1	0.94 ± 0.20	−6.0	−10.5 ± 2.6	4.43
Tetrasaccharide	6.1 ± 0.2	0.94 ± 0.01	−7.11	−5.60 ± 0.03	−1.51
	5.0 ± 0.2	0.98 ± 0.01	−7.11	−5.74 ± 0.03	−1.37

3.6. Structure of Ligand-Loaded cGRIFIN

Our attempts to crystallize rGal-5 bound to the blood-group B tetrasaccharide were unsuccessful. Thus, we decided to test chicken GRIFIN (cGRIFIN) as a model for the binding of this compound. This very stable protein has been previously crystallized in several conditions [32]. We were able to obtain crystals of cGRIFIN in the presence of the blood-group B tetrasaccharide. These crystals diffracted up to a resolution of 1.14 Å (Table 1). This high-resolution data allowed us to build the sugar structure in the electron density in both carbohydrate-binding sites of the dimer. A comparison of the lactose-bound (PDB 5NLE) and the tetrasaccharide-bound cGRIFIN structures shows the absence of any significant structural change between these two structures, the RMSD value being 0.382 Å for all Cα atoms. The GalB moiety fully superposes with the galactose moiety of lactose, forming H-bonds with His46, Asn48, Arg50, Asn59 and Glu69. On the other hand, the GlcNacB moiety is rotated in the tetrasaccharide compared to the glucose moiety of lactose. Despite this change in the conformation, the H-bond with Glu69 is conserved. The acetamido group is exposed to the solvent as it is the FucA moiety. The GalA moiety establishes two additional H-bonds, one of them linking the 6'-hydroxyl group with the NE atom of Trp66. The second one extends the binding site beyond the S4 strand, linking the 2'-hydroxyl group with Glu32 (Figure 5A).

The superposition of the lactose-bound rGal-5 structure with the blood-group B tetrasaccharide-bound cGRIFIN gave an RMSD of 0.69 Å for all Cα atoms, showing the similarity of both complexes. This similarity allows us to analyze the interactions that could be established between this ligand and rGal-5 (Figure 5B). The GalB and GlcNacB moieties of the tetrasaccharide could stablish the same interactions as those observed for lactose, including the one with Arg82. The GalA moiety is properly placed to interact with the NE atom of Trp66 and with the side chain of Gln45, a residue from the S3 strand. In addition, Arg43 faces the FucA ring and could interact with this moiety, expanding the ligand-protein surface of contact (Figure 5C). This last residue belongs to the loop connecting the S3 and F2 strands, the region interacting with the N-terminal residues in the ligand bound rGal-5 structure.

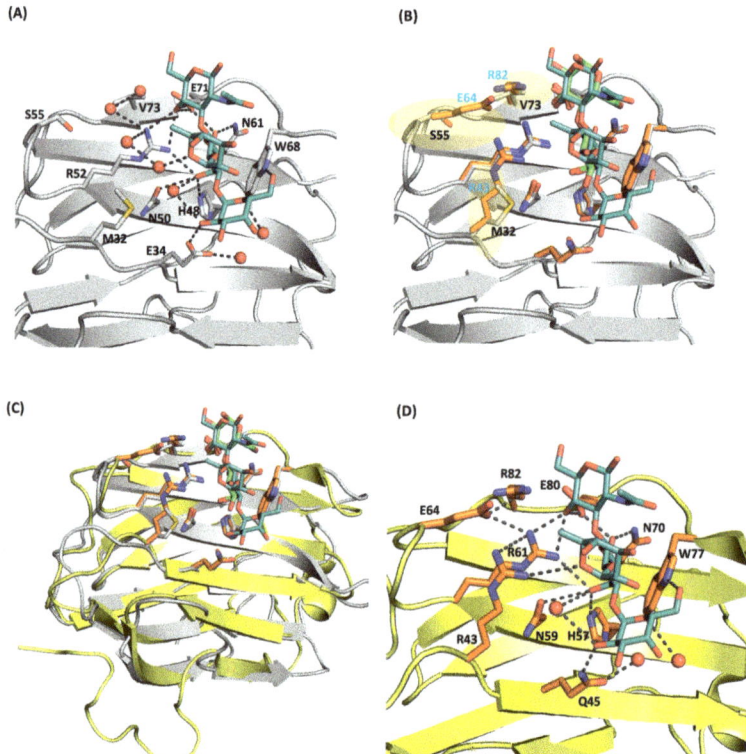

Figure 5. (**A**) Close-up view of the carbohydrate-binding site (CBD) of cGRIFIN showing the interactions between the histo-blood-group B tetrasaccharide and the active site residues [Met32 (M32), Glu34 (E34), His48 (H48), Asn50 (N50), Arg52 (R52), Ser55 (S55), Asn61 (N61), Trp68 (W68), Glu71 (E71) and Val73 (V73)]. (**B**) Structural comparison between cGRIFIN bound to the blood-group B tetrasaccharide complex (carbons in white) with the rGal-5/lactose complex (lactose in green, carbons of the interacting residues in orange). While residues such as His57, Asn59, Arg61, Asn70, Glu80 and Arg82 occupy almost the same position than residues in the former structure, the presence of charged residues (Arg53, Glu64 and Arg82) (highlighted in semitransparent yellow color) may lead to direct or water-mediated interactions with the ligand. (**C**) Superposition of the cGRIFIN/tetrasaccharide structure (grey; the histo-blood-group B tetrasaccharide in cyan) with the rGal-5–lactose complex (yellow; lactose in green) highlighting the differences in the active site residues between cGRIFIN and rGal-5 [Met32 (M32) to Arg43 (R43), Ser55 (S55) to Glu64 (E64), and Val73 (V73) to Arg82 (R82)]. (**D**) The rGal-5/histo-blood-group B tetrasaccharide modeled by superposition with the cGRIFIN/tetrasaccharide structure. The active site residues involved in bindig of the ligand are: Arg43 (R43), Gln45 (Q45), His57 (H57), Asn59 (N59), Arg61 (R61), Glu64 (E64), Asn70 (N70), Trp77 (W77), Glu80 (E80) and Arg82 (R82). Residues Arg43, Glu64 and Arg82 might explain the affinity of rGal-5 for the histo-blood-group B (type 2) tetrasaccharide.

3.7. Oligomerization State of Full-Length rGal-5

Despite the disparity in packing inside the asymmetric unit of rGal-5 crystals without and with a ligand, intramolecular interactions, computed using the PISA Web server [41], did not appear to be sufficient to promote oligomerization, in both cases.

We experimentally confirmed by analytical ultracentrifugation and small-angle X-ray scattering (SAXS) in solution the absence of any oligomerization. The range of protein concentrations from 0.45 mg/mL up to 6 mg/mL was covered to trace any tendency to form oligomers at high non-physiological concentrations. In sedimentation velocity experiments in the absence or presence of 0.1 M lactose, rGal-5 (at 0.45 mg/mL) appeared as a single peak with a sedimentation coefficient of 1.8 ± 0.1 S (s20, w value of 1.9 ± 0.1 S, after correcting for the effect of solvent density and viscosity) (Supplementary Material, Figure S1). This result is fully in line with a globular protein of a mass of 13.8 kDa, as calculated from its amino acid sequence. Small-angle X-ray scattering (SAXS) data for the rGal-5–lactose complex at the concentrations of 4 mg/mL and 6 mg/mL yielded a particle distribution that is also attributed to a molecular mass of approximately 13 kDa (Figure 6). These data sets further substantiate that rGal-5 is a monomer in solution (under these conditions), as it is in the obtained crystals.

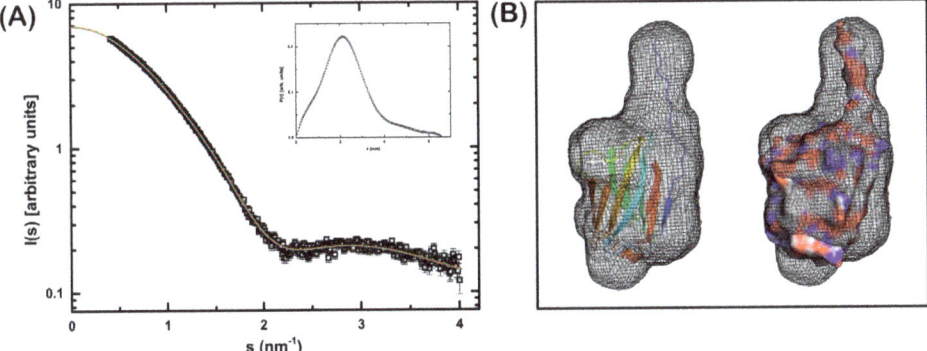

Figure 6. Small-angle X-ray scattering (SAXS) experiment supports a monomeric quaternary structure of rGal-5 in solution. (**A**) SAXS scattering profile of rGal-5. Black squares represent the experimental data, the red line the theoretical fitting obtained with the program GNOM. Inset: pair distances distribution function. (**B**) Ab initio SAXS model generated with the program DAMMIN (grey mesh). The crystallographic structure of the rGal-5 CRD domain with additional N-terminal residues modelled in an extended conformation is shown inside the envelope.

4. Discussion

Gal-5 is an N-tailed proto-type-like galectin present exclusively in rats and exists in solution as a monomer. To fill the gap of the structural characterization of this particular protein and obtain information of a full-length N-tailed proto-type-like galectin, the three-dimensional structures of the *apo* form and in complex with lactose were determined by X-ray crystallography. The protein crystallizes as a monomer in the absence and in the presence of ligand. In solution, analytical ultracentrifugation and SAXS experiments extended the available evidence for the lack of any intermolecular association.

Within the ligand-binding site, the side chains of the residues of the signature sequence for sugar recognition do not adopt their position by ligand binding. In the ligand-free structure, a network of water molecules or the presence of a single glycerol molecule takes the place of the core of a cognate glycan in the preformed contact site. The validity of the concept for such an intimate preorganization that can also accommodate a compound with a sugar-like constellation of hydroxyls, such as glycerol, has been thoroughly documented for the CRD of human Gal-3 [48–50], also described for murine Gal-4's N-terminal CRD [51]. In order to define cognate compounds, glycan array testing revealed preferential affinity of rGal5 for the histo-blood-group B determinant and its fucose-less trisaccharide as well as LacNAc-based tetrasaccharides among the tested set of mammalian glycans. The calorimetric titrations reflect the affinity gain for LacNAc and the blood-group B epitope relative to lactose (Table 2 and Figure S2).

Reflecting its proposed origin, rGal-5 shares structural features with Gal-9C (respective data available for the human protein) to a great extent. The availability of individual crystallographic information for human (h) Gal-9N and Gal-9C [52–55] made it possible to superimpose the rGal-5 structure to both Gal-9 CRDs. The calculated RMSD value is smaller for hGal-9C (0.37 Å) than for hGal-9N (0.56 Å). rGal-5 and hGal-9C could be overlaid almost perfectly, with loops occupying similar positions around the ligand-binding site (Figure 7A). As observed in the hGal-9N structure, a short β-strand is formed in the ligand-free rGal-5. This extension of the β-sandwich involves residues Pro9 and Tyr10 (Figure 7B). The N-terminal CRD of the human tandem-repeat-type Gal-9 thus mimics rGal-5's tendency for gaining some order in the N-terminal extension.

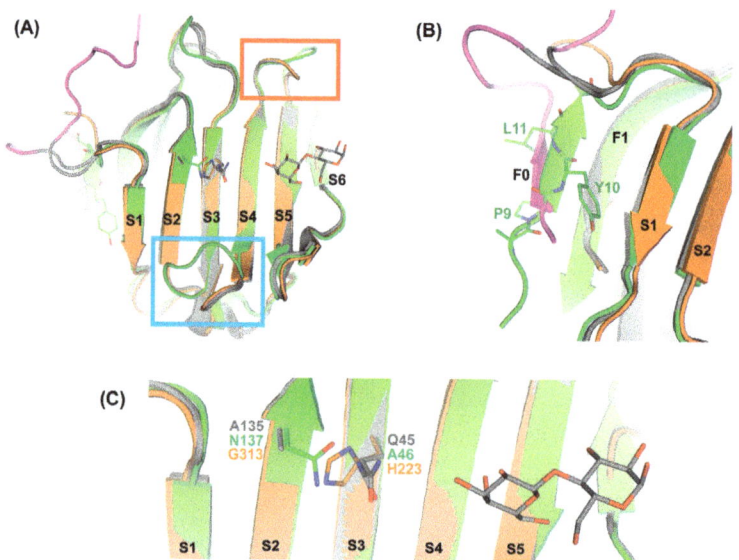

Figure 7. (**A**) Structural comparison of rGal-5 (grey) with the N-terminal (green) and the C-terminal (orange) CRD domains of hGal-9 showing the closer relationship between the first and the last structure. Differences in loops connecting strands in the S-face (labelled S1 to S6) are highlighted with colored squares. (**B**) Close view of N-terminal residues showing the formation at the F0 strand in rGal5, as observed in the structure of the N-terminal domain for hGal-9. (**C**) Specificity for LacNAc motif repeats in hGal-9N involves a hydrogen bond with Asn137 (N137) in the hGal-9N, while the corresponding residues in hGal-9C and rGal5 are Gly313 (G313) and Ala135 (A135), respectively. Furthermore, His223 (H223, in hGal-9C) and Gln45 (Q45, in rGal5) occupy the equivalent position of Ala46 (A46) in hGal-9N, causing steric impediment.

Structural differences between rGal-5 and hGal-9C relative to the N-terminal CRD of hGal-9 were found in the loop regions by insertion or deletion of residues (Figure 7A). These differences can be linked to shifts in specificities in glycan-binding between both CRDs in hGal-9, such as a reduction in affinity towards the LacNAc oligomers for the C-terminal with respect to the N-terminal CRD [38,56,57]. Loops connecting the F2-S3 and S3-S4 strands have additional residues in hGal-9N, covering the S2-S4 β-strands and forming a highly favorable binding site for LacNAc and its oligomers (polyLacNAc repeats) in N- and O-glycans and keratan sulphate, interacting with both CRDs [56–58]. Binding of these ligands involves a hydrogen bond with Asn137 in the hGal-9N [38]. The corresponding residues in hGal-9C and rGal5 are Gly313 and Ala135, respectively, hindering this interaction. In addition, His223 (in hGal-9C) and Gln45 (in rGal5) occupy the equivalent position of Ala46 in hGal-9N, causing a steric impediment for ligand accommodation within this region of the S3 β-strand (Figure 7B). This residue was linked to the specificity of hGal-9N for polyLacNAc repeats, the Forssman pentasaccharide and the histo-blood-group A hexasaccharide [52,53]. The loops connecting the S4-S5 and the S5-S6 strands are shorter in rGal5 and hGal-9C than in hGal-9N. These loops form the entrance for the ligand-binding site, their similar shape letting rGal-5 and the C-terminal CRD of hGal-9 share affinity.

The distinctive characteristic of rGal-5 (and GRP) is the N-terminal extension to the canonical CRD. Since the galectin CRD can interact with binding partners beyond the site for accommodating lactose and can engage in two types of contact at the same time (e.g., Gal-3 binding glycan and the chemokine CXCL12 [59]), changes between flexible and ordered arrangements of a tail may establish a molecular switch to let a protein ligand dock or not onto this region. For example, the considerably longer tail in Gal-3 has been shown by ESI MS and NMR [60,61] to backfold. This move results in blocking the access to a region of the S-face of the CRD. The systematic design of the Gal-3 variants with truncated versions attached to the CRD facilitated the possibility of generating a new double-stranded antiparallel β-sheet at the F-face [62,63]. In that case, the obtained information indicated the potential for forming an ordered structural element in the distal section that may have a bearing on the presentation of the Ser acceptor for phosphorylation [62]. Identification of respective counterreceptors for rGal-5 with a contact at this site will be required to support such an idea, giving further work a clear direction.

Interestingly, the monomeric C-type lectin RegIIIγ (HIP/PAP in mice) express a flexible N-terminal extension, which is a prosegment maintaining the protein in a biologically inactive state and is proteolytically removed to let the CRD become antibacterial via peptidoglycan binding [64]. Equally important, functional assays with an engineered variant to establish a protein pair (such as lectin with and without the N-terminal tail) can be informative for rGal-5 and GRP to trace the physiological significance for the extension, such as for the N-tailed proto-type-like galectins. The detection of the relevance of the isomer state of the Pro4 or Pro5 peptide bond in two galectins, chicken galectin-1B and human galectin-7, in the quaternary structure, illustrates the apparent fine-tuning of galectin activity by in-built molecular switches [65,66]. More work on the N-terminal tail is encouraged as a result of this.

Supplementary Materials: The following are available online at https://www.mdpi.com/article/10.3390/biom11121854/s1, Figure S1: Sedimentation velocity of rGal5 in the absence and in the presence of lactose. Figure S2: Binding of lactose, LacNAc and histo-blood-group B (type 2) tetrasaccharide to rGal5 studied by Isothermal Titration Calorimetry. Table S1: Results of the glycan array.

Author Contributions: All authors contributed to this work. H.-J.G. and A.R. conceived and designed the study; F.M.R. and A.R. obtained the crystal structures; F.J.M. performed SAXS and ITC experiments; N.V.S. and N.V.B. performed glycan array experiments; A.-K.L., methodology; writing—original draft, F.M.R., A.R. and H.-J.G.; writing—review and editing, F.M.R., F.J.M., H.K., N.V.B., H.-J.G. and A.R.; supervision, H.K., N.V.B., H.-J.G. and A.R. All authors have read and approved the published version of the manuscript.

Funding: This research was funded by the Ministerio de Economía, Industria y Competitividad, Gobierno de España, grant number BFU2016-77835-R.

Institutional Review Board Statement: Not applicable.

Informed Consent Statement: Not applicable.

Data Availability Statement: The structures have been deposited at the Protein Data Bank and are available with the identifications listed in Table 1.

Acknowledgments: Hans-Joachim Gabius died unexpectedly on August 2 while writing this manuscript. This work contains many of his ideas and is a testimony to his tireless creative energy. We want to honor his memory by including him in the list of authors in a privileged place. We cordially thank Juan Roman Luque Ortega from the Analytical Ultracentrifugation and Light Scattering Facility of the CIB as well as the staff of beamlines BM14, BM29 (ESRF Synchrotron) and BL13-XALOC (Alba Synchrotron) for their generous and much appreciated support. We thank Ivan Ryzhov (IBCh RAS, Moscow) for synthesis of the blood-group B (type 2) tetrasaccharide.

Conflicts of Interest: The authors declare no conflict of interest.

References

1. Gabius, H.-J.; Roth, J. An introduction to the sugar code. *Histochem. Cell Biol.* **2017**, *147*, 111–117. [CrossRef] [PubMed]
2. Corfield, A.P. Eukaryotic protein glycosylation: A primer for histochemists and cell biologists. *Histochem. Cell Biol.* **2017**, *147*, 119–147. [CrossRef]
3. Kopitz, J. Lipid glycosylation: A primer for histochemists and cell biologists. *Histochem. Cell Biol.* **2017**, *147*, 175–198. [CrossRef] [PubMed]
4. Suzuki, N. Glycan diversity in the course of vertebrate evolution. *Glycobiology* **2018**, *29*, 625–644. [CrossRef] [PubMed]
5. Cummings, R.D. Stuck on sugars: How carbohydrates regulate cell adhesion, recognition, and signaling. *Glycoconj. J.* **2019**, *36*, 241–257. [CrossRef] [PubMed]
6. Kaltner, H.; Abad-Rodríguez, J.; Corfield, A.P.; Kopitz, J.; Gabius, H.-J. The sugar code: Letters and vocabulary, writers, editors and readers and biosignificance of functional glycan-lectin pairing. *Biochem. J.* **2019**, *476*, 2623–2655. [CrossRef]
7. Kilpatrick, D.C. Animal lectins: A historical introduction and overview. *Biochim. Biophys. Acta* **2002**, *1572*, 187–197. [CrossRef]
8. Manning, J.C.; Romero, A.; Habermann, F.A.; García Caballero, G.; Kaltner, H.; Gabius, H.-J. Lectins: A primer for histochemists and cell biologists. *Histochem. Cell Biol.* **2017**, *147*, 199–222. [CrossRef] [PubMed]
9. Gunther, G.R.; Wang, J.L.; Yahara, I.; Cunningham, B.A.; Edelman, G.M. Concanavalin A derivatives with altered biological activities. *Proc. Natl. Acad. Sci. USA* **1973**, *70*, 1012–1016. [CrossRef]
10. Ludwig, A.-K.; Kaltner, H.; Kopitz, J.; Gabius, H.-J. Lectinology 4.0: Altering modular (ga)lectin display for functional analysis and biomedical applications. *Biochim. Biophys. Acta* **2019**, *1863*, 935–940. [CrossRef]
11. Kasai, K.-I.; Hirabayashi, J. Galectins: A family of animal lectins that decipher glycocodes. *J. Biochem.* **1996**, *119*, 1–8. [CrossRef]
12. Cooper, D.N.W. Galectinomics: Finding themes in complexity. *Biochim. Biophys. Acta* **2002**, *1572*, 209–231. [CrossRef]
13. Kaltner, H.; Toegel, S.; García Caballero, G.; Manning, J.C.; Ledeen, R.W.; Gabius, H.-J. Galectins: Their network and roles in immunity/tumor growth control. *Histochem. Cell Biol.* **2017**, *147*, 239–256. [CrossRef] [PubMed]
14. García Caballero, G.; Kaltner, H.; Kutzner, T.J.; Ludwig, A.-K.; Manning, J.C.; Schmidt, S.; Sinowatz, F.; Gabius, H.-J. How galectins have become multifunctional proteins. *Histol. Histopathol.* **2020**, *35*, 509–539. [PubMed]
15. Hughes, R.C. Mac-2: A versatile galactose-binding protein of mammalian tissues. *Glycobiology* **1994**, *4*, 5–12. [CrossRef] [PubMed]
16. Gao, X.; Liu, J.; Liu, X.; Li, L.; Zheng, J. Cleavage and phosphorylation: Important post-translational modifications of galectin-3. *Cancer Metastasis Rev.* **2017**, *36*, 367–374. [CrossRef]
17. Romero, A.; Gabius, H.-J. Galectin-3: Is this member of a large family of multifunctional lectins (already) a therapeutic target? *Expert Opin. Ther. Targets* **2019**, *23*, 819–828. [CrossRef] [PubMed]
18. Zhou, D.; Ge, H.; Sun, J.; Gao, Y.; Teng, M.; Niu, L. Crystal structure of the C-terminal conserved domain of human GRP, a galectin-related protein, reveals a function mode different from those of galectins. *Proteins* **2008**, *71*, 1582–1588. [CrossRef] [PubMed]
19. García Caballero, G.; Flores-Ibarra, A.; Michalak, M.; Khasbiullina, N.; Bovin, N.V.; André, S.; Manning, J.C.; Vértesy, S.; Ruiz, F.M.; Kaltner, H.; et al. Galectin-related protein: An integral member of the network of chicken galectins. 1. From strong sequence conservation of the gene confined to vertebrates to biochemical characteristics of the chicken protein and its crystal structure. *Biochim. Biophys. Acta* **2016**, *1860*, 2285–2297. [CrossRef] [PubMed]
20. Cerra, R.F.; Gitt, M.A.; Barondes, S.H. Three soluble rat β-galactoside-binding lectins. *J. Biol. Chem.* **1985**, *260*, 10474–10477. [CrossRef]
21. Jung, S.K.; Fujimoto, D. A novel β-galactoside-binding lectin in adult rat kidney. *J. Biochem.* **1994**, *116*, 547–553. [CrossRef]
22. Gitt, M.A.; Wiser, M.F.; Leffler, H.; Herrmann, J.; Xia, Y.; Massa, S.M.; Cooper, D.N.W.; Lusis, A.J.; Barondes, S.H. Sequence and mapping of galectin-5, a β-galactoside-binding lectin, found in rat erythrocytes. *J. Biol. Chem.* **1995**, *270*, 5032–5038. [CrossRef] [PubMed]
23. Wada, J.; Kanwar, Y.S. Identification and characterization of galectin-9, a novel β-galactoside-binding mammalian lectin. *J. Biol. Chem.* **1997**, *272*, 6078–6086. [CrossRef] [PubMed]
24. Lensch, M.; Lohr, M.; Russwurm, R.; Vidal, M.; Kaltner, H.; André, S.; Gabius, H.-J. Unique sequence and expression profiles of rat galectins-5 and -9 as a result of species-specific gene divergence. *Int. J. Biochem. Cell Biol.* **2006**, *38*, 1741–1758. [CrossRef]
25. Kaltner, H.; Raschta, A.-S.; Manning, J.C.; Gabius, H.-J. Copy-number variation of functional galectin genes: Studying animal galectin-7 (p53-induced gene 1 in man) and tandem-repeat-type galectins-4 and -9. *Glycobiology* **2013**, *23*, 1152–1163. [CrossRef] [PubMed]
26. André, S.; Kaltner, H.; Lensch, M.; Russwurm, R.; Siebert, H.-C.; Fallsehr, C.; Tajkhorshid, E.; Heck, A.J.R.; von Knebel-Döberitz, M.; Gabius, H.-J.; et al. Determination of structural and functional overlap/divergence of five proto-type galectins by analysis of the growth-regulatory interaction with ganglioside GM1 in silico and in vitro on human neuroblastoma cells. *Int. J. Cancer* **2005**, *114*, 46–57. [CrossRef]
27. Leffler, H.; Barondes, S.H. Specificity of binding of soluble rat lung lectins to substituted and unsubstituted mammalian β-galactosides. *J. Biol. Chem.* **1986**, *261*, 10119–10126. [CrossRef]
28. Dam, T.K.; Gabius, H.-J.; André, S.; Kaltner, H.; Lensch, M.; Brewer, C.F. Galectins bind to the multivalent glycoprotein asialofetuin with enhanced affinities and a gradient of decreasing binding constants. *Biochemistry* **2005**, *44*, 12564–12571. [CrossRef]

29. Wu, A.M.; Singh, T.; Wu, J.H.; Lensch, M.; André, S.; Gabius, H.-J. Interaction profile of galectin-5 with free saccharides and mammalian glycoproteins: Probing its fine-specificity and the effect of naturally clustered ligand presentation. *Glycobiology* **2006**, *16*, 524–537. [CrossRef]
30. Beer, A.; André, S.; Kaltner, H.; Lensch, M.; Franz, S.; Sarter, K.; Schulze, C.; Gaipl, U.S.; Kern, P.; Herrmann, M.; et al. Human galectins as sensors for apoptosis/necrosis-associated surface changes of granulocytes and lymphocytes. *Cytom. A* **2008**, *73*, 139–147. [CrossRef] [PubMed]
31. Barrès, C.; Blanc, L.; Bette-Bobillo, P.; André, S.; Mamoun, R.; Gabius, H.-J.; Vidal, M. Galectin-5 is bound onto the surface of rat reticulocyte exosomes and modulates vesicle uptake by macrophages. *Blood* **2010**, *115*, 696–705. [CrossRef]
32. Ruiz, F.M.; Gilles, U.; Ludwig, A.-K.; Sehad, C.; Shiao, T.C.; García Caballero, G.; Kaltner, H.; Lindner, I.; Roy, R.; Reusch, D.; et al. Chicken GRIFIN: Structural characterization in crystals and in solution. *Biochimie* **2018**, *146*, 127–138. [CrossRef] [PubMed]
33. Gabius, H.-J. Influence of type of linkage and spacer on the interaction of β-galactoside-binding proteins with immobilized affinity ligands. *Anal. Biochem.* **1990**, *189*, 91–94. [CrossRef]
34. Kutzner, T.J.; Gabba, A.; FitzGerald, F.G.; Shilova, N.V.; García Caballero, G.; Ludwig, A.-K.; Manning, J.C.; Knospe, C.; Kaltner, H.; Sinowatz, F.; et al. How altering the modular architecture affects aspects of lectin activity: Case study on human galectin-1. *Glycobiology* **2019**, *29*, 593–607. [CrossRef] [PubMed]
35. García Caballero, G.; Beckwith, D.; Shilova, N.V.; Gabba, A.; Kutzner, T.J.; Ludwig, A.-K.; Manning, J.C.; Kaltner, H.; Sinowatz, F.; Cudic, M.; et al. Influence of protein (human galectin-3) design on aspects of lectin activity. *Histochem. Cell Biol.* **2020**, *154*, 135–153. [CrossRef] [PubMed]
36. Kabsch, W. XDS. *Acta Crystallogr.* **2010**, *D66*, 125–132. [CrossRef] [PubMed]
37. Winn, M.D.; Ballard, C.C.; Cowtan, K.D.; Dodson, E.J.; Emsley, P.; Evans, P.R.; Keegan, R.M.; Krissinel, E.B.; Leslie, A.G.; McCoy, A.; et al. Overview of the CCP4 suite and current developments. *Acta Crystallogr.* **2011**, *D67*, 235–242.
38. Yoshida, H.; Teraoka, M.; Nishi, N.; Nakakita, S.; Nakamura, T.; Hirashima, M.; Kamitori, S. X-Ray structures of human galectin-9 C-terminal domain in complexes with a biantennary oligosaccharide and sialyllactose. *J. Biol. Chem.* **2010**, *285*, 36969–36976. [CrossRef]
39. Adams, P.D.; Afonine, P.V.; Bunkoczi, G.; Chen, V.B.; Davis, I.W.; Echols, N.; Headd, J.J.; Hung, L.W.; Kapral, G.J.; Grosse-Kunstleve, R.W.; et al. PHENIX: A comprehensive Python-based system for macromolecular structure solution. *Acta Crystallogr.* **2010**, *D66*, 213–221. [CrossRef]
40. Emsley, P.; Lohkamp, B.; Scott, W.G.; Cowtan, K. Features and development of Coot. *Acta Crystallogr.* **2010**, *D66*, 486–501.
41. Krissinel, E.; Henrick, K. Inference of macromolecular assemblies from crystalline state. *J. Mol. Biol.* **2007**, *372*, 774–797. [CrossRef] [PubMed]
42. DeLano, W. *The PyMOL Molecular Graphics System*; Delano Scientifics: Palo Alto, CA, USA, 2002.
43. Pernot, P.; Round, A.; Barrett, R.; De Maria Antolinos, A.; Gobbo, A.; Gordon, E.; Huet, J.; Kieffer, J.; Lentini, M.; Mattenet, M.; et al. Upgraded ESRF BM29 beamline for SAXS on macromolecules in solution. *J. Synchrotron Radiat.* **2013**, *20*, 660–664. [CrossRef] [PubMed]
44. Franke, D.; Svergun, D.I. DAMMIF, a program for rapid ab-initio shape determination in small-angle scattering. *J. Appl. Crystallogr.* **2009**, *42*, 342–346. [CrossRef]
45. Volkov, V.V.; Svergun, D.I. Uniqueness of ab initio shape determination in small-angle scattering. *J. Appl. Crystallogr.* **2003**, *36*, 860–864. [CrossRef]
46. Ryzhov, I.M.; Korchagina, E.Y.; Popova, I.S.; Tyrtysh, T.V.; Paramonov, A.S.; Bovin, N.V. Block synthesis of A (type 2) and B (type 2) tetrasaccharides related to the human ABO blood group system. *Carbohydr. Res.* **2016**, *430*, 59–71. [CrossRef] [PubMed]
47. Ludwig, A.-K.; Michalak, M.; Xiao, Q.; Gilles, U.; Medrano, F.J.; Ma, H.; FitzGerald, F.G.; Hasley, W.D.; Melendez-Davila, A.; Liu, M.; et al. Design-functionality relationships for adhesion/growth-regulatory galectins. *Proc. Natl. Acad. Sci. USA* **2019**, *116*, 2837–2842. [CrossRef]
48. Collins, P.M.; Hidari, K.I.; Blanchard, H. Slow diffusion of lactose out of galectin-3 crystals monitored by X-ray crystallography: Possible implications for ligand-exchange protocols. *Acta Crystallogr.* **2007**, *D63*, 415–419. [CrossRef] [PubMed]
49. Saraboji, K.; Håkansson, M.; Genheden, S.; Diehl, C.; Qvist, J.; Weininger, U.; Nilsson, U.J.; Leffler, H.; Ryde, U.; Akke, M.; et al. The carbohydrate-binding site in galectin-3 is preorganized to recognize a sugar-like framework of oxygens: Ultra-high-resolution structures and water dynamics. *Biochemistry* **2012**, *51*, 296–306. [CrossRef]
50. Manzoni, F.; Wallerstein, J.; Schrader, T.E.; Ostermann, A.; Coates, L.; Akke, M.; Blakeley, M.P.; Oksanen, E.; Logan, D.T. Elucidation of hydrogen bonding patterns in ligand-free, lactose- and glycerol-bound galectin-3c by neutron crystallography to guide drug design. *J. Med. Chem.* **2018**, *61*, 4412–4420. [CrossRef]
51. Krejciríková, V.; Pachl, P.; Fábry, M.; Maly, P.; Rezáčová, P.; Brynda, J. Structure of the mouse galectin-4 N-terminal carbohydrate-recognition domain reveals the mechanism of oligosaccharide recognition. *Acta Crystallogr.* **2011**, *D67*, 204–211. [CrossRef] [PubMed]
52. Nagae, M.; Nishi, N.; Nakamura-Tsuruta, S.; Hirabayashi, J.; Wakatsuki, S.; Kato, R. Structural analysis of the human galectin-9 N-terminal carbohydrate recognition domain reveals unexpected properties that differ from the mouse orthologue. *J. Mol. Biol.* **2008**, *375*, 119–135. [CrossRef] [PubMed]

53. Nagae, M.; Nishi, N.; Murata, T.; Usui, T.; Nakamura, T.; Wakatsuki, S.; Kato, R. Structural analysis of the recognition mechanism of poly-N-acetyllactosamine by the human galectin-9 N-terminal carbohydrate recognition domain. *Glycobiology* **2009**, *19*, 112–117. [CrossRef] [PubMed]
54. Solís, D.; Maté, M.J.; Lohr, M.; Ribeiro, J.P.; López-Merino, L.; André, S.; Buzamet, E.; Cañada, F.J.; Kaltner, H.; Lensch, M.; et al. N-Domain of human adhesion/growth-regulatory galectin-9: Preference for distinct conformers and non-sialylated N-glycans and detection of ligand-induced structural changes in crystal and solution. *Int. J. Biochem. Cell Biol.* **2010**, *42*, 1019–1029. [CrossRef] [PubMed]
55. Yoshida, H.; Nishi, N.; Wada, K.; Nakamura, T.; Hirashima, M.; Kuwabara, N.; Kato, R.; Kamitori, S. X-ray structure of a protease-resistant mutant form of human galectin-9 having two carbohydrate recognition domains with a metal-binding site. *Biochem. Biophys. Res. Commun.* **2017**, *490*, 1287–1293. [CrossRef]
56. Hirabayashi, J.; Hashidate, T.; Arata, Y.; Nishi, N.; Nakamura, T.; Hirashima, M.; Urashima, T.; Oka, T.; Futai, M.; Müller, W.E.G.; et al. Oligosaccharide specificity of galectins: A search by frontal affinity chromatography. *Biochim. Biophys. Acta* **2002**, *1572*, 232–254. [CrossRef]
57. Iwaki, J.; Tateno, H.; Nishi, N.; Minamisawa, T.; Nakamura-Tsuruta, S.; Itakura, Y.; Kominami, J.; Urashima, T.; Nakamura, T.; Hirabayashi, J. The Galβ-(syn)-gauche configuration is required for galectin-recognition disaccharides. *Biochim. Biophys. Acta* **2011**, *1810*, 643–651. [CrossRef]
58. Miller, M.C.; Cai, C.; Wichapong, K.; Bhaduri, S.; Pohl, N.L.B.; Linhardt, R.J.; Gabius, H.-J.; Mayo, K.H. Structural insight into the binding of human galectins to corneal keratan sulfate, its desulfated form and related saccharides. *Sci. Rep.* **2020**, *10*, 15708. [CrossRef]
59. Eckardt, V.; Miller, M.C.; Blanchet, X.; Duan, R.; Leberzammer, J.; Duchene, J.; Soehnlein, O.; Megens, R.T.A.; Ludwig, A.-K.; Dregni, A.; et al. Chemokines and galectins form heterodimers to modulate inflammation. *EMBO Rep.* **2020**, *21*, e47852. [CrossRef]
60. Kopitz, J.; André, S.; von Reitzenstein, C.; Versluis, K.; Kaltner, H.; Pieters, R.J.; Wasano, K.; Kuwabara, I.; Liu, F.-T.; Cantz, M.; et al. Homodimeric galectin-7 (p53-induced gene 1) is a negative growth regulator for human neuroblastoma cells. *Oncogene* **2003**, *22*, 6277–6288. [CrossRef]
61. Ippel, H.; Miller, M.C.; Vértesy, S.; Zheng, Y.; Canada, F.J.; Suylen, D.; Umemoto, K.; Romanò, C.; Hackeng, T.; Tai, G.; et al. Intra- and intermolecular interactions of human galectin-3: Assessment by full-assignment-based NMR. *Glycobiology* **2016**, *26*, 888–903. [CrossRef]
62. Kopitz, J.; Vértesy, S.; André, S.; Fiedler, S.; Schnölzer, M.; Gabius, H.-J. Human chimera-type galectin-3: Defining the critical tail length for high-affinity glycoprotein/cell surface binding and functional competition with galectin-1 in neuroblastoma cell growth regulation. *Biochimie* **2014**, *104*, 90–99. [CrossRef] [PubMed]
63. Flores-Ibarra, A.; Vértesy, S.; Medrano, F.J.; Gabius, H.-J.; Romero, A. Crystallization of a human galectin-3 variant with two ordered segments in the shortened N-terminal tail. *Sci. Rep.* **2018**, *8*, 9835. [CrossRef] [PubMed]
64. Mukherjee, S.; Partch, C.L.; Lehotzky, R.E.; Whitham, C.V.; Chu, H.; Bevins, C.L.; Gardner, K.H.; Hooper, L.V. Regulation of C-type lectin antimicrobial activity by a flexible N-terminal prosegment. *J. Biol. Chem.* **2009**, *284*, 4881–4888. [CrossRef] [PubMed]
65. López-Lucendo, M.F.; Solís, D.; Sáiz, J.L.; Kaltner, H.; Russwurm, R.; André, S.; Gabius, H.-J.; Romero, A. Homodimeric chicken galectin CG-1B (C-14): Crystal structure and detection of unique redox-dependent shape changes involving inter- and intrasubunit disulfide bridges by gel filtration, ultracentrifugation, site-directed mutagenesis, and peptide mass fingerprinting. *J. Mol. Biol.* **2009**, *386*, 366–378.
66. Miller, M.C.; Nesmelova, I.V.; Daragan, V.A.; Ippel, H.; Michalak, M.; Dregni, A.; Kaltner, H.; Kopitz, J.; Gabius, H.-J.; Mayo, K.H. Pro4 prolyl peptide bond isomerization in human galectin-7 modulates the monomer-dimer equilibrium to affect function. *Biochem. J.* **2020**, *477*, 3147–3165. [CrossRef]

Review

Galectins in Endothelial Cell Biology and Angiogenesis: The Basics

Victor L. Thijssen

Cancer Center Amsterdam, Department of Radiation Oncology, Amsterdam UMC Location VUmc, De Boelelaan 1118, 1081 HV Amsterdam, The Netherlands; v.thijssen@amsterdamumc.nl

Abstract: Angiogenesis, the growth of new blood vessels out of existing vessels, is a complex and tightly regulated process. It is executed by the cells that cover the inner surface of the vasculature, i.e., the endothelial cells. During angiogenesis, these cells adopt different phenotypes, which allows them to proliferate and migrate, and to form tube-like structures that eventually result in the generation of a functional neovasculature. Multiple internal and external cues control these processes and the galectin protein family was found to be indispensable for proper execution of angiogenesis. Over the last three decades, several members of this glycan-binding protein family have been linked to endothelial cell functioning and to different steps of the angiogenesis cascade. This review provides a basic overview of our current knowledge regarding galectins in angiogenesis. It covers the main findings with regard to the endothelial expression of galectins and highlights their role in endothelial cell function and biology.

Keywords: vasculature; gene expression; tube formation; sprouting; VEGF; integrins; galectin; extracellular matrix; microenvironment

1. Introduction

With an estimated length of at least 100,000 kilometers (±60,000 miles), the adult human vasculature forms an immense infrastructure encompassing all blood vessels, ranging from the large arteries and veins to the countless number of small capillaries. This vast vascular bed ensures that all organs, tissues, and cells in the body have access to sufficient amounts of oxygen and nutrients and that waste materials can be disposed of. In addition, platelets and blood-borne cells, such as leukocytes, are able to travel to all parts of the body via the vasculature. The key players that are involved in building, maintaining, and providing functionality to the blood vessel system are the endothelial cells. These cells are of mesodermal origin and they cover the inner surface of all blood vessels. As such, the vascular endothelium serves as the main interface between all components in the blood and the underlying tissues. Consequently, endothelial cells participate in several biological processes, e.g., coagulation, inflammation, and transendothelial transport/migration [1]. Moreover, in case of a demand for new vessels, the endothelial cells can be triggered to start the formation of new blood vessels, a process referred to as angiogenesis. Angiogenesis is not only an intricate part of different physiological processes, e.g., the menstrual cycle, embryogenesis, or wound healing, it is also involved in different pathologies, including cancer [2]. In fact, 50 years ago it was shown that tumor tissues, once they reached a few cubic millimeters, depended on activation of angiogenesis in order to maintain growth [3,4]. If tumor cells fail to induce angiogenesis, the growing tumor mass is provided with insufficient oxygen and nutrients and remains dormant [5]. Consequently, activation of tumor angiogenesis is considered a hallmark of cancer and targeting this process has been recognized as a potent strategy for cancer therapy [6,7].

Angiogenesis, both in the physiological and pathological context, is a complex and multistep process during which endothelial cells respond to a multitude of external and internal signals [7]. These signals trigger endothelial cells to adopt different phenotypes that ultimately result in the formation of new blood vessels. It is now well recognized that several galectins are involved in facilitating different endothelial activities during angiogenesis [8–10]. The current review will provide a basic overview of the current knowledge regarding the role of galectins in endothelial cell biology and angiogenesis.

2. Endothelial Galectin Expression

The mammalian galectin family comprises 15 members, 11 of which were also found expressed at the protein level in humans. These glycan-binding proteins share a so-called carbohydrate recognition domain (CRD) of approximately 130 amino acids, which is composed of two antiparallel beta-sheets that fold in a beta-sandwich. The beta-sandwich structure is slightly curved forming a groove in which carbohydrate binding occurs. Main interactions with beta-galactoside containing glycans involve a core-binding site inside the groove that contains several evolutionary conserved amino acids (Figure 1a). Glycan-binding specificity and affinity of each galectin are further mediated through small structural differences within (and outside) the binding groove. Based on the number and structural arrangement of the CRDs, galectins can be classified into three subgroups (Figure 1b), i.e., prototype galectins (single CRD; gal-1/-2/-7/-10/-13/-14), chimeric galectins (single CRD with an N-terminal non-lectin domain; gal-3), and tandem repeat galectins (two linked CRDs; gal-4/-8/-9/-12). Their ability to engage in glycan-dependent, as well as glycan-independent interactions, allows them to exert multiple functions in many different biological processes. Indeed, galectins are expressed by many different cell types where they can be found intracellularly and/or extracellularly (Figure 1c). For an extensive and more in-depth background on the structure, glycan-binding, and function of galectins, see [11–13]).

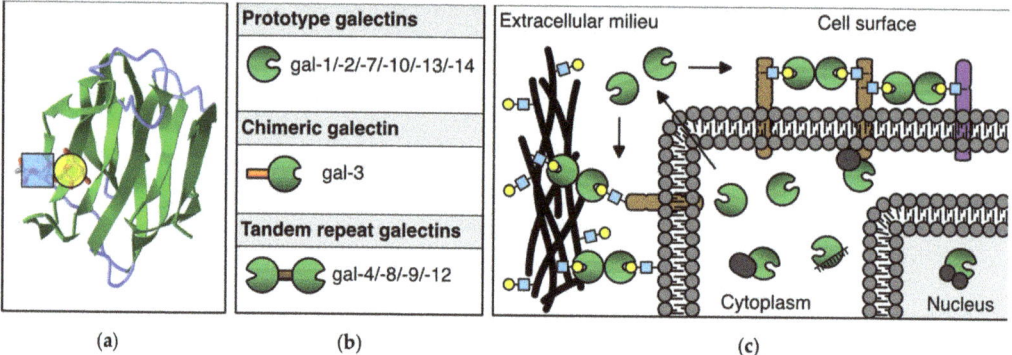

Figure 1. The galectin protein family. (**a**) Cartoon of the anti-parallel beta-sheet structure forming the carbohydrate recognition domain of galectin-1. On the left, the interaction of a LacNAc(N-acetyllactosamine) moiety in the binding groove is shown. (**b**) Overview of the 11 mammalian galectins that are expressed in humans. See text for explanation of the subgroups. (**c**) Schematic representation of the (extra)cellular location of galectins. In the extracellular environment and on the cell surface, galectins can interact with glycoconjugates to facilitate, e.g., cell–ECM and cell–cell interactions. In addition, galectins can mediate interactions between molecules in the cell membrane. In the cytosol and nucleus, galectins can engage in (mostly) glycan-independent protein/protein interactions involved in, e.g., signaling and mRNA splicing.

With regard to the expression of galectins in endothelial cells, we performed a broad galectin-profiling study in 2008 showing that the endothelial expression of galectins is mainly restricted to galectin-1, -3, -8, and -9 (Figure 2a) [14]. The mRNA expression levels of three other galectins (galectin-2, -4, -12) were low and not confirmed at the protein level, while mRNA expression of the remaining galectins (galectin-7, -10, -13, -14) could not be detected at all. The findings of this extensive galectin profiling study corroborated previous observations [15–18] and were later confirmed by different research groups in endothelial cells from variable origins [19–24]. Only recently, it was reported that galectin-2 protein expression was also detectable in endothelial cells, albeit in a specific context, i.e., in fetal endothelial cells in the placenta of women with gestational diabetes mellitus [25]. While this needs further confirmation, it indicates that the cellular context can control specific endothelial galectin expression. In line with this, galectin-8 expression appears to be higher in primary isolated lymphatic endothelial cells when compared to regular endothelial cells [19,26]. Likewise, galectin-3 expression was reported to be higher in endothelial progenitor cells as compared to normal endothelial cells [27]. Furthermore, we (and others) have shown that several cytokines, growth factors, and other molecules can alter galectin expression levels in endothelial cells (Table 1) [9]. For example, galectin-8 and galectin-9 expressions are reduced in serum-activated primary isolated endothelial cells as compared to non-activated counterparts [22]. At the same time, treatment with interferon gamma (IFNγ) can trigger the endothelial expression of galectin-9 [23,24,28,29]. IFNγ, as well as other cytokines, was also shown to increase endothelial galectin-1 expression [16,30], and more recently, cathepsin L was found to induce endothelial expression of galectin-1 [31]. The expression of galectin-3 by endothelial cells was shown to be induced by, e.g., matrix component fibronectin [32], advanced glycosylation end products [33], and interacting neutrophils [34]. It is important to note that the overview presented in Table 1 is likely far from complete, as it is still poorly understood which and how environmental triggers affect endothelial galectin expression. This is illustrated by the observation that many cancer tissues—often characterized by an aberrant microenvironment—display altered expression of endothelial galectins (for overview see [9]). For example, we have shown that galectin-9 expression, which is reduced in activated endothelial cells [14], is significantly increased in the tumor endothelium of different cancer types [35]. Since tumor cells secrete many different factors to modulate their microenvironment, including the immune infiltrate and the vasculature, it can be anticipated that the list of proteins provided here merely represents the tip of the iceberg when it comes to regulation of endothelial galectin expression.

To further complicate matters, it has been shown that extracellular triggers, such as cytokines, growth factors, or hypoxia, can alter the glycosylation patterns on the endothelial cell surface, which in turn affects the binding of galectins to the cells [36,37]. For example, hypoxia was shown to increase the presence of β1-6GlcNAc-branched N-glycans, poly-LacNAc structures, and fucosylated glycans on the endothelial cell surface, while α2-6 sialylation and α2,3-sialylated moieties are reduced [36,38]. Such alterations change the permissiveness of the endothelial cell towards specific galectin binding, which in turn affects how galectins control endothelial cell functionality.

At the same time, galectins can trigger the endothelial expression and release of cytokines and growth factors [21,39–41]. All of this points towards a complex relationship between the microenvironment and endothelial galectin expression. It is an ongoing challenge to unravel this relationship, in particular in the in vivo context where multiple triggers can simultaneously influence the endothelial cell phenotype.

It is also important to realize that endothelial galectin expression is not only controlled at the transcriptional level, but also at the post-transcriptional and post-translational level (Figure 2b). For example, alternative splicing has been shown to occur for tandem repeat galectin-8 and galectin-9. In endothelial cells, the alternative splicing can give rise to up to 3 different protein isoforms by affecting the length of the linker region between the two CRDs [14,20,22,29,42,43]. While the regulatory mechanisms that control the alternative splicing remain elusive, we observed that cytokines and growth factors could affect the mRNA expression levels of the different endothelial galectin-9 splice variants in vitro [22]. Whether it is also true for endothelial galectin-8, and how this is regulated in endothelial cells in vivo, requires further investigation.

The need for more research also applies to the role of post-translational modifications of galectins in endothelial cells. Different protein modifications with different functional effects on galectins have been reported, including proteolytic cleavage, phosphorylation, and S-nitrosylation [44–51]. With regard to angiogenesis, it was shown that proteolytic processing of galectin-3 influences the angioregulatory activity of the protein (see also next section) [52]. We also found that the two distinct galectin-9 CRDs, which can be generated upon proteolytic cleavage [51], have different effects on endothelial cell function [23]. However, a comprehensive insight in the post-translational modifications of endothelial galectins, how it is regulated, and how it affects endothelial cell function is still lacking.

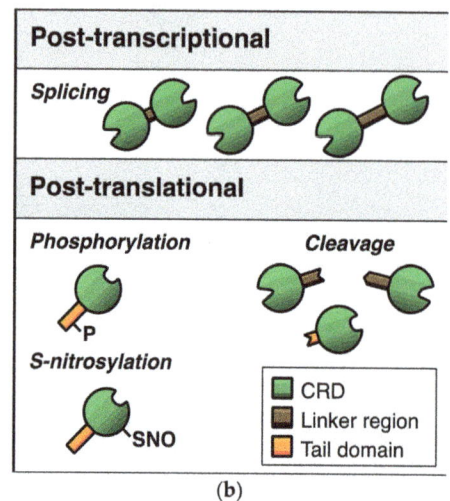

Figure 2. Endothelial galectins. (a) Schematic representation of the four dominant galectins that are expressed by endothelial cells. (b) Overview of the main post-transcriptional and post-translational modifications that occur in endothelial galectins. Note that the modifications shown here are illustrative and do not represent the actual location of modification in the respective proteins.

Table 1. Regulators of endothelial galectin expression.

Galectin	Expression Induced by	Expression Reduced by
Galectin-1	IL-1β, IFNγ, TNFα, LDL, LPS, Cathepsin L, High serum [b]	-
Galectin-3	IL-1β, fibronectin, AGEs, asialofetuin, neutrophil adhesion/transmigration	-
Galectin-8 [a]	-	High serum [b]
Galectin-9	IFNγ, IFNβ, IL-10, viral RNA	VEGF, IL-1, High serum [b,c]

IL = Interleukin, IFN = Interferon, LDL = low-density lipoprotein, LPS = lipopolysaccharide, AGE = Advanced glycosylation end products, VEGF = vascular endothelial growth factor. [a] higher expression in lymphatic EC as compared to normal EC. [b] Cells cultured in 20% serum. [c] Differential effects on specific splice variants.

Finally, an important aspect that should be taken in consideration when studying endothelial galectin expression is the cellular localization. As briefly mentioned before, galectins can be located intracellularly and extracellularly. In fact, galectins can be found in specific compartments of a cell, including the nucleus, the cytoplasm, and the cell membrane [14,20,53,54]. In addition, galectins can be secreted into the extracellular milieu [21,39,41]. With regard to the cellular localization and secretion of galectins, again, environmental clues that regulate endothelial activity appear to play an important role. For example, activation of cultured endothelial cells by high serum conditions or by a tumor conditioned medium was shown to increase cell surface exposure of galectin-1, -8, and -9 [14,30]. In line with this, the surface translocation of endothelial galectin-9 can be triggered by IFNγ [28,29] while secretion of galectin-8 has been linked to treatment with LPS [21]. Furthermore, endothelial cells in tumor tissues show altered cellular location of galectins as compared to normal endothelium [14,20]. All of this further supports the concept that the microenvironment is a key regulator of endothelial galectin expression as it provides most of the signals to which the endothelial cells respond.

3. Galectins in Endothelial Cell Function and Angiogenesis

As evident from the previous section, the expression of endothelial galectins is controlled by many different factors. Such a complex level of regulation suggests that galectins are involved in different aspects of endothelial cell function. Indeed, research over the last three decades has shown that adequate expression and function of galectins is required in multiple endothelial cell activities related to, e.g., inflammation and immunomodulation [55], coagulation [56], and angiogenesis [9,57–59]. While there is increasing interest in the immunomodulatory functions of endothelial galectins [24,60], this review will solely focus on the angioregulatory role. In particular, the role of the individual endothelial galectins in cellular functions related to angiogenesis will be highlighted.

3.1. Galectin-1

The first studies reporting on endothelial galectin-1 expression appeared around the 1990s in the previous century [15,17,61,62]. While these findings hinted towards a possible function in angiogenesis, the first clear evidence that directly linked galectin-1 to endothelial cell biology was provided by us in 2006. Using a protein–protein interaction screen, we identified galectin-1 as the target protein of a synthetic peptide inhibitor of angiogenesis [63]. Subsequent research showed that galectin-1 was essential for different endothelial cell functions during angiogenesis, in particular cell proliferation and migration [63]. Nowadays, it is well established that endothelial cells in vitro as well as in vivo angiogenesis rely on galectin-1 [31,64–70]. Only recently, the importance of galectin-1 in endothelial cell biology was again confirmed in a study that explored vascular remodeling after cerebral ischemia [71]. This study also reiterated the important link between galectin-1 and the angiostimulatory protein vascular endothelial growth factor (VEGF) by showing an association between galectin-1 and VEGF/VEGF receptor expression in endothelial cells [71]. Previous work had already shown that galectin-1 can delay endocytosis of VEGFR2 [72] and that binding of galectin-1 to the VEGFR2 co-receptor neuropilin-1 enhanced receptor phosphorylation and downstream signaling [64]. Moreover, Croci et al. found that altered glycosylation of VEGFR2 allows galectin-1 to activate receptor signaling in VEGF refractory tumors [36]. Importantly, this was linked to interactions of galectin-1 with non-sialylated N-linked glycans on the VEGF receptor [36], indicative of an important role of endothelial cell glycosylation in the sensitivity to galectins. Interestingly, it was also recently suggested that galectin-1 might interact with VEGF mRNA transcripts, which might interfere with VEGF translation and/or secretion [73]. This could be related to the possible role of galectin-1 in splicing [74] but needs further validation. Nevertheless, it is evident that galectin-1 can induce endothelial cell activation and control or even replace the angiostimulatory activity of VEGF. Altogether, these findings show that the galectin-1/VEGF/VEGFR2 axis represents an important route for inducing and maintaining endothelial cell activation. Of note, galectin-1 was also shown to regulate vascular permeability involving neuropilin-1/VEGFR1 mediated signaling [75].

Apart from the interactions with VEGF (co)receptors, galectin-1 was also shown to bind to CD146 (melanoma cell adhesion molecule; MCAM), which resides on the endothelial cell surface. It was suggested that this interaction prevented galectin-1-induced apoptosis with CD146 serving as a galectin-1 scavenger molecule [76]. While the observation that galectin-1 can induce endothelial cell apoptosis appears contradictory to its angiostimulatory role, it is important to realize that high concentrations of galectin-1 were used in this particular study (millimolar range). Indeed, the activity of galectin-1, in particular in relation to glycan-binding functionality, is dependent on the ability to form homodimers [77–80]. At too low concentrations, insufficient numbers of dimers will be formed, while at too high concentrations, an excess of dimers might interfere with effective crosslinking of glycoproteins. Therefore, the effects of galectin-1 on endothelial cells (and on other cells) were found to be concentration-dependent with inhibitory effects at high concentrations [65,81,82]. This biphasic activity should always be taken into account when studying the function of galectin-1 in endothelial cell biology and angiogenesis.

Finally, many other proteins were identified that engage in either protein-carbohydrate or protein-protein interactions with galectin-1 [83]. While some of these proteins have known functions in endothelial biology, e.g., thrombospondin, integrins $\alpha1\beta1$ and $\alpha5\beta1$, it is still poorly understood whether and how these interactions contribute to endothelial cell function or angiogenesis.

3.2. Galectin-3

The first compelling evidence that galectin-3 is involved in angiogenesis was provided by Nangia-Makker et al. The authors observed increased endothelial cell tube formation in the presence of galectin-3 as well as a higher number of blood vessels in Matrigel plugs in mice that contained galectin-3 compared to Matrigel alone [84,85]. The angiostimulatory activity, including enhanced migration, proliferation, and in vivo angiogenesis, was later confirmed by others [72,86–89]. However, the context is important, as galectin-3 was also shown to increase endothelial cell dysfunction in the presence of oxidized low-density lipoprotein [90] and to inhibit endothelial cell proliferation [91].

Similar to galectin-1, galectin-3 can delay VEGFR2 endocytosis and stimulate VEGFR2 phosphorylation and signaling [72,86,87]. In fact, when applied together, galectin-1 and galectin-3 were also shown to trigger VEGFR1 signaling in endothelial cells [72]. Next to increased VEGFR2 signaling, the angiostimulatory activity of galectin-3 was recently also linked to the protein's ability to interact with JAG1, a NOTCH1 ligand. Dos Santos et al. described that binding of galectin-3 to JAG1 increased the half-life of the latter resulting in enhanced JAG1/NOTCH1 signaling and stimulation of sprouting angiogenesis [92]. Interestingly, a galectin-3/JAG1/NOTCH1 signaling axis has also been linked to transdifferentiation of pulmonary artery-derived endothelial cells into a smooth muscle cell-like phenotype [91]. In line with this, it has recently been described that galectin-3 can regulate endothelial-to-mesenchymal transition of human lung micro-endothelial cells [93]. Although it requires further investigations, it is tempting to speculate that galectin-3 plays a role in regulating the balance between a migratory and proliferative phenotype of endothelial cells, which is key during angiogenesis.

Of note, it was reported that the angiogenic activity of galectin-3 is lost upon removal of the N-terminal tail [87,94]. In line with this, it was shown that proteolytic removal of a large part of the tail by PSA hampers the angiogenic activity of galectin-3 [48]. At the same time, proteolytic cleavage by MMP within the N-terminal tail of galectin-3 can stimulate the angiogenic activity [52]. Moreover, aminopeptidase N (CD13) was suggested to enhance the angiostimulatory activity of galectin-3 by proteolytic processing [95]. Apparently, galectin-3, and in particular the non-CRD tail, is susceptible to proteolytic cleavage, which is important for the activity of the protein during angiogenesis. Whether galectin-3 processing affects both VEGFR2- and JAG1/NOTCH1-mediated signaling requires further investigation. It is however tempting to speculate that modifications of the tail region control the ability of galectin-3 to oligomerize, which might differentially affect both signaling pathways.

The ability of galectin-3 to stimulate endothelial migration and tube formation has been linked to integrin-$\alpha V/\beta3$. Galectin-3 was found to induce glycosylation-dependent clustering of integrin-$\alpha V/\beta3$, resulting in enhanced FAK signaling [86]. Consequently, blocking integrin-$\alpha V/\beta3$ hampered the VEGF-induced migration and tube formation by galectin-3 [86]. Galectin-3 was also found to form a complex with integrin-$\alpha3/\beta1$ and the proteoglycan NG2, which could be involved in mediating the angiostimulatory activity of the latter [96]. More recently, a study by Sedláǿ et al. also suggested a role for glycan-independent effects of galectin-3/integrin interactions. The authors described that blocking antibodies targeting integrin-$\alpha V/\beta3$, integrin-$\alpha5/\beta1$, or integrin-$\alpha2/\beta1$ could hamper endothelial cell adhesion to a galectin-3-coated surface [97]. All of these findings suggest an important link between galectin-3 and integrins in controlling endothelial cell biology, in particular with regard to endothelial cell migration and adhesion.

Apart from integrins, galectin-3 was shown to interact with other proteins on the endothelial cell surface, including CD31 (PECAM-1), CD146 (MCAM), CD144 (VE-cadherin), CD106 (endoglin) [41,98]. As described above, CD146 could serve as a galectin-1 scavenger molecule to hamper galectin-1-induced apoptosis [76]. The interaction of galectin-3 with CD146 was shown to activate AKT signaling and stimulate the release of cytokines by endothelial cells [41]. In addition, the CD146 interaction reduced endothelial cell migration [99]. Whether CD146 also serves as a galectin-3 scavenger or whether galectin-1/CD146 interactions affect cytokine release and migration remains to be studied. In addition, the functional consequences of the other galectin-3/protein interactions in endothelial cell biology are still largely unknown and should be further explored.

3.3. Galectin-8

In contrast to galectin-1 and galectin-3, research on the role of galectin-8 in endothelial cell biology and angiogenesis is relatively sparse. Nevertheless, it was shown that this tandem repeat galectin is also involved in regulation of endothelial cell function (for excellent review see [57]). In part, the regulatory activity appears to be dependent on the endothelial cell phenotype and the presence of galectin-8-binding proteins that are associated with that phenotype. For example, in lymph endothelial cells, which display high expression of galectin-8, a glycosylated transmembrane protein called podoplanin was identified as an important binding partner. Podoplanin is a specific marker of lymphatic vessels and indeed, lymph endothelial cells were able to bind to surface-immobilized galectin-8 while regular endothelial cells were not [19]. At the same time, in a tube formation assay on collagen, galectin-8 was inhibitory towards lymph endothelial cells [19] while stimulatory towards regular endothelial cells [20]. In the latter, CD166 (ALCAM) was identified as a binding partner suggesting that the interaction of galectin-8 with specific endothelial cell surface molecules determines the angioregulatory function of the protein. In line with this, Hadari et al. showed that endothelial cell binding to vitronectin was hardly enhanced by galectin-8 since this adhesion is mediated through integrin$\alpha V\beta 3$, which only shows limited interaction with galectin-8 [100]. Instead, galectin-8 was shown to interact with other integrin subunits present in endothelial cells, including $\alpha 3$, $\alpha 5$, and $\beta 1$ [100,101]. As such, galectin-8 can be expected to differentially mediate cell adhesion and migration, depending on the presence of specific extracellular matrix components as well as certain endothelial cell surface molecules. The complexity of such interactions was shown by Chen et al., again, in the context of lymphangiogenesis. The authors not only confirmed the interaction between podoplanin and galectin-8, but also presented elegant data that supported a model in which galectin-8 clustered podoplanin with integrins $\alpha 1\beta 1/\alpha 5\beta 1$ in order to activate signaling pathways in lymphangiogenesis. By additional clustering of this complex with VEGR3, signaling was further potentiated [26]. Interestingly, galectin-8 was also shown to interact with CD44 [102], a surface molecule that is associated with angiogenesis [103,104] and that also binds podoplanin to promote tumor cell migration [105]. To what extent potential protein clusters of galectin-8/podoplanin/CDD44/integrins contribute to, e.g., endothelial cell migration, is currently not known.

Apart from a role in (lymph)endothelial cell adhesion and migration, galectin-8 has also been shown to induce a pro-inflammatory phenotype in endothelial cells, which was characterized by increased secretion of proinflammatory cytokines and increased binding of platelets [21]. More recently, it was suggested that galectin-8 enhanced the stimulatory effects of VEGF on endothelial cell proliferation and migration, but these effects were small. Moreover, the effects were only observed at the lowest concentration and galectin-8 alone did not stimulate proliferation and migration [106]. The clearest stimulatory effect of galectin-8 on angiogenesis, both with or without VEGF, was observed in the in vivo chorioallantoic membrane assay [106]. Whether these findings hint towards a function of galectin-8 in VEGF-signaling, similar as galectin-1 and galectin-3 needs further confirmation. In that regard, it has been shown that galectin-8 can bind to VEGFR2 [101].

Finally, galectin-8 was also shown to increase vascular permeabilization, similar as reported for galectin-1 [75]. However, the galectin-8 induced permeabilization appears to be triggered by a different pathway, i.e., activation of eNOS and disruption of adherens junctions through S-nitrosylation of p120 (Catenin Delta-1). Moreover, here, β1 integrins appeared to be involved [101].

3.4. Galectin-9

With regard to the regulation of endothelial cell function, galectin-9 is relatively a new kid on the block. Indeed, because endothelial galectin-9 expression was identified as an eosinophil chemoattractant [107] and was found induced by, e.g., IFNγ and viral RNA [29,108,109], the protein has been mainly studied in the context of immunomodulation [110,111]. With regard to angiogenesis, we have shown that exogenous application of galectin-9M, the dominant isoform in endothelial cells, hampers in vivo angiogenesis in the chicken chorioallantoic membrane assay [22,23]. In contrast, O'Brien et al. reported increased in vivo angiogenesis using a Matrigel plug assay in mice [112]. Although these appear as opposite findings, the Matrigel plug experiments actually confirmed the findings that galectin-9M serves as a chemoattractant for endothelial cells [22,112], while the CAM experiments confirmed the inhibitory effects of galectin-9M on proliferation and migration [22,23]. However, it should be noted that the effects of galectin-9M are concentration dependent and often show a biphasic effect, similar as described for galectin-1. In addition, the activity depends on the cellular activation status [23] as well as on the origin of the cells with primary endothelial cells being more sensitive (low nM range) compared to immortalized endothelial cells (high nM range) [22]. While this already indicates a complex regulatory role of galectin-9M in angiogenesis, matters are further complicated by the fact that multiple galectin-9 isoform exist and that the protein is subject to proteolytic cleavage as described previously. Indeed, when exploring the effects of the separate galectin-9 CRDs, we observed neutralization or even reversal of activity compared to galectin-9M. These effects again depended on endothelial cell activation status [23]. Thus, galectin-9 clearly regulates multiple aspects of endothelial cell biology and angiogenesis, but regulation is complex and the ultimate outcome depends on many intrinsic and extrinsic factors.

4. Summary and Outstanding Questions

Over the last thirty years, it has become evident that multiple galectin family members are expressed by endothelial cells and that these multifunctional proteins play key roles in endothelial cell biology and angiogenesis. As described here, endothelial galectin expression appears to involve primarily galectin-1, -3, -8, and -9. Importantly, different environmental conditions and triggers were found to affect the following: (i) the galectin expression level; (ii) the presence of (processed) galectin isoforms; and (iii) the cellular localization of galectins. Deciphering the mechanisms and pathways that control these aspects of endothelial galectin expression represents an important challenge for future research. This is particularly relevant in the context of disease as alterations in (vascular) galectin expression have been associated with different pathologies, including cancer [9,35].

With regard to galectins as regulators of angiogenesis, many insights have been gained. It has become clear that galectins can regulate vessel permeability and vessel growth and that they contribute to multiple endothelial cell functions, including activation, proliferation, migration, tube formation, and sprouting (summarized in Table 2 and Figure 3).

Table 2. Effect of galectins on endothelial cell function and angiogenesis *.

Process	Galectin-1	Galectin-3	Galectin-8	Galectin-9M [f]
Activation	↑	↑	↑ [a]	UNK
Proliferation	↑/↓ [b]	↑	=	↓/= [d]
Migration	↑/↓ [b]	↑	↑	↑/↓ [b,d]
Tube formation	↑	↑	↑/↓ [c]	↑/= [d]
Sprouting	↑	↑	↑	=
Permeability	↑	UNK	↑	UNK
Angiogenesis in vivo	↑	↑	↑	↓ [e]

* Some effects are based on single studies and require additional confirmation. In addition, as described in the text, effects might be dependent on specific experimental conditions, galectin isoforms, or on the endothelial cell phenotype. UNK, unknown; [a] inflammatory activation; [b] concentration dependent, i.e., stimulation in low nM range and inhibition in high nM/low µM range. [c] Dependent on lymph (↓) vs. regular (↑) endothelial cell phenotype; [d] dependent on cell activation status; [e] only at high dose (500 nM). [f] Different effects were found for the separate domains of galectin-9 (see [23]).

In general, most galectins appear to be stimulatory but it is also evident that the effects are dependent on many different aspects, including the source and environmental context of the endothelial cell, the activation status, and importantly, the local concentration of the galectins. In that regard, it is important to realize that the findings described in this review were obtained using endothelial cells from a multitude of origins. This includes primary cells obtained from different tissues, e.g., umbilical, dermal, omental, pulmonary, aortic, as well as different endothelial cell lines from human or mouse origin. Moreover, different culture conditions were used with regard to, e.g., serum conditions, growth factors, matrix proteins. Since all of these differences can affect galectin expression and functionality, it is important to emphasize that the generalizations described here might be different for specific endothelial cells under specific conditions. In fact, deciphering the interplay between these aspects remains a future challenge since much of our current knowledge relies on in vitro findings or on studies using a single galectin. As it was shown that combined application of galectins can enhance the angiostimulatory activity [72], and it was suggested that galectins might engage in heterodimer formation [113], it is relevant to further explore the role of multiple galectins simultaneously in appropriate in vitro and in vivo models. In addition, while this review focused on the main endothelial galectins, recent findings have shown that other galectins can also play a role in regulating angiogenesis. For example, it was shown that galectin-12 expression is increased in adipose tissue under hypoxic conditions. Interestingly, hypoxia was also found to change the glycan-repertoire of endothelial cells, making them more permissive towards galectin-12 binding [38]. Subsequently, the authors showed that galectin-12 could act as a chemoattractant for endothelial cells and that the proteins stimulated tube formation in vitro. In addition, galectin-12 was required for adequate vascularization of adipose tissue in vivo [38]. In addition, galectin-13 has been linked to vascular remodeling, specifically of the uterine vasculature [114]. Although the latter awaits confirmation in humans, all these findings highlight the need for further research into the angioregulatory role of different galectins and specifically the relation with altered endothelial cell glycosylation patterns. As already recognized by Croci et al., a key future challenge will be to obtain a comprehensive insight in the endothelial cell glycome under physiological and pathophysiological conditions vessels, both in the preclinical and clinical settings. This will help to understand how galectins (as well as other glycan-binding proteins) are able to regulate vascular signaling programs and how to interfere with such programs in the context of therapy [37].

It is important to recognize that apart from the direct effects on angiogenesis, galectins can also trigger blood vessel growth indirectly. For example, galectins that are secreted or presented on the surface of endothelial cells can serve as chemoattractants for immune cells or as platelet activators [115]. This can trigger the release of angioregulatory molecules, such as cytokines and growth factors, which influence endothelial cell function and activity [115]. In addition, galectins in the extracellular milieu can serve as scavenger molecules for cytokines [116,117]. All of this further contributes to the angiomodulatory role of galectins. In particular, the width of galectin-cytokine interactions requires further research, as both protein families can exert angioregulatory as well as immunomodulatory functions [118–120]. As such, endothelial galectins hold a key position in the interface between the vasculature and immune cells, which should be further explored.

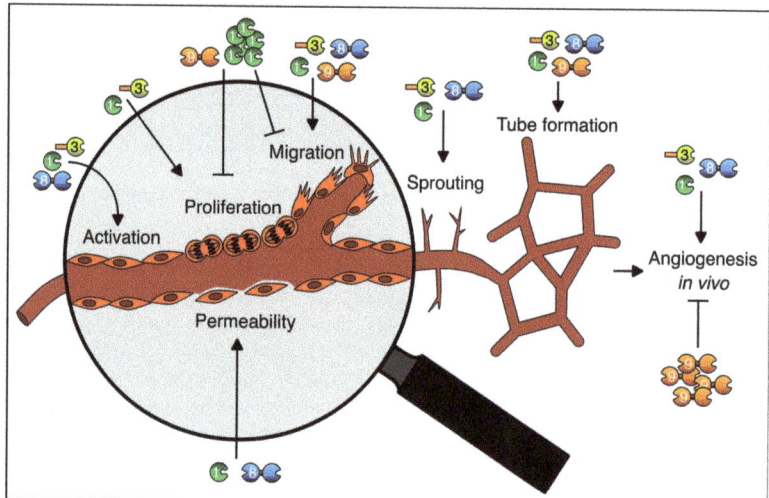

Figure 3. Graphical abstract of the roles of galectins in endothelial cell function and angiogenesis. See text and Table 2 for further explanation.

5. Future Perspectives

As described above, the research community has made considerable steps forward in understanding how galectins contribute to endothelial cell biology and angiogenesis. Nevertheless, many questions remain unanswered and there are sufficient challenges and questions that should be addressed in order to fully grasp the complex functions of galectins in vascular biology. At the same time, increasing insights in the function of galectins in the vasculature also provides opportunities, especially in the context of pathologies or diseases that are associated with aberrant vascular functionality, e.g., cardiovascular disease and cancer. Indeed, many galectin-targeting agents were developed by us and others, ranging from peptides, small molecules, and glycan-based ligands, to blocking antibodies, and have been shown to interfere with galectin functions during angiogenesis [63,66,70,84,121–124]. While a major future challenge is to translate these preclinical findings to clinical applications, it can be anticipated that such galectin-targeting molecules can be used for direct therapeutic applications as well as for indirect applications, including drug delivery and diagnostic imaging. Ultimately, this could help to develop novel and better treatment modalities for patients suffering from diseases that are associated with deregulated vascular galectin expression and or galectin dysfunction.

Funding: This research received no external funding.

Data Availability Statement: Not applicable.

Conflicts of Interest: The author declares no conflict of interest.

References

1. Galley, H.F.; Webster, N.R. Physiology of the endothelium. *Br. J. Anaesth.* **2004**, *93*, 105–113. [CrossRef]
2. Potente, M.; Gerhardt, H.; Carmeliet, P. Basic and therapeutic aspects of angiogenesis. *Cell* **2011**, *146*, 873–887. [CrossRef]
3. Folkman, J. Tumor angiogenesis: Therapeutic implications. *N. Engl. J. Med.* **1971**, *285*, 1182–1186. [PubMed]
4. Folkman, J.; Merler, E.; Abernathy, C.; Williams, G. Isolation of a tumor factor responsible for angiogenesis. *J. Exp. Med.* **1971**, *133*, 275–288. [CrossRef]
5. Folkman, J. Anti-Angiogenesis: New concept for therapy of solid tumors. *Ann. Surg.* **1972**, *175*, 409–416. [CrossRef]
6. Hanahan, D.; Weinberg, R.A. Hallmarks of cancer: The next generation. *Cell* **2011**, *144*, 646–674. [CrossRef] [PubMed]
7. Carmeliet, P.; Jain, R.K. Molecular mechanisms and clinical applications of angiogenesis. *Nature* **2011**, *473*, 298–307. [CrossRef] [PubMed]
8. Griffioen, A.W.; Thijssen, V.L. Galectins in tumor angiogenesis. *Ann. Transl. Med.* **2014**, *2*, 90. [CrossRef] [PubMed]
9. Thijssen, V.L.; Rabinovich, G.A.; Griffioen, A.W. Vascular galectins: Regulators of tumor progression and targets for cancer therapy. *Cytokine Growth Factor Rev.* **2013**, *24*, 547–558. [CrossRef]
10. Elola, M.T.; Ferragut, F.; Méndez-Huergo, S.P.; Croci, D.O.; Bracalente, C.; Rabinovich, G.A. Galectins: Multitask signaling molecules linking fibroblast, endothelial and immune cell programs in the tumor microenvironment. *Cell Immunol.* **2018**, *333*, 34–45. [CrossRef] [PubMed]
11. Johannes, L.; Jacob, R.; Leffler, H. Galectins at a glance. *J. Cell Sci.* **2018**, *131*. [CrossRef]
12. Vasta, G.R. Galectins as pattern recognition receptors: Structure, function, and evolution. *Adv. Exp. Med. Biol.* **2012**, *946*, 21–36. [CrossRef] [PubMed]
13. Hirabayashi, J.; Hashidate, T.; Arata, Y.; Nishi, N.; Nakamura, T.; Hirashima, M.; Urashima, T.; Oka, T.; Futai, M.; Muller, W.E.; et al. Oligosaccharide specificity of galectins: A search by frontal affinity chromatography. *Biochim. Biophys. Acta* **2002**, *1572*, 232–254. [CrossRef]
14. Thijssen, V.L.; Hulsmans, S.; Griffioen, A.W. The galectin profile of the endothelium: Altered expression and localization in activated and tumor endothelial cells. *Am. J. Pathol.* **2008**, *172*, 545–553. [CrossRef] [PubMed]
15. Lotan, R.; Belloni, P.N.; Tressler, R.J.; Lotan, D.; Xu, X.C.; Nicolson, G.L. Expression of galectins on microvessel endothelial cells and their involvement in tumour cell adhesion. *Glycoconj. J.* **1994**, *11*, 462–468. [CrossRef] [PubMed]
16. Baum, L.G.; Seilhamer, J.J.; Pang, M.; Levine, W.B.; Beynon, D.; Berliner, J.A. Synthesis of an endogenous lectin, galectin-1, by human endothelial cells is up-regulated by endothelial cell activation. *Glycoconj. J.* **1995**, *12*, 63–68. [CrossRef] [PubMed]
17. Clausse, N.; van den Brule, F.; Waltregny, D.; Garnier, F.; Castronovo, V. Galectin-1 expression in prostate tumor-associated capillary endothelial cells is increased by prostate carcinoma cells and modulates heterotypic cell-cell adhesion. *Angiogenesis* **1999**, *3*, 317–325. [CrossRef]
18. Shekhar, M.P.; Nangia-Makker, P.; Tait, L.; Miller, F.; Raz, A. Alterations in galectin-3 expression and distribution correlate with breast cancer progression: Functional analysis of galectin-3 in breast epithelial-endothelial interactions. *Am. J. Pathol.* **2004**, *165*, 1931–1941. [CrossRef]
19. Cueni, L.N.; Detmar, M. Galectin-8 interacts with podoplanin and modulates lymphatic endothelial cell functions. *Exp. Cell Res.* **2009**, *315*, 1715–1723. [CrossRef]
20. Cardenas Delgado, V.M.; Nugnes, L.G.; Colombo, L.L.; Troncoso, M.F.; Fernandez, M.M.; Malchiodi, E.L.; Frahm, I.; Croci, D.O.; Compagno, D.; Rabinovich, G.A.; et al. Modulation of endothelial cell migration and angiogenesis: A novel function for the "tandem-repeat" lectin galectin-8. *FASEB J.* **2011**, *25*, 242–254. [CrossRef]
21. Cattaneo, V.; Tribulatti, M.V.; Carabelli, J.; Carestia, A.; Schattner, M.; Campetella, O. Galectin-8 elicits pro-inflammatory activities in the endothelium. *Glycobiology* **2014**, *24*, 966–973. [CrossRef]
22. Heusschen, R.; Schulkens, I.A.; van Beijnum, J.; Griffioen, A.W.; Thijssen, V.L. Endothelial LGALS9 splice variant expression in endothelial cell biology and angiogenesis. *Biochim. Biophys. Acta* **2014**, *1842*, 284–292. [CrossRef]
23. Aanhane, E.; Schulkens, I.A.; Heusschen, R.; Castricum, K.; Leffler, H.; Griffioen, A.W.; Thijssen, V.L. Different angioregulatory activity of monovalent galectin-9 isoforms. *Angiogenesis* **2018**, *21*, 545–555. [CrossRef] [PubMed]
24. Chakraborty, A.; Staudinger, C.; King, S.L.; Erickson, F.C.; Lau, L.S.; Bernasconi, A.; Luscinskas, F.W.; Perlyn, C.; Dimitroff, C.J. Galectin-9 bridges human B cells to vascular endothelium while programming regulatory pathways. *J. Autoimmun.* **2021**, *117*, 102575. [CrossRef] [PubMed]
25. Hepp, P.; Unverdorben, L.; Hutter, S.; Kuhn, C.; Ditsch, N.; Groß, E.; Mahner, S.; Jeschke, U.; Knabl, J.; Heidegger, H.H. Placental Galectin-2 Expression in Gestational Diabetes: A Systematic, Histological Analysis. *Int. J. Mol. Sci.* **2020**, *21*, E2404. [CrossRef] [PubMed]
26. Chen, W.S.; Cao, Z.; Sugaya, S.; Lopez, M.J.; Sendra, V.G.; Laver, N.; Leffler, H.; Nilsson, U.J.; Fu, J.; Song, J.; et al. Pathological lymphangiogenesis is modulated by galectin-8-dependent crosstalk between podoplanin and integrin-associated VEGFR-3. *Nat. Commun.* **2016**, *7*, 11302. [CrossRef] [PubMed]
27. Furuhata, S.; Ando, K.; Oki, M.; Aoki, K.; Ohnishi, S.; Aoyagi, K.; Sasaki, H.; Sakamoto, H.; Yoshida, T.; Ohnami, S. Gene expression profiles of endothelial progenitor cells by oligonucleotide microarray analysis. *Mol. Cell Biochem.* **2007**, *298*, 125–138. [CrossRef] [PubMed]
28. Alam, S.; Li, H.; Margariti, A.; Martin, D.; Zampetaki, A.; Habi, O.; Cockerill, G.; Hu, Y.; Xu, Q.; Zeng, L. Galectin-9 protein expression in endothelial cells is positively regulated by histone deacetylase 3. *J. Biol. Chem.* **2011**, *286*, 44211–44217. [CrossRef]

29. Imaizumi, T.; Kumagai, M.; Sasaki, N.; Kurotaki, H.; Mori, F.; Seki, M.; Nishi, N.; Fujimoto, K.; Tanji, K.; Shibata, T.; et al. Interferon-{gamma} stimulates the expression of galectin-9 in cultured human endothelial cells. *J. Leukoc. Biol.* **2002**, *72*, 486–491.
30. He, J.; Baum, L.G. Endothelial cell expression of galectin-1 induced by prostate cancer cells inhibits T-cell transendothelial migration. *Lab. Invest.* **2006**, *86*, 578–590. [CrossRef]
31. Pranjol, M.Z.I.; Zinovkin, D.A.; Maskell, A.R.T.; Stephens, L.J.; Achinovich, S.L.; Los', D.M.; Nadyrov, E.A.; Hannemann, M.; Gutowski, N.J.; Whatmore, J.L. Cathepsin L-induced galectin-1 may act as a proangiogenic factor in the metastasis of high-grade serous carcinoma. *J. Transl. Med.* **2019**, *17*, 216. [CrossRef]
32. Ahrens, I.; Domeij, H.; Topcic, D.; Haviv, I.; Merivirta, R.M.; Agrotis, A.; Leitner, E.; Jowett, J.B.; Bode, C.; Lappas, M.; et al. Successful in vitro expansion and differentiation of cord blood derived CD34+ cells into early endothelial progenitor cells reveals highly differential gene expression. *PLoS ONE* **2011**, *6*, e23210. [CrossRef]
33. Deo, P.; Glenn, J.V.; Powell, L.A.; Stitt, A.W.; Ames, J.M. Upregulation of oxidative stress markers in human microvascular endothelial cells by complexes of serum albumin and digestion products of glycated casein. *J. Biochem. Mol. Toxicol.* **2009**, *23*, 364–372. [CrossRef]
34. Gil, C.D.; La, M.; Perretti, M.; Oliani, S.M. Interaction of human neutrophils with endothelial cells regulates the expression of endogenous proteins annexin 1, galectin-1 and galectin-3. *Cell Biol. Int.* **2006**, *30*, 338–344. [CrossRef] [PubMed]
35. Thijssen, V.L.; Heusschen, R.; Caers, J.; Griffioen, A.W. Galectin expression in cancer diagnosis and prognosis: A systematic review. *Biochim. Biophys. Acta* **2015**, *1855*, 235–247. [CrossRef] [PubMed]
36. Croci, D.O.; Cerliani, J.P.; Dalotto-Moreno, T.; Méndez-Huergo, S.P.; Mascanfroni, I.D.; Dergan-Dylon, S.; Toscano, M.A.; Caramelo, J.J.; García-Vallejo, J.J.; Ouyang, J.; et al. Glycosylation-Dependent Lectin-Receptor Interactions Preserve Angiogenesis in Anti-VEGF Refractory Tumors. *Cell* **2014**, *156*, 744–758. [CrossRef] [PubMed]
37. Croci, D.O.; Cerliani, J.P.; Pinto, N.A.; Morosi, L.G.; Rabinovich, G.A. Regulatory role of glycans in the control of hypoxia-driven angiogenesis and sensitivity to anti-angiogenic treatment. *Glycobiology* **2014**, *24*, 1283–1290. [CrossRef] [PubMed]
38. Maller, S.M.; Cagnoni, A.J.; Bannoud, N.; Sigaut, L.; Pérez Sáez, J.M.; Pietrasanta, L.I.; Yang, R.Y.; Liu, F.T.; Croci, D.O.; Di Lella, S.; et al. An adipose tissue galectin controls endothelial cell function via preferential recognition of 3-fucosylated glycans. *FASEB J.* **2020**, *34*, 735–753. [CrossRef]
39. Chen, C.; Duckworth, C.A.; Fu, B.; Mark Pritchard, D.; Rhodes, J.M.; Yu, L.-G. Circulating galectins -2, -4 and -8 in cancer patients make important contributions to the increased circulation of several cytokines and chemokines that promote angiogenesis and metastasis. *Br. J. Cancer* **2014**. [CrossRef] [PubMed]
40. Chen, C.; Duckworth, C.A.; Zhao, Q.; Pritchard, D.M.; Rhodes, J.M.; Yu, L.-G. Increased circulation of galectin-3 in cancer induces secretion of metastasis-promoting cytokines from blood vascular endothelium. *Clin. Cancer Res.* **2013**, *19*, 1693–1704. [CrossRef] [PubMed]
41. Colomb, F.; Wang, W.; Simpson, D.; Zafar, M.; Beynon, R.; Rhodes, J.M.; Yu, L.G. Galectin-3 interacts with the cell-surface glycoprotein CD146 (MCAM, MUC18) and induces secretion of metastasis-promoting cytokines from vascular endothelial cells. *J. Biol. Chem.* **2017**, *292*, 8381–8389. [CrossRef]
42. Spitzenberger, F.; Graessler, J.; Schroeder, H.E. Molecular and functional characterization of galectin 9 mRNA isoforms in porcine and human cells and tissues. *Biochimie* **2001**, *83*, 851–862. [CrossRef]
43. Friedel, M.; André, S.; Goldschmidt, H.; Gabius, H.J.; Schwartz-Albiez, R. Galectin-8 enhances adhesion of multiple myeloma cells to vascular endothelium and is an adverse prognostic factor. *Glycobiology* **2016**, *26*, 1048–1058. [CrossRef] [PubMed]
44. Balan, V.; Nangia-Makker, P.; Jung, Y.S.; Wang, Y.; Raz, A. Galectin-3: A novel substrate for c-Abl kinase. *Biochim. Biophys. Acta* **2010**, *1803*, 1198–1205. [CrossRef] [PubMed]
45. Gao, X.; Liu, J.; Liu, X.; Li, L.; Zheng, J. Cleavage and phosphorylation: Important post-translational modifications of galectin-3. *Cancer Metastasis Rev.* **2017**, *36*, 367–374. [CrossRef]
46. Davis, C.M.; Hiremath, G.; Wiktorowicz, J.E.; Soman, K.V.; Straub, C.; Nance, C.; Quintanilla, N.; Pazdrak, K.; Thakkar, K.; Olive, A.P.; et al. Proteomic Analysis in Esophageal Eosinophilia Reveals Differential Galectin-3 Expression and S-Nitrosylation. *Digestion* **2016**, *93*, 288–299. [CrossRef]
47. Berbís, M.Á.; André, S.; Cañada, F.J.; Pipkorn, R.; Ippel, H.; Mayo, K.H.; Kübler, D.; Gabius, H.J.; Jiménez-Barbero, J. Peptides derived from human galectin-3 N-terminal tail interact with its carbohydrate recognition domain in a phosphorylation-dependent manner. *Biochem. Biophys. Res. Commun.* **2014**, *443*, 126–131. [CrossRef]
48. Balan, V.; Nangia-Makker, P.; Kho, D.H.; Wang, Y.; Raz, A. Tyrosine-phosphorylated galectin-3 protein is resistant to prostate-specific antigen (PSA) cleavage. *J. Biol. Chem.* **2012**, *287*, 5192–5198. [CrossRef]
49. Prudova, A.; auf dem Keller, U.; Butler, G.S.; Overall, C.M. Multiplex N-terminome analysis of MMP-2 and MMP-9 substrate degradomes by iTRAQ-TAILS quantitative proteomics. *Mol. Cell. Proteomics.* **2010**, *9*, 894–911. [CrossRef]
50. Mazurek, N.; Conklin, J.; Byrd, J.C.; Raz, A.; Bresalier, R.S. Phosphorylation of the beta-galactoside-binding protein galectin-3 modulates binding to its ligands. *J. Biol. Chem.* **2000**, *275*, 36311–36315. [CrossRef]
51. Nishi, N.; Itoh, A.; Shoji, H.; Miyanaka, H.; Nakamura, T. Galectin-8 and galectin-9 are novel substrates for thrombin. *Glycobiology* **2006**, *16*, 15C–20C. [CrossRef]
52. Nangia-Makker, P.; Wang, Y.; Raz, T.; Tait, L.; Balan, V.; Hogan, V.; Raz, A. Cleavage of galectin-3 by matrix metalloproteases induces angiogenesis in breast cancer. *Int. J. Cancer* **2010**, *127*, 2530–2541. [CrossRef]

53. Gabius, H.J.; Brehler, R.; Schauer, A.; Cramer, F. Localization of endogenous lectins in normal human breast, benign breast lesions and mammary carcinomas. *Virchows Arch. B Cell Pathol. Incl. Mol. Pathol.* **1986**, *52*, 107–115. [CrossRef]
54. Patterson, R.J.; Wang, W.; Wang, J.L. Understanding the biochemical activities of galectin-1 and galectin-3 in the nucleus. *Glycoconj. J.* **2004**, *19*, 499–506. [CrossRef]
55. Juszczynski, P.; Ouyang, J.; Monti, S.; Rodig, S.J.; Takeyama, K.; Abramson, J.; Chen, W.; Kutok, J.L.; Rabinovich, G.A.; Shipp, M.A. The AP1-dependent secretion of galectin-1 by Reed Sternberg cells fosters immune privilege in classical Hodgkin lymphoma. *Proc. Natl. Acad. Sci. USA* **2007**, *104*, 13134–13139. [CrossRef]
56. Schattner, M.; Rabinovich, G.A. Galectins: New agonists of platelet activation. *Biol. Chem.* **2013**, *394*, 857–863. [CrossRef] [PubMed]
57. Troncoso, M.F.; Ferragut, F.; Bacigalupo, M.L.; Cárdenas Delgado, V.M.; Nugnes, L.G.; Gentilini, L.; Laderach, D.; Wolfenstein-Todel, C.; Compagno, D.; Rabinovich, G.A.; et al. Galectin-8: A matricellular lectin with key roles in angiogenesis. *Glycobiology* **2014**. [CrossRef]
58. Funasaka, T.; Raz, A.; Nangia-Makker, P. Galectin-3 in angiogenesis and metastasis. *Glycobiology* **2014**, *24*, 886–891. [CrossRef] [PubMed]
59. Thijssen, V.L.; Griffioen, A.W. Galectin-1 and galectin-9 in angiogenesis; A sweet couple. *Glycobiology* **2014**. [CrossRef] [PubMed]
60. Nambiar, D.K.; Aguilera, T.; Cao, H.; Kwok, S.; Kong, C.; Bloomstein, J.; Wang, Z.; Rangan, V.S.; Jiang, D.; von Eyben, R.; et al. Galectin-1-driven T cell exclusion in the tumor endothelium promotes immunotherapy resistance. *J. Clin. Invest.* **2019**. [CrossRef]
61. Whitney, P.; Maxwell, S.; Ryan, U.; Massaro, D. Synthesis and binding of lactose-specific lectin by isolated lung cells. *Am. J. Physiol.* **1985**, *248*, C258–C264. [CrossRef]
62. Allen, H.J.; Sucato, D.; Gottstine, S.; Kisailus, E.; Nava, H.; Petrelli, N.; Castillo, N.; Wilson, D. Localization of endogenous beta-galactoside-binding lectin in human cells and tissues. *Tumour Biol.* **1991**, *12*, 52–60. [CrossRef]
63. Thijssen, V.L.; Postel, R.; Brandwijk, R.J.; Dings, R.P.; Nesmelova, I.; Satijn, S.; Verhofstad, N.; Nakabeppu, Y.; Baum, L.G.; Bakkers, J.; et al. Galectin-1 is essential in tumor angiogenesis and is a target for antiangiogenesis therapy. *Proc. Natl. Acad. Sci. USA* **2006**, *103*, 15975–15980. [CrossRef] [PubMed]
64. Hsieh, S.H.; Ying, N.W.; Wu, M.H.; Chiang, W.F.; Hsu, C.L.; Wong, T.Y.; Jin, Y.T.; Hong, T.M.; Chen, Y.L. Galectin-1, a novel ligand of neuropilin-1, activates VEGFR-2 signaling and modulates the migration of vascular endothelial cells. *Oncogene* **2008**, *27*, 3746–3753. [CrossRef] [PubMed]
65. Thijssen, V.L.; Barkan, B.; Shoji, H.; Aries, I.M.; Mathieu, V.; Deltour, L.; Hackeng, T.M.; Kiss, R.; Kloog, Y.; Poirier, F.; et al. Tumor cells secrete galectin-1 to enhance endothelial cell activity. *Cancer Res.* **2010**, *70*, 6216–6224. [CrossRef] [PubMed]
66. van Beijnum, J.R.; Thijssen, V.L.; Läppchen, T.; Wong, T.J.; Verel, I.; Engbersen, M.; Schulkens, I.A.; Rossin, R.; Grüll, H.; Griffioen, A.W.; et al. A key role for galectin-1 in sprouting angiogenesis revealed by novel rationally designed antibodies. *Int. J. Cancer* **2016**. [CrossRef] [PubMed]
67. Laderach, D.J.; Gentilini, L.D.; Giribaldi, L.; Delgado, V.C.; Nugnes, L.; Croci, D.O.; Al Nakouzi, N.; Sacca, P.; Casas, G.; Mazza, O.; et al. A Unique Galectin Signature in Human Prostate Cancer Progression Suggests Galectin-1 as a Key Target for Treatment of Advanced Disease. *Cancer Res.* **2013**, *73*, 86–96. [CrossRef]
68. Croci, D.O.; Salatino, M.; Rubinstein, N.; Cerliani, J.P.; Cavallin, L.E.; Leung, H.J.; Ouyang, J.; Ilarregui, J.M.; Toscano, M.A.; Domaica, C.I.; et al. Disrupting galectin-1 interactions with N-glycans suppresses hypoxia-driven angiogenesis and tumorigenesis in Kaposi's sarcoma. *J. Exp. Med.* **2012**, *209*, 1985–2000. [CrossRef] [PubMed]
69. Tang, D.; Gao, J.; Wang, S.; Ye, N.; Chong, Y.; Huang, Y.; Wang, J.; Li, B.; Yin, W.; Wang, D. Cancer-associated fibroblasts promote angiogenesis in gastric cancer through galectin-1 expression. *Tumour Biol.* **2015**. [CrossRef] [PubMed]
70. Pérez Sáez, J.M.; Hockl, P.F.; Cagnoni, A.J.; Méndez Huergo, S.P.; García, P.A.; Gatto, S.G.; Cerliani, J.P.; Croci, D.O.; Rabinovich, G.A. Characterization of a neutralizing anti-human galectin-1 monoclonal antibody with angioregulatory and immunomodulatory activities. *Angiogenesis* **2021**, *24*, 1–5. [CrossRef]
71. Cheng, Y.H.; Jiang, Y.F.; Qin, C.; Shang, K.; Yuan, Y.; Wei, X.J.; Xu, Z.; Luo, X.; Wang, W.; Qu, W.S. Galectin-1 Contributes to Vascular Remodeling and Blood Flow Recovery After Cerebral Ischemia in Mice. *Transl. Stroke Res.* **2021**. [CrossRef] [PubMed]
72. D'Haene, N.; Sauvage, S.; Maris, C.; Adanja, I.; Le Mercier, M.; Decaestecker, C.; Baum, L.; Salmon, I. VEGFR1 and VEGFR2 Involvement in Extracellular Galectin-1- and Galectin-3-Induced Angiogenesis. *PLoS ONE* **2013**, *8*, e67029. [CrossRef]
73. Wei, J.; Li, D.K.; Hu, X.; Cheng, C.; Zhang, Y. Galectin-1-RNA interaction map reveals potential regulatory roles in angiogenesis. *FEBS Lett.* **2021**, *595*, 623–636. [CrossRef]
74. Park, J.W.; Voss, P.G.; Grabski, S.; Wang, J.L.; Patterson, R.J. Association of galectin-1 and galectin-3 with Gemin4 in complexes containing the SMN protein. *Nucleic Acids Res.* **2001**, *29*, 3595–3602. [CrossRef]
75. Wu, M.H.; Ying, N.W.; Hong, T.M.; Chiang, W.F.; Lin, Y.T.; Chen, Y.L. Galectin-1 induces vascular permeability through the neuropilin-1/vascular endothelial growth factor receptor-1 complex. *Angiogenesis* **2014**, *17*, 839–849. [CrossRef]
76. Jouve, N.; Despoix, N.; Espeli, M.; Gauthier, L.; Cypowyj, S.; Fallague, K.; Schiff, C.; Dignat-George, F.; Vély, F.; Leroyer, A.S. The involvement of CD146 and its novel ligand Galectin-1 in apoptotic regulation of endothelial cells. *J. Biol. Chem.* **2013**, *288*, 2571–2579. [CrossRef]
77. Leffler, H.; Carlsson, S.; Hedlund, M.; Qian, Y.; Poirier, F. Introduction to galectins. *Glycoconj. J.* **2004**, *19*, 433–440. [CrossRef] [PubMed]

78. Perillo, N.L.; Pace, K.E.; Seilhamer, J.J.; Baum, L.G. Apoptosis of T cells mediated by galectin-1. *Nature* **1995**, *378*, 736–739. [CrossRef]
79. Cho, M.; Cummings, R.D. Characterization of monomeric forms of galectin-1 generated by site-directed mutagenesis. *Biochemistry* **1996**, *35*, 13081–13088. [CrossRef]
80. Salomonsson, E.; Larumbe, A.; Tejler, J.; Tullberg, E.; Rydberg, H.; Sundin, A.; Khabut, A.; Frejd, T.; Lobsanov, Y.D.; Rini, J.M.; et al. Monovalent interactions of galectin-1. *Biochemistry* **2010**, *49*, 9518–9532. [CrossRef]
81. Adams, L.; Scott, G.K.; Weinberg, C.S. Biphasic modulation of cell growth by recombinant human galectin-1. *Biochim. Biophys. Acta* **1996**, *1312*, 137–144. [CrossRef]
82. Vas, V.; Fajka-Boja, R.; Ion, G.; Dudics, V.; Monostori, E.; Uher, F. Biphasic effect of recombinant galectin-1 on the growth and death of early hematopoietic cells. *Stem Cells* **2005**, *23*, 279–287. [CrossRef] [PubMed]
83. Camby, I.; Le Mercier, M.; Lefranc, F.; Kiss, R. Galectin-1: A small protein with major functions. *Glycobiology* **2006**, *16*, 137R–157R. [CrossRef]
84. Nangia-Makker, P.; Honjo, Y.; Sarvis, R.; Akahani, S.; Hogan, V.; Pienta, K.J.; Raz, A. Galectin-3 induces endothelial cell morphogenesis and angiogenesis. *Am. J. Pathol.* **2000**, *156*, 899–909. [CrossRef]
85. Nangia-Makker, P.; Hogan, V.; Honjo, Y.; Baccarini, S.; Tait, L.; Bresalier, R.; Raz, A. Inhibition of human cancer cell growth and metastasis in nude mice by oral intake of modified citrus pectin. *J. Natl. Cancer Inst.* **2002**, *94*, 1854–1862. [CrossRef] [PubMed]
86. Markowska, A.I.; Liu, F.T.; Panjwani, N. Galectin-3 is an important mediator of VEGF- and bFGF-mediated angiogenic response. *J. Exp. Med.* **2010**, *207*, 1981–1993. [CrossRef] [PubMed]
87. Markowska, A.I.; Jefferies, K.C.; Panjwani, N. Galectin-3 protein modulates cell surface expression and activation of vascular endothelial growth factor receptor 2 in human endothelial cells. *J. Biol. Chem.* **2011**, *286*, 29913–29921. [CrossRef] [PubMed]
88. Wesley, U.V.; Vemuganti, R.; Ayvaci, E.R.; Dempsey, R.J. Galectin-3 enhances angiogenic and migratory potential of microglial cells via modulation of integrin linked kinase signaling. *Brain Res.* **2013**, *1496*, 1–9. [CrossRef] [PubMed]
89. Wan, S.Y.; Zhang, T.F.; Ding, Y. Galectin-3 enhances proliferation and angiogenesis of endothelial cells differentiated from bone marrow mesenchymal stem cells. *Transplant. Proc.* **2011**, *43*, 3933–3938. [CrossRef] [PubMed]
90. Ou, H.C.; Chou, W.C.; Hung, C.H.; Chu, P.M.; Hsieh, P.L.; Chan, S.H.; Tsai, K.L. Galectin-3 aggravates ox-LDL-induced endothelial dysfunction through LOX-1 mediated signaling pathway. *Environ. Toxicol.* **2019**, *34*, 825–835. [CrossRef] [PubMed]
91. Zhang, L.; Li, Y.M.; Zeng, X.X.; Wang, X.Y.; Chen, S.K.; Gui, L.X.; Lin, M.J. Galectin-3- Mediated Transdifferentiation of Pulmonary Artery Endothelial Cells Contributes to Hypoxic Pulmonary Vascular Remodeling. *Cell Physiol. Biochem.* **2018**, *51*, 763–777. [CrossRef]
92. Dos Santos, S.N.; Sheldon, H.; Pereira, J.X.; Paluch, C.; Bridges, E.M.; El-Cheikh, M.C.; Harris, A.L.; Bernardes, E.S. Galectin-3 acts as an angiogenic switch to induce tumor angiogenesis via Jagged-1/Notch activation. *Oncotarget* **2017**, *8*, 49484–49501. [CrossRef]
93. Jia, W.; Wang, Z.; Gao, C.; Wu, J.; Wu, Q. Trajectory modeling of endothelial-to-mesenchymal transition reveals galectin-3 as a mediator in pulmonary fibrosis. *Cell Death Dis.* **2021**, *12*, 327. [CrossRef] [PubMed]
94. Mirandola, L.; Yu, Y.; Chui, K.; Jenkins, M.R.; Cobos, E.; John, C.M.; Chiriva-Internati, M. Galectin-3C inhibits tumor growth and increases the anticancer activity of bortezomib in a murine model of human multiple myeloma. *PLoS ONE* **2011**, *6*, e21811. [CrossRef]
95. Yang, E.; Shim, J.S.; Woo, H.J.; Kim, K.W.; Kwon, H.J. Aminopeptidase N/CD13 induces angiogenesis through interaction with a pro-angiogenic protein, galectin-3. *Biochem. Biophys. Res. Commun.* **2007**, *363*, 336–341. [CrossRef] [PubMed]
96. Fukushi, J.; Makagiansar, I.T.; Stallcup, W.B. NG2 proteoglycan promotes endothelial cell motility and angiogenesis via engagement of galectin-3 and alpha3beta1 integrin. *Mol. Biol. Cell* **2004**, *15*, 3580–3590. [CrossRef]
97. Sedlář, A.; Trávníčková, M.; Bojarová, P.; Vlachová, M.; Slámová, K.; Křen, V.; Bačáková, L. Interaction between Galectin-3 and Integrins Mediates Cell-Matrix Adhesion in Endothelial Cells and Mesenchymal Stem Cells. *Int. J. Mol. Sci.* **2021**, *22*, 5144. [CrossRef] [PubMed]
98. Gallardo-Vara, E.; Ruiz-Llorente, L.; Casado-Vela, J.; Ruiz-Rodríguez, M.J.; López-Andrés, N.; Pattnaik, A.K.; Quintanilla, M.; Bernabeu, C. Endoglin Protein Interactome Profiling Identifies TRIM21 and Galectin-3 as New Binding Partners. *Cells* **2019**, *8*, E1082. [CrossRef]
99. Zhang, Z.; Zheng, Y.; Wang, H.; Zhou, Y.; Tai, G. CD146 interacts with galectin-3 to mediate endothelial cell migration. *FEBS Lett.* **2018**, *592*, 1817–1828. [CrossRef]
100. Hadari, Y.R.; Arbel-Goren, R.; Levy, Y.; Amsterdam, A.; Alon, R.; Zakut, R.; Zick, Y. Galectin-8 binding to integrins inhibits cell adhesion and induces apoptosis. *J. Cell Sci.* **2000**, *113*, 2385–2397. [CrossRef]
101. Zamorano, P.; Koning, T.; Oyanadel, C.; Mardones, G.A.; Ehrenfeld, P.; Boric, M.P.; González, A.; Soza, A.; Sánchez, F.A. Galectin-8 induces endothelial hyperpermeability through the eNOS pathway involving S-nitrosylation-mediated adherens junction disassembly. *Carcinogenesis* **2019**, *40*, 313–323. [CrossRef]
102. Eshkar Sebban, L.; Ronen, D.; Levartovsky, D.; Elkayam, O.; Caspi, D.; Aamar, S.; Amital, H.; Rubinow, A.; Golan, I.; Naor, D.; et al. The involvement of CD44 and its novel ligand galectin-8 in apoptotic regulation of autoimmune inflammation. *J. Immunol.* **2007**, *179*, 1225–1235. [CrossRef] [PubMed]
103. Forster-Horvath, C.; Meszaros, L.; Raso, E.; Dome, B.; Ladanyi, A.; Morini, M.; Albini, A.; Timar, J. Expression of CD44v3 protein in human endothelial cells in vitro and in tumoral microvessels in vivo. *Microvasc. Res.* **2004**, *68*, 110–118. [CrossRef]

104. Griffioen, A.W.; Coenen, M.J.; Damen, C.A.; Hellwig, S.M.; van Weering, D.H.; Vooys, W.; Blijham, G.H.; Groenewegen, G. CD44 is involved in tumor angiogenesis; an activation antigen on human endothelial cells. *Blood* **1997**, *90*, 1150–1159. [CrossRef]
105. Martín-Villar, E.; Fernández-Muñoz, B.; Parsons, M.; Yurrita, M.M.; Megías, D.; Pérez-Gómez, E.; Jones, G.E.; Quintanilla, M. Podoplanin associates with CD44 to promote directional cell migration. *Mol. Biol. Cell* **2010**, *21*, 4387–4399. [CrossRef] [PubMed]
106. Varinská, L.; Fáber, L.; Petrovová, E.; Balážová, L.; Ivančová, E.; Kolář, M.; Gál, P. Galectin-8 Favors VEGF-Induced Angiogenesis: In Vitro Study in Human Umbilical Vein Endothelial Cells and In Vivo Study in Chick Chorioallantoic Membrane. *Anticancer. Res.* **2020**, *40*, 3191–3201. [CrossRef]
107. Matsumoto, R.; Matsumoto, H.; Seki, M.; Hata, M.; Asano, Y.; Kanegasaki, S.; Stevens, R.L.; Hirashima, M. Human ecalectin, a variant of human galectin-9, is a novel eosinophil chemoattractant produced by T lymphocytes. *J. Biol. Chem.* **1998**, *273*, 16976–16984. [CrossRef] [PubMed]
108. Imaizumi, T.; Yoshida, H.; Nishi, N.; Sashinami, H.; Nakamura, T.; Hirashima, M.; Ohyama, C.; Itoh, K.; Satoh, K. Double-stranded RNA induces galectin-9 in vascular endothelial cells: Involvement of TLR3, PI3K, and IRF3 pathway. *Glycobiology* **2007**, *17*, 12C–15C. [CrossRef]
109. Ishikawa, A.; Imaizumi, T.; Yoshida, H.; Nishi, N.; Nakamura, T.; Hirashima, M.; Satoh, K. Double-stranded RNA enhances the expression of galectin-9 in vascular endothelial cells. *Immunol Cell Biol.* **2004**, *82*, 410–414. [CrossRef]
110. Hirashima, M.; Kashio, Y.; Nishi, N.; Yamauchi, A.; Imaizumi, T.A.; Kageshita, T.; Saita, N.; Nakamura, T. Galectin-9 in physiological and pathological conditions. *Glycoconj. J.* **2004**, *19*, 593–600. [CrossRef]
111. Heusschen, R.; Griffioen, A.W.; Thijssen, V.L. Galectin-9 in tumor biology: A jack of multiple trades. *Biochim. Biophys. Acta* **2013**, *1836*, 177–185. [CrossRef]
112. O'Brien, M.J.; Shu, Q.; Stinson, W.A.; Tsou, P.S.; Ruth, J.H.; Isozaki, T.; Campbell, P.L.; Ohara, R.A.; Koch, A.E.; Fox, D.A.; et al. A unique role for galectin-9 in angiogenesis and inflammatory arthritis. *Arthritis Res. Ther.* **2018**, *20*, 31. [CrossRef]
113. Miller, M.C.; Ludwig, A.K.; Wichapong, K.; Kaltner, H.; Kopitz, J.; Gabius, H.J.; Mayo, K.H. Adhesion/growth-regulatory galectins tested in combination: Evidence for formation of hybrids as heterodimers. *Biochem. J.* **2018**, *475*, 1003–1018. [CrossRef] [PubMed]
114. Drobnjak, T.; Jónsdóttir, A.M.; Helgadóttir, H.; Runólfsdóttir, M.S.; Meiri, H.; Sammar, M.; Osol, G.; Mandalà, M.; Huppertz, B.; Gizurarson, S. Placental protein 13 (PP13) stimulates rat uterine vessels after slow subcutaneous administration. *Int. J. Womens Health* **2019**, *11*, 213–222. [CrossRef] [PubMed]
115. Etulain, J.; Negrotto, S.; Tribulatti, M.V.; Croci, D.O.; Carabelli, J.; Campetella, O.; Rabinovich, G.A.; Schattner, M. Control of angiogenesis by galectins involves the release of platelet-derived proangiogenic factors. *PLoS ONE* **2014**, *9*, e96402. [CrossRef]
116. Eckardt, V.; Miller, M.C.; Blanchet, X.; Duan, R.; Leberzammer, J.; Duchene, J.; Soehnlein, O.; Megens, R.T.; Ludwig, A.K.; Dregni, A.; et al. Chemokines and galectins form heterodimers to modulate inflammation. *EMBO Rep.* **2020**, e47852. [CrossRef] [PubMed]
117. Gordon-Alonso, M.; Hirsch, T.; Wildmann, C.; van der Bruggen, P. Galectin-3 captures interferon-gamma in the tumor matrix reducing chemokine gradient production and T-cell tumor infiltration. *Nat. Commun.* **2017**, *8*, 793. [CrossRef]
118. Nagarsheth, N.; Wicha, M.S.; Zou, W. Chemokines in the cancer microenvironment and their relevance in cancer immunotherapy. *Nat. Rev. Immunol.* **2017**, *17*, 559–572. [CrossRef] [PubMed]
119. Zlotnik, A.; Yoshie, O. The chemokine superfamily revisited. *Immunity* **2012**, *36*, 705–716. [CrossRef]
120. Rabinovich, G.A.; Croci, D.O. Regulatory circuits mediated by lectin-glycan interactions in autoimmunity and cancer. *Immunity* **2012**, *36*, 322–335. [CrossRef] [PubMed]
121. Schulkens, I.A.; Griffioen, A.W.; Thijssen, V.L. *ACS Symposium Series: Galectins and Disease Implications for Targeted Therapeutics*; Klyosov, A., Ed.; ACS Publications: Washington, DC, USA, 2012; p. 233.
122. Rabinovich, G.A.; Cumashi, A.; Bianco, G.A.; Ciavardelli, D.; Iurisci, I.; D'Egidio, M.; Piccolo, E.; Tinari, N.; Nifantiev, N.; Iacobelli, S. Synthetic lactulose amines: Novel class of anticancer agents that induce tumor-cell apoptosis and inhibit galectin-mediated homotypic cell aggregation and endothelial cell morphogenesis. *Glycobiology* **2006**, *16*, 210–220. [CrossRef] [PubMed]
123. Ito, K.; Scott, S.A.; Cutler, S.; Dong, L.F.; Neuzil, J.; Blanchard, H.; Ralph, S.J. Thiodigalactoside inhibits murine cancers by concurrently blocking effects of galectin-1 on immune dysregulation, angiogenesis and protection against oxidative stress. *Angiogenesis* **2011**, *14*, 293–307. [CrossRef] [PubMed]
124. Dings, R.P.M.; Miller, M.C.; Nesmelova, I.; Astorgues-Xerri, L.; Kumar, N.; Serova, M.; Chen, X.; Raymond, E.; Hoye, T.R.; Mayo, K.H. Antitumor agent calixarene 0118 targets human galectin-1 as an allosteric inhibitor of carbohydrate binding. *J. Med. Chem.* **2012**, *55*, 5121–5129. [CrossRef] [PubMed]

Review

The Diagnostic and Therapeutic Potential of Galectin-3 in Cardiovascular Diseases

Grażyna Sygitowicz *, Agata Maciejak-Jastrzębska and Dariusz Sitkiewicz

Department of Clinical Chemistry and Laboratory Diagnostics, Medical University of Warsaw, 02-097 Warsaw, Poland; agata.maciejak@wum.edu.pl (A.M.-J.); dariusz.sitkiewicz@gmail.com (D.S.)
* Correspondence: gsygitowicz@poczta.onet.pl

Abstract: Galectin-3 plays a prominent role in chronic inflammation and has been implicated in the development of many disease conditions, including heart disease. Galectin-3, a regulatory protein, is elevated in both acute and chronic heart failure and is involved in the inflammatory pathway after injury leading to myocardial tissue remodelling. We discussed the potential utility of galectin-3 as a diagnostic and disease severity/prognostic biomarker in different cardio/cerebrovascular diseases, such as acute ischemic stroke, acute coronary syndromes, heart failure and arrhythmogenic cardiomyopathy. Over the last decade there has been a marked increase in the understanding the role of galectin-3 in myocardial fibrosis and inflammation and as a therapeutic target for the treatment of heart failure and myocardial infarction.

Keywords: galectin-3; cardiac fibrosis; heart failure; atrial fibrillation; chronic inflammation; MMPs; microRNAs; lncRNAs

1. Introduction

Galectins are a family of glycan-binding proteins, which were named and classified in 1994 considering their affinity for β-galactosides and significant sequence similarity in the carbohydrate-binding domains (CRDs) [1,2]. To date, 16 protein-coding galectin genes have been identified including 12 members in human tissues encoding galectins 1, 2, 3, 4, 7, 8, 9, 10, 12, 13, 14 and 16 [2–4]. Galectins 5 and 6 are present in rodents, while galectins 11 and 15 are found in sheep and goats [2].

Galectins are synthesized in cytosol but they can be secreted from the cells in not yet fully elucidated mechanisms. Functioning both outside and inside the cell—galectins participate in various cell processes, including transport of glycoprotein vesicles, chemotaxis, proliferation, pre-RNA splicing and apoptosis. Extracellular galectins can act through binding to cell surface glycans, while intracellular effectors, mediating galectin function, still remain, to a great extent, unknown.

The members of the galectin family are classified into three types according to their molecular architecture: (1) prototype, which are usually homodimers containing one carbohydrate recognition domain in each subunit; (2) tandem, which are monomers containing two CRDs joined by a linker sequence; (3) chimeric, containing C-terminal CRD joined to a large repeating sequence and N-terminal domain [2].

2. Galectin-3 Structure and Functions

Galectin-3 (Gal-3) is widespread and present in various organs including: the lungs, heart, stomach, colon, adrenals, uterus and ovaries [5]. Galectin-3 is the only galectin of chimeric type in the galectin family. Galectin-3, also known as a binding protein for IgE, Mac2, CBP30 and CP35, is encoded by the *LGALS3* gene present on chromosome 14, locus q21-22. It consists of six exons and five introns involving about 17 kilobases. Galectin-3 expression is regulated by promotor methylation status of *LGALS3*, and such elements as: CRE motifs, region similar to nuclear factor kappa B (NF-κB), GC regions located in galectin-3 promoter [6]. The galectin-3 gene contains also a special regulatory element called galig (galectin-3 internal gene) located in the second intron of the *LGALS3* gene [7].

Galectin-3 consists of 251 amino acid residues of relative molecular mass 29–35 kDa and it has been identified for the first time in murine peritoneal macrophages [1]. It contains three domains: (1) short N-terminal constituting a unique region of 12 amino acids and containing a site of serine 6 phosphorylation for controlling its nuclear location and abolishing its affinity to ligand [8,9]; (2) a 100-amino acid sequence similar to collagen, containing proline, glycine and tyrosine tandem repetitions and containing a fissionable domain of collagenase H, in which histidine 64 is the site of action of matrix metalloproteinases (MMPs), such as MMP-9 and MMP-2; (3) spherical C-terminal CRD containing an Asp-Trp-Gly-Arg motif (NWGR), similar to those described in the anti-apoptotic protein Bcl-2 [1].

Galectin-3 can form dimers or pentamers in specific circumstances, when galectin-3 concentration is high or when ligands are present [10]. Monomeric galectin-3 undergoes physicochemical modifications, which increase the range of its biological functionality, particularly extracellular activity. The most important mechanisms leading to galectin-3 "bioactivation" is multimerization and formation of galectin-3 lattice. Galectin-3 has various functions, depending on its cellular location (Table 1).

Table 1. Various functions of galectin-3 in relation to cellular location.

Location	Function	References
Cytoplasm	Anti-apoptotic effect Proliferation	[6,7,11–13]
Nucleus	Gene transcription regulation pre-mRNA splicing promotion	[12,13]
Cell-surface	Diffusion and compartmentalization regulation Kinase and membrane receptors signalling	[14,15]
Extracellular environment	Cell adhesion Migration Growth regulation Pro-apoptotic effect	[16,17]

Galectin-3 participates in various pathophysiological processes, including apoptosis [6], adhesion [16], angiogenesis [17], cell migration [17], proliferation [16], and differentiation [18], but its main function is induction of inflammatory condition and fibrotic process [6] (Figure 1).

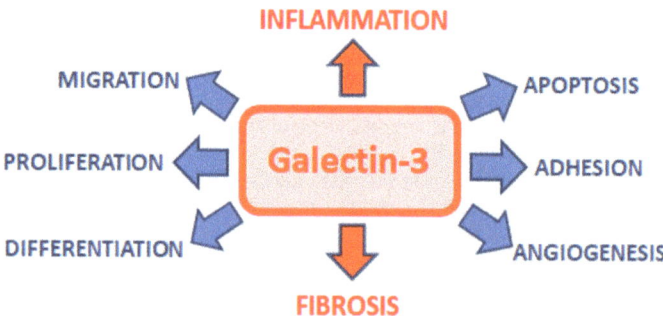

Figure 1. Biological functions of galectin-3.

3. Galectin-3 in Cardiovascular Diseases

In view of its multidirectional activity, galectin-3 plays an important role in many various clinical conditions and disease entities. Increased galectin-3 expression has been documented in cardiovascular diseases (CVDs), such as atherosclerosis, acute ischemic stroke, acute coronary syndrome (ACS) and heart failure (HF), arterial hypertension, cardiomyopathies or atrial fibrillation (AF) [19] (Figure 2).

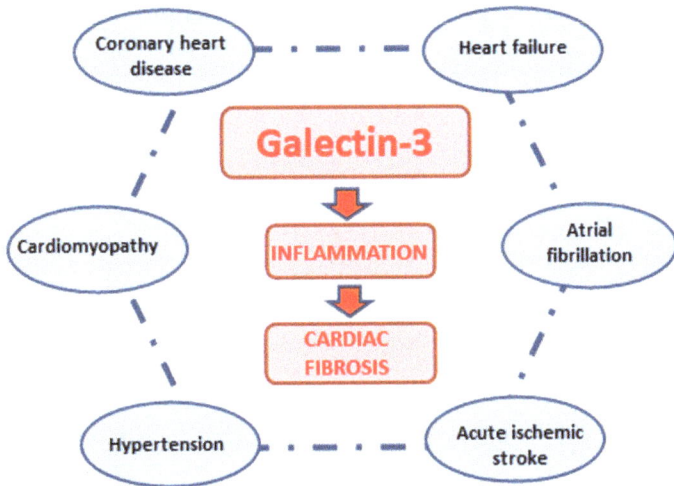

Figure 2. Galectin-3 and cardiovascular diseases.

3.1. Atherosclerosis

Atherosclerosis is a complex inflammatory process, initiated by changed permeability of arterial wall cells and focal subendothelial accumulation of LDL lipoproteins, leading to formation of atheromatous plaques characterised by inflammatory condition and increased oxidative stress [20,21]. The later inflammatory reaction includes a massive participation of monocytes and macrophages changing the structure of the vascular wall [10,22]. Endothelial cells become activated and vascular smooth muscle cells (VSMCs) actively proliferate to produce extracellular matrix. Over the years, the participation of many inflammatory condition markers in the atherosclerotic process has been studied [21–25] and then a potential role of galectin-3 as a mediator of atherosclerosis has been suggested. Many studies have shown that galectin-3 contributes to macrophage differentiation [26], formation of foam cells [27], endothelial dysfunction [27,28] and VSMC proliferation and migration in atherogenesis [29]. The amplification of cardiovascular system inflammatory condition and accumulation of lipids in macrophages caused by galectin-3 are the most important mechanisms of atherosclerosis development, which are stimulated by local or circulating galectin-3.

Study performed by Ou et al. [30], using human umbilical vein endothelial cells (HUVECs), showed that exogenous addition of galectin-3 and oxidised low-density lipoprotein (oxLDL) to cell cultures increased the expression of lectin-like oxLDL receptor 1 (LOX-1) and promoted endothelial dysfunction via LOX-1/ROS/p38/NF-kB-mediated signalling pathway. It has been suggested by authors that galectin-3 enhances LOX-1 expression and induced pro-inflammatory response [30].

3.2. Acute Myocardial Infarction

Galectin-3 has been described as a factor contributing to the development and destabilisation of atheromatous plaques through propagation of inflammatory condition, interaction with lipopolysaccharides (LPSs) and promotion of VSMC phenotypic transformation [31]. Therefore, galectin-3 can be involved in the pathogenesis of ACSs caused by atherosclerosis. A significantly increased galectin-3 expression has been demonstrated in the early phase of acute myocardial infarction (AMI) and ACS [32,33]. In in vivo studies, using recombinant galectin-3 in rat experimental model, it has been observed higher collagen deposition and thick collagen content in the infarct region. This study presented potential application of exogenous galectin-3 to assess the myocardial remodelling process after MI [34]. Both experimental and clinical studies have demonstrated that galectin-3 is an independent predictor of mortality for any cause, death for cardiovascular causes and development of HF [35,36]. In the later phase of ACS, galectin-3 enhances the transition from acute to chronic inflammatory condition and causes myocardial fibrosis, leading to unfavourable ventricular remodelling [35].

3.3. Heart Failure

In recent decades, cardiac remodelling and myocardial fibrosis have been accepted as the main heart failure-inducing factors. Galectin-3, exogenously added, is closely related to myocardial fibrosis and is strongly expressed in cardiac myofibroblasts, which can be used as an independent predictor of myocardial fibrosis [37]. Sharma et al. [37] revealed that recombinant galectin-3, exogenously injected into the pericardial sac of healthy rats over a long term, led to LV dysfunction and deterioration of cardiac function. Galectin-3 induced collagen deposition and fibroblast proliferation. Liu et al. [38] demonstrated that when the transforming growth factor β (TGFβ)/Smad3 signalling protein pathway was inhibited by N-acetyl-seryl-aspartyl-lysyl-proline (Ac-SDKP), the expression of profibrotic and inflammatory factors induced by intrapericardial infusion of galectin-3 was significantly reduced and cardiac remodelling and dysfunction were less pronounced. That fact suggests that galectin-3-induced myocardial remodelling may be associated with an activation of TGFβ/Smad3 signal transduction pathway. On the other hand, another experiment [39] has proved that excessive galectin-3 expression favours type I collagen synthesis, while galectin-3 secretion inhibition reduces the synthesis and deposition of that collagen.

3.4. Hypertrophic Cardiomyopathy

Hypertrophic cardiomyopathy (HCM), a primary myocardial disease, is characterised by myocardial hypertrophy. The development of cardiac fibrosis and irreversible change of ventricular structure significantly contribute to sudden cardiac death in patients with HCM [40]. In their case-control study, Yakar et al. [41] analysed and compared the relationship between serum galectin-3 concentration and left ventricular (LV) function in 40 patients with HCM and in 35 age-matched healthy volunteers. That study demonstrated that concentration of serum galectin-3 was significantly elevated in HCM patients compared with the control group. Furthermore, levels of serum galectin-3 were positively correlated with interventricular septum thickness and LV mass index. The concentration of serum galectin-3 was, however, not related to the degree of LV outflow tract obstruction. For that reason, it was accepted that serum galectin-3 concentration was related to the degree of LV hypertrophy but not to the LV diastolic and systolic dysfunction [41]. However, whether galectin-3 plays a major causative role in discussed process is not completely justified because there are studies showing that up-regulation of cardiac galectin-3 is not a critical disease modulator of cardiomyopathy induced by β_2-adrenoceptor over-expression [42]. These findings raise a possibility that role of galectin-3 in the development of cardiomyopathy might not be universal but rather dependent to disease aetiology. On the other hand, another experiment [43] has revealed that galectin-3 is not a critical mediator of the fibrotic cardiomyopathy associated with pressure overload. Myocardial galectin-3 expression did not affect the survival, systolic and diastolic dysfunction, and cardiac fibrosis.

3.5. Atrial Fibrillation

Atrial fibrillation (AF) is a cardiac arrhythmia characterised by a rapid and irregular heart rate and is the most frequent and most serious arrhythmia. AF is closely related to high mortality rate, cerebral stroke and HF [44]. Particularly, atrial interstitial fibrosis seems to be the key factor contributing to AF development [45]. It is known that left atrial (LA) interstitial fibrosis plays a significant role in initiating and maintaining atrial fibrillation [45,46].

Galectin-3 enhances cardiac fibrosis and remodelling and is a well-documented cause of arrhythmia. Most studies have focused on the relationship between levels of serum galectin-3 and cardiac fibrosis in HF. The role of galectin-3 in the pathophysiology of atrial fibrillation has not been fully elucidated as yet. Sonmez et al. [47] were the first to study the levels of new biomarkers of inflammation in serum in patients with atrial fibrillation compared with those with sinus rhythm. The results of the study demonstrated that the levels of the new circulating markers of remodelling, such as galectin-3, MMP-9 and PIIINP (amino terminal peptide of type III procollagen) were significantly higher in patients with atrial fibrillation. Moreover, galectin-3, MMP-9 and PIIINP concentrations in serum were strongly positively correlated with LA volume and LA volume index [47]. The observational study by Gurses et al. [48] demonstrated also that serum galectin-3 concentration and LA volume index were significantly higher in patients with atrial fibrillation, which could have suggested that concentration of serum galectin-3 in AF can be correlated with atrial remodelling. The study also demonstrated that serum galectin-3 concentration was significantly higher in patients with persistent atrial fibrillation than in those with paroxysmal atrial fibrillation [48]. The study conducted by Yalcin et al. [49] revealed that the LA volume index and serum galectin-3 concentration were independently correlated with the range of LA fibrosis detected by means of delayed-enhancement magnetic resonance imaging (DE-MRI) in patients with paroxysmal atrial fibrillation with preserved LV function. These results suggest that concentration of serum galectin-3 is significantly correlated with atrial remodelling in patients with paroxysmal atrial fibrillation with preserved LV function [49].

3.6. Arterial Hypertension

Arterial hypertension is usually a chronic pathological condition characterised by elevated arterial pressure. Hypertensive cardiac remodelling starts with inflammation, increased deposition of extracellular matrix proteins leading then to myocardial fibrosis and, finally, to heart dysfunction [50]. It has been demonstrated that serum galectin-3 concentration increases in hypertensive patients, but that phenomenon is more pronounced in patients with LV hypertrophy. Therefore, galectin-3, as an important biomarker of early cardiac remodelling, is independently correlated with left ventricular remodelling (LVR). Moreover, LV mass was independently correlated with concentration of serum galectin-3 in hypertensive patients [51].

Primary aldosteronism is regarded as the most frequent cause of secondary arterial hypertension. Patients with primary aldosteronism show an increased susceptibility to heart muscle inflammation and fibrosis [52]. It has been demonstrated that aldosterone can induce galectin-3 secretion [50]. Galectin-3 is one of the most important mediators between macrophage activation and myocardial fibrosis. In a prospective pilot clinical study, serum galectin-3 concentration was significantly higher in the group of patients with aldosterone-producing adenoma. Moreover, both the degree of myocardial fibrosis and serum galectin-3 concentration returned to normal after adrenalectomy [53]. Azibani et al. [54] observed for the first time that hyperaldosteronism increased the number of inflammatory factors, including galectin-3, and accelerated the hypertension-induced fibrosis. Many studies in this aspect were conducted. However, the role of galectin-3 in hyperaldosteronism-induced inflammation remains unclear [54].

3.7. Acute Ischemic Stroke (AIS)

AIS is a cerebral tissue infarction caused by occlusion of cerebral arteries, with injury of the neurons, astrocytes and oligodendrocytes. It is the most important vascular event in the central nervous system, leading to death or disability. Dong et al. [55] assessed the role of galectin-3 in patients with AIS. They demonstrated that serum galectin-3 concentration was significantly higher in AIS patients compared with healthy individuals, what increased the intensity of AIS and infarction volume. Besides that, higher levels of serum galectin-3 were independently correlated with increased risk of death, significant disability, recurrent stroke and vascular events [56,57]. In in vitro studies, performed using human cortical neuronal (HCN) cell lines culture, a knockdown of galectin-3 expression with siRNA dramatically increased neuronal cell viability and simultaneously reduced apoptosis and serum levels of proinflammatory cytokines, including interleukin-1, -6 (IL-1, -6) and NF-κB and also caspase-3, a protein associated with apoptosis [55]. As a result of ischaemia, galectin-3 is released by microglia in the injured cerebral tissue. Then, the released galectin-3 contributes to microglia activation through binding to toll-like receptor 4 (TLR4), what causes an exacerbation of the inflammatory response [58]. Furthermore, the effect of serum galectin-3 concentration of AIS patients may be associated with serum lipid concentration regulation. AIS is reversely proportionally related to HDL cholesterol concentration, which, in turn, is reversely related to mortality from ischemic stroke and vascular events [59]. Earlier studies have demonstrated a reverse relationship between galectin-3 and HDL-C in serum, what suggests that AIS regulation, in which galectin-3 participates, is possibly associated with dyslipidaemia and inflammatory condition [60,61]. In a prospective cohort study involving 2970 patients with AIS, increased galectin-3 and decreased HDL-C levels of serum were observed, what could have intensified inflammatory condition and oxidative stress after ischemic stroke [57]. Moreover, it seems that concentrations of serum galectin-3 and HDL-C exerta combined effect on AIS prognosis.

Furthermore, a neuroprotective effect in cerebral stroke has been described. Galectin-3 is of key importance for activation, migration and proliferation of microglia after ischemic stroke [62]. Galectin-3 deficiency is associated with a significant increase of the size of ischemic lesions and number of apoptotic neurons. Microglial activation and proliferation mediated by galectin-3 are associated with Gal-3/IGF-R1 (insulin-like growth factor receptor 1) interaction in response to ischemic injury [62]. The data suggest that galectin-3 plays a neuroprotective role in injured brain.

In summary, galectin-3 can simultaneously exert a negative effect and show a neuroprotective action in AIS. One of the possible causes is that circulating galectin-3 serum levels are associated with various stages of AIS and a coincidence of two unfavourable effects or the neuroprotective role are not necessarily associated with the circulating galectin-3.

4. Diagnostic Usefulness of Serum Galectin-3 Concentration

The inflammatory process and especially its acute phase revealed as a result of bacterial infection and in certain types of tumours, is closely related to the circulating neutrophils. The recruitment of the cells is a characteristic feature of acute inflammation and they are the first cells to migrate towards the inflammation site. Galectin-3 is strongly involved in the modulations of inflammatory processes and disorders underlying the inflammatory condition. The proinflammatory effect of galectin-3 is associated with activation of NF-kB transcription factor, induction of tumour necrosis factor α (TNF-α) and interleukin 6, regulation of cell adhesion, promotion of cell activation and chemotaxis and regulation of cell growth or apoptosis [63].

Inflammatory condition underlies many atherosclerosis-related diseases. Galectin-3 seems to be an important and, at the same time, useful biomarker of atherosclerosis and its special role has been observed in the process of atheromatous plaque destabilisation. A positive relationship has been demonstrated between serum galectin-3 concentration and the number and calcification area of atheromatous plaques [64]. Moreover, high concentrations of serum galectin-3 are a harbinger of clinical failures associated with higher risk of overall mortality or mortality due to cardiovascular reasons and HF [65]. The authors of another paper [66] have stressed that higher galectin-3 serum concentrations determined in patients on admission to hospital are associated with severe course of stroke and frequently with a poor prognosis at discharge from hospital. An unfavourable prognosis is frequently even poorer when a close relationship is present between levels of serum galectin-3 and markers of lipid metabolism [60], carbohydrate metabolism [67,68], renal function [69,70] or echocardiographic parameters serving for assessment of myocardial function and structure [71,72]. In individuals, in whom no metabolic disorders or excessive body mass or increased triglyceride and total cholesterol concentrations have been found, it has been demonstrated that the age itself can distinguish subjects with higher and lower serum galectin-3 concentrations (individuals <40 years of age: 11.5 (9.5–13.60) ng/mL vs. those aged ≥40 years: 12.4 (10.6–14.4) ng/mL) [73].

In the assessment of the cardiovascular risk of primary importance are disorders of lipid–carbohydrate profile or renal function. It was found [60] that in individuals aged <40 years after myocardial infarction, serum galectin-3 concentration was strongly positively related to non-HDL-cholesterol concentration and negatively related to HDL-cholesterol level. Patients with type 2 diabetes mellitus had higher serum galectin-3 concentrations than healthy individuals [67,68]. In patients with chronic and acute heart failure a relationship was observed between higher serum galectin-3 concentrations and renal failure parameters assessed by cystatin C or uric acid concentrations or by reduced estimated glomerular filtration rate (eGFR) [74]. The studies conducted by Mueller et al. [75] demonstrated that patients with eGFR >90 mL/min/1.73 m^2 had lower levels of serum galectin-3 (median: 10.7 (9.3–12.4) ng/mL) compared with patients with eGFR <90 mL/min/1.73 m^2, in whom higher levels of serum galectin-3 were found (median: 12.1 (10.2–14.1) ng/mL). The diverse clinical involvement of galectin-3 in many diseases is due to its role as a regulator of acute and chronic inflammation that links inflammation-related macrophages and the promotion of fibrosis. This makes galectin-3 not an organ-specific marker, but a marker specific to the pathogenesis of inflammatory and/or fibrotic disorders. This is because the primary sources of circulating galectin-3 are not always identifiable. A patient with heart disease may have varying degrees of inflammation and progression of the fibrotic process. Thus, the serum galectin-3 concentration may reflect different stages of the pathophysiological state. This is because the level of circulating galectin-3 in patients with different stages of heart disease does not differentiate between myocarditis and fibrosis and therefore it does not specifically reflect these conditions. Moreover, there are gender-related differences in serum galectin-3 concentrations. In women, higher serum galectin-3 concentrations were observed compared to men, as well as a stronger correlation between concentration of serum galectin-3 and other cardiovascular disease risk factors [61].

An increased secretion of galectin-3 promotes a release of inflammatory mediators, including TGF-β or interleukin 1 and 2, and also intensifies cardiac fibroblast proliferation. Activated cardiac fibroblasts are the main source of extracellular matrix (ECM) proteins, particularly collagen. Galectin-3 affects in the first place synthesis of type I collagen, leading to an impairment of the homeostasis between type I/III collagen content and thus to impairment of the systolic/diastolic function of the myocardium. These disorders contribute to progression of myocardial failure [76].

Galectins, apart from modulating inflammatory processes, play the dominant role in fibrotic processes. In heart failure the pathophysiological element of key importance is the progressing fibrosis of the myocardial tissue. In the myocardium, the level of galectin-3 expression is almost undetectable in cardiomyocytes, but in cardiac fibroblasts that lectin reaches a significantly higher concentration. Chronic kidney disease (CKD) is one of the risk factors of cardiovascular diseases, hence cardiac biomarkers play a significant role in the development of kidney diseases. An elevated serum galectin-3 concentration is associated with myofibroblast proliferation or intense fibrogenesis and may be a harbinger of kidney fibrosis process or even of CKD development [70].

In heart failure galectin-3 is released into extracellular space, promoting fibrotic process through activation of fibroblasts. The fibroblast activation is characterised by increased expression of cytoskeletal protein—alpha smooth muscle actin (α-SMA)—an intracellular fibrosis marker, and collagen type 1 (COL1α1)—an extracellular fibrosis marker. Both α-SMA and COL1α1 regulation in tissues affected by fibrosis is mediated by galectin-3. That process is mediated by activation of cyclin-dependent kinase inhibitor 1A (CDKN1A), inhibin beta A, fibronectin 1, as well as extracellular signal-regulated kinase (ERK) and phosphatidylinositol 3-kinase (PI3K). Thus, the central place in the regulation of fibrotic process development is taken by galectin-3, which participates in the regulation of expression of fibrotic matrix components (α-SMA and COL1α1) and in extracellular matrix turnover through a number of tissue inhibitors of metalloproteinases (TIMPs) and matrix metalloproteinases. Galectin-3 influences the development of myocardial fibrosis through an effect on the important intermediates of that regulation: phosphatase and tensin homolog (PTEN) and protein tyrosine kinase 2 (PTK2). Exerting an effect on PTEN, it inhibits MMP-14 activity, contributing to the development of fibrosis, and inhibits MMP-9 activity, preventing the cardiac fibrosis process. On the other hand, the promoting of myocardial fibrosis with participation of MMP-9 is mediated by PTK2 [77].

The effect of galectin-3 is closely related to the heart failure markers useful in clinical practice, namely natriuretic peptides. The studies conducted by Felker et al. [78] have confirmed the relationship between increased galectin-3 serum concentration and the intensity of heart failure in individual NYHA classes, monitored by NT-proBNP concentration. In other studies a positive correlation between serum galectin-3 concentration and a negative correlation between left ventricular ejection fraction (LVEF%) and the degree of heart failure progression (max. NYHA IV class) have been demonstrated. A relationship has also been found between plasma galectin-3 concentration and the change of the left ventricular structure and function, what has confirmed that galectin-3 can participate in the left ventricular remodelling process in patients with HF [79,80]. Moreover, the studies have confirmed a higher specificity of galectin-3 in predicting the occurrence and development of HF than that in the case of determination of NT-proBNP concentration alone. However, this has not been shown with respect to galectin-3 sensitivity in the prognosis of the occurrence and development of HF [80].

Our studies have unequivocally stressed the importance of galectin-3 as a predictor of death in one-year follow-up [81]. The research has also shown a positive correlation between concentrations of serum NT-proBNP and galectin-3 ($r = 0.565$, $p = 0.035$). In the studies conducted in patients with HF in NYHA III class [82] it has been observed that galectin-3 is an important predictor of the risk of death, taking into account the age and gender and also HF intensity (based on NT-proBNP concentration) and renal function disorders (acc. to eGFR value). In the studies by Tang et al. [83], increased plasma galectin-3 concentration was associated with advanced age and poor renal function. That correlation, revealed in a group of patients with chronic systolic heart failure, demonstrated that high plasma galectin-3 concentration was associated with renal failure and shorter survival of the patients.

On the other hand, Fermann et al. [84] demonstrated in their study higher serum galectin-3 concentrations in patients with myocardial failure and with confirmed renal dysfunction. It is worth to stress here, that in the case of heart or kidney failure it is recommended to analyse in serum galectin-3 concentration together with natriuretic peptide concentration values. A joint analysis of changes of these markers offers a possibility of a more accurate prognostication and, at the same time, confirmation of organ failure development.

A group of researchers [85] made an attempt to find a correlation between levels of serum galectin-3 and NT-proBNP and the inflammatory condition assessed by serum hsCRP concentration in patients with the first myocardial infarction, treated by percutaneous coronary intervention (PCI). In the paper by Szadkowska et al. [85] a detailed analysis of the above-mentioned markers in serum, depending on galectin-3 level, demonstrated three times higher NT-proBNP concentrations and two times higher hsCRP levels in patients presenting higher galectin-3 concentrations (>16 ng/mL), than in those with galectin-3 concentrations <16 ng/mL. In the paper, a positive correlation was also demonstrated between NT-proBNP and hsCRP concentrations ($r = 0.45$, $p < 0.001$) and galectin-3 and hsCRP concentrations ($r = 0.20$, $p < 0.05$).

In clinical practice, apart from commonly used heart failure markers—natriuretic peptides, increasingly frequently new tools are tested for the assessment of the degree of heart failure—and that concerns not only myocardial failure. They include, among other markers: GDF-15 (growth and differentiation factor-15) and ST2 (suppression of tumourigenicity 2), which belongs to the interleukin 1 family. GDF-15 is regarded as a prognostic marker in cardiovascular diseases and is frequently determined in combination with other prognostic factors, such as NT-proBNP or hsTnT. High GDF-15 concentrations were noted in hypertrophic and dilated cardiomyopathies, after volume overload, ischaemia and heart failure [86]. The ST2 receptor has two isoforms: membrane-bound receptor type 2 (ST2L) and soluble form (sST2), present in serum and most frequently used in diagnostic procedures [70,87]. The form sST2 is a "bait" for IL-33 and thus it counteracts its interaction with the membrane-bound ST2L, blocking the paracrine fibroblast-cardiomyocyte communication system and reducing the cardioprotective effect of IL-33 [87].

A good clinical practice is, however, a joint assessment of the changes of concentrations of several parameters. An assessment of the cardiovascular risk is possible with a higher sensitivity with simultaneous measurements of the concentrations of: sST2, hsTnI, hsCRP or GDF-15 [70]. The addition of serum galectin-3 concentration to that group is extremely useful in the case of disorders associated with myocardial dysfunction. It should be also mentioned that the biological variability of galectin-3 is low in comparison to other established and novel cardiovascular biomarkers (e.g., NT-proBNP, GDF-15, ST2) [19,88]. Apart from this, galectin-3 had 8.1% within-individual coefficient of variation in healthy controls and chronic HF patients [88]. Schindler et al. [89] presented that galectin-3 shown relatively low biological variability in healthy individuals and stable HF patients. Furthermore, among healthy subjects, galectin-3 had minimal biological variation in both the short- and long-term without sex differences.

Galectin-3 is a biomarker of ventricular remodelling and myocardial fibrosis [90]. It is also an important regulator of chronic and acute inflammatory condition and inflammation leading to fibrosis in various tissues [91]. The study by Tuegel et al. [92] demonstrates that simultaneously determined high concentrations of three biomarkers in serum: galectin-3, sST2 and GDF-15 in patients with chronic kidney disease are associated with a higher mortality of such patients. At the same time, it has been observed that only GDF-15 serum concentration is associated with development of heart failure. Serum galectin-3 concentration shows a low tissue specificity and, therefore, a multimarker strategy of cardiovascular risk assessment is needed [70]. Two studies involving patients with HF (CORONA and COACH trials) analysed the kinetics of changes in levels of serum galectin-3 after 0, 3 and 6 months [19,93]. The increase in serum galectin-3 concentration by $\geq 15\%$ was found to indicate a 50% higher risk of mortality and hospitalization due to HF compared to patients with stable values of galectin-3 in the same time range, regardless of age, sex, diabetes, renal function, ejection fraction (LVEF) and NT-proBNP concentration. Moreover, the results of our study [81] demonstrated that patients with acute heart failure who died within 1 year of follow-up had significantly higher levels of serum galectin-3 at baseline compared to those who survived (55.6 ± 37.6 ng/mL vs. 15.0 ± 7.04 ng/mL; $p = 0.005$), which is in agreement with some previous reports. Our study was only a preliminary pilot research and the size of the study group is small.

In summary, it should be stressed that the proinflammatory effect of galectin-3 is not only limited to its participation in the pathogenesis of cardiovascular diseases. Recently, a significant importance of that marker has been demonstrated in the first place in SARS-CoV-2 infections. In patients with a severe COVID-19 course a systemic inflammatory condition develops, with intense cytokine storm, being the cause of respiratory failure and other multiple organ injuries [94,95]. It seems that serum galectin-3 concentration reflects the severity of COVID-19 course. In patients infected with SARS-CoV-2, in whom an unfavourable course of the disease forced administration of intensive therapy in view of respiratory failure, an almost three times higher serum galectin-3 concentrations were found, compared with patients not requiring treatment at intensive care units (ICUs) (23.46 (15.51–27.80) ng/mL vs. 8.93 (7.58–12.97) ng/mL, respectively) [63].

The study conducted by Kuśnierz-Cabala et al. [63] also called attention to the use of galectin-3 in the diagnosis of pneumonia, particularly in situations of severe COVID-19 course. In patients with pneumonia, an almost twice higher serum galectin-3 concentrations were found compared with those, in whom no pneumonia developed (13.30 (8.93–17.38) ng/mL vs. 8.55 (6.73–10.98) ng/mL, respectively). In that group of patients also higher concentrations were revealed of other proinflammatory markers: IL-6, pentraxin-3 (PTX-3), endothelial damage marker–soluble fms-like tyrosine kinase-1 (sFlt-1) and a number of tissue damage markers [63].

5. Therapeutic Potential of Galectin-3

A galectin-3-targeted therapy requires an understanding of the biology and pathologies associated with that molecule. Galectin-3 plays an important role in chronic inflammatory condition and is involved in the development of many diseases, including diseases of the heart [96–98], kidneys [99,100], viral infections [101], autoimmune diseases [102], neurodegenerative disorders [103–105] and many neoplastic diseases. In recent years a significant intensification has been observed of the studies on the role of galectin-3 as a central regulator of fibrosis and on therapeutic strategies aimed at inhibition of galectin-3 function in profibrotic diseases. The strategy of galectin-3 expression or function inhibition may be effective, not disturbing at the same time its normal main function.

Preclinical studies, in which a deletion of galectin-3-encoding gene was used, demonstrated the role of galectin-3 on the long list of cardiovascular diseases, including those focused on heart remodelling [106,107], hyperaldosteronism and arterial hypertension [108,109], acute myocardial infarction [110], myocardial ischemic/reperfusion injuries [111] and dilated cardiomyopathy [112]. That led to putting forward the hypothesis that galectins are potential targets for new anti-tumour and anti-inflammatory compounds. This hypothesis has been supported in several in vivo studies with the use of amino acid-derived lactulose-amine [113] and modified citrus pectin (MCP) [114]. The mentioned derivatives inhibit the expression of galectin-3. A development of small and large molecules blocking galectin-3 function may pave the way for a novel and exciting therapy with inhibitors, the action of which is, however, limited in healthy tissues.

5.1. Galectin-3 Inhibitors

5.1.1. Monosaccharides

Pharmacological inhibition of galectin-3 with N-acetyllactosamine (N-Lac) prevented left ventricular dysfunction in heart-failure-susceptible REN2 rats [107] and also demonstrated a protective effect against hypertensive nephropathy in REN2 rats [115]. Low molecular mass saccharides, such as lactose or N-Lac, cannot be, however, used as "drugs" since they are rapidly absorbed and metabolised. Low molecular mass organic compounds, galactose derivatives, have also been synthesised and tested in respect of binding to the carbohydrate-binding domain of various galectins [116]. An inhibitor, 3,3′-ditriazolylthiodigalactoside exerted the strongest effect on galectin-3, with low values of the inhibition constant Kd = 29 nM [116].

5.1.2. Galactomannans and Modified Citrus Pectins

Galactomannans (GMs) are plant-derived galectin-3 antagonists. GM-CT-01, known under proprietary name Davanat® is a galactomannan with molecular mass 50 kDa and half-life from 12 h to 18 h [117,118]. Various types of modified citrus pectins with masses over 1000 kDa, e.g., GCS-100 [118] and PectaSol-C® are already available on the market. The galectin-3-inhibiting effect of MCP was tested in various cell and animal models. The tests included: inhibition of haemaglutination by galectin-3, reduction of heart inflammatory condition, suppression of organ fibrosis and reduction of atherosclerosis in mice with apolipoprotein E deficiency [25,108,119]. The inhibiting effect of MCP on *LGALS3* gene expression was tested in cardiac fibroblasts in a rat experimental model of heart failure [120]. It was demonstrated [120] that MCP alleviated heart dysfunction, decreased the degree of myocardial damage and reduced collagen deposition. A reduction of *LGALS3*, TLR4, MyD88 gene expression and NF-κB-p65 factor inhibition were also described. A reduction was also observed of expression of IL-1β, IL-18, TNF-α– the proinflammatory cytokines involved in the pathogenesis of heart failure. The use of MCP as a galectin-3 antagonist exerted a favourable effect on myocardial dysfunction process through inhibition of inflammation and fibrosis. The use of galectin-3 as a potential therapeutic target in the treatment of heart fibrosis after infarction may bring many benefits [120]. It should also be said that in many in vivo studies no specificity of MCPs has been established, and it is possible that their inhibitory activity is caused by an effect also on other therapeutic targets.

5.1.3. Thiodigalactosides

Recently, thiodigalactoside derivatives have been developed, with action targeted at new CRD sites, other than the canonical binding site. TD-139 is a thiodigalactoside analogue approved by the FDA for treatment of idiopathic pulmonary fibrosis, in the form of inhalation powder, and it has been speculated that it shows an effect antagonistic to galectin-3 through binding to B and E subsites [118,121]. TD-139 is a small molecule ($C_{28}H_{30}F_2N_6O_8S$) of about 648 g/mol molecular mass and it can bind with a high affinity to both galectin-3 and galectin-1. Although the mechanism of action still remains unclear, it is supposed that the molecule allosterically modulates the CRD of galectin-3. Some research groups have also reported that thiodigalactosides can be preferentially adjusted to form more specific galectin-3 inhibitors [122].

5.1.4. Heparin-Based Inhibitors

Heparin-based inhibitors are a relatively new and attractive group of galectin-3 inhibitors, which are sulphated or acetylated heparin derivatives. The results of in vitro studies have demonstrated that they are non-cytotoxic and selective for galectin-3 (i.e., they do not inhibit galectin-1, -4 and -8). The experimental in vivo studies with nude mice demonstrated that compounds induced by galectin-3 significantly inhibited the metastasizing of human melanoma and colonic cancer cells to the lungs. Moreover, the compounds showed no detectable anti-thrombotic activity and they seemed to be promising therapeutic agents [123]. They were, however, only tested in vivo models of metastases and, in the future, studies should be conducted in order to estimate their potential as anti-fibrotic agents.

5.1.5. Neoglycoconjugates

Galectin-3 binds to branched saccharides with increased binding power, and large dendrimers connected with lactose as a functional group, provide an "excess of ligands" for galectin-3 binding. Michel et al. [124] studied the effect of various types of dendrimers, functionally bound to lactose, on tumour cell aggregation. They found that smaller dendrimers inhibited cell aggregation, possibly through competitive inhibition, while larger dendrimers containing several terminal lactose groups increased aggregation, providing thus many sites for galectin-3 binding [124]. Recently, yet other chemically modified glycoproteins (neoglycoproteins) have been developed, which show a potential to be used as new therapeutic molecules against fibrosis, through effective targeting at galectin-3. They serve not only as ligands with a high affinity, but can be also modulated in order to achieve selectivity for galectin-3, compared with other galectins. Their use in clinical context has not been assessed yet [125].

5.1.6. Peptide Based Compounds

Amino-terminally truncated galectin-3 (Gal-3C) has been studied in the therapy of galectin-3-related tumours and seems to be a promising therapeutic target, showing a low toxicity profile [126,127]. There have been, however, no sufficient studies conducted, assessing its potential in the treatment of other galectin-3-associated disorders. Recently, Sun et al. [128] used the galectin-3-binding peptide i.e., G3-C12 for inhibition of intracellular galectin-3 in tumour cells. Since G3-C12 is highly selective for galectin-3 compared with other galectins, it acts as a selective galectin-3 target ligand. Then, when that peptide is coupled with the drug by means of a universal drug carrier, such as copolymer of N-(2-hydroxypropyl) methacrylamide (HPMA), the created G3-C12-HPMA-drug conjugate can easily penetrate the cells with galectin-3 over-expression [128]. The above-described concept of selective intracellular galectin-3 supplying can also find use in other scenarios of galectin-3-related diseases, such as organ fibrosis and HF.

Galectin-3 inhibitors are promising therapeutic agents, but still not much is known about their critical features, such as: in vivo power of action, absorption, metabolism, pharmacokinetics and toxicology. Although it has been demonstrated that they are active in some disease models, further studies are needed on the mechanism of their action, to establish whether they are active in a therapeutically acceptable model of supplying and dosing of the drugs for patients.

6. NcRNAs as the Modulators of *LGALS3* Gene Expression and a New Potential Therapeutic Target

The learning of the role of galectin-3 in the processes of myocardial fibrosis, inflammation and postinfarction dysfunction has also started studies, the aim of which was the search, at molecular level, for post-transcriptional mechanisms regulating *LGALS3* gene expression. The effect has been studied of short, non-coding RNAs, microRNAs (miRNAs), which participate in the regulation of many genes involved in such processes as: cell differentiation, division, proliferation, apoptosis and angiogenesis [129]. The mechanism of miRNAs action is based on the inhibition of protein translation process or direct degradation of mRNA of the target genes through binding of miRNAs to the complementary region of the target mRNA molecule [130]. Much data is available confirming the participation of miRNAs in the process of physiological regulation of the heart function and also in the progression of cardiovascular diseases [131–134].

The increased interest in circulating miRNAs as potential biomarkers of cardiovascular diseases has caused that successive studies started to appear, concerning the question whether miRNAs molecules can reflect the changes occurring at various stages of a pathological process. The correlation was studied between miRNAs specific to acute heart failure and the serum concentration of well-documented biochemical biomarkers including galectin-3 [135]. A negative correlation was demonstrated between miR-199a-3p expression and galectin-3 serum concentration in the 48th hour of hospitalisation in patients with impaired heart function and poor prognosis ($r = -0.73$; $p < 0.001$). The relationship observed between microRNA and established biomarkers, including galectin-3, can contribute to better elucidation of the role of unfavourable heart remodelling processes and fibrosis in the pathogenesis of acute heart failure. In study by Song et al. [136] *LGALS3* gene expression is increased in cardiac fibrosis process and over-expression of miR-199a expression has also been observed with hypertrophy heart failure. The biochemical and molecular biomarkers analysed can serve for predicting and better identification of patients with unfavourable prognosis in the case of impaired heart function [136,137].

The study by Zhang et al. [138] presented, in vivo and in vitro, using a murine experimental model, the role of miR-27b and explained some mechanisms underlying myocardial hypertrophy. In the case of miR-27b over-expression the heart function in the course of hypertrophy was restored, what suggested a protective role of miR-27b against the pathological process. The study demonstrated and experimentally confirmed that *LGALS3* was the target gene for miR-27b. Experimental *LGALS3* gene inactivation significantly suppressed the myocardial hypertrophy process. Both *LGALS3* and miR-27b have a potential as therapeutic and diagnostic targets in the treatment of cardiovascular diseases, including myocardial hypertrophy [138].

Increased *LGALS3* expression was also noted in the experimental murine model, in damaged cardiac tissue subjected to myocardial ischemic reperfusion (I/R) procedure [139]. On the other hand, an experimental silencing of *LGAL3* gene expression alleviated myocardial damage resulting from the I/R procedure. A bioinformatic prediction of the interactions between mRNA and miRNA demonstrated that the *LGALS3* was regulated by miR-204-5p. A reduced miR-204-5p expression level was observed in the cardiac tissue subjected to the I/R procedure. Moreover, a negative correlation was demonstrated between miR-204-5p expression and *LGALS* expression in the studied experimental system.

Another important observation from the cited study [139] concerns the interaction between miR-204-5p and long non-coding RNA (lncRNA) *KCNQ1OT1*. A reduced *KCNQ1OT1* expression can provide protection against heart injury in the course of I/R after myocardial infarction, and *KCNQ1OT1* can modulate the expression of genes through a network of lncRNA/miRNA/mRNA interactions. *KCNQ1OT1*, binding to miR-204-5p, modulates the process of heart injury in the course of I/R through interaction with *LGALS3*. An increased *KCNQ1OT1* expression was noted in the heart tissue subjected to I/R, and a negative correlation was demonstrated between *KCNQ1OT1* expression and miR-204-5p expression in the studied experimental model. From the therapeutic point of view, the observation seems important, that experimental reduction of *KCNQ1OT1* expression and increase of miR-204-5p expression suppress cardiac injury in the course of I/R through a reduction of *LGALS3* expression. The presence of the network of *KCNQ1OT1*/miR-204-5p/*LGALS3* interactions may constitute a potential therapeutic target in the case of myocardial damage after I/R procedure [139].

LGALS3 expression and influence of the lncRNA *SNHG20* and miR-335 regulatory factors were assessed in the process of myocardial fibrosis and hypertrophy induced by angiotensin II (Ang II) in a murine experimental model [140]. An increased *SNHG20* expression was noted, which was reflected in an increase of *LGALS3* expression, but a decreased miR-335 expression was observed in the cardiac tissue. An experimental loss of *SNHG20* function, which resulted in *SNHG20* expression reduction, caused a reduction of expression of the proteins involved in heart fibrosis and apoptosis processes and also increased the viability of the cells. An experimental over-expression or silencing of *SNHG20* through interaction with the miR-335/*LGALS3* system can modulate the cardiac fibrosis process induced by Ang II. *SNHG20* lncRNA can be another therapeutic target in the process of myocardial fibrosis and hypertrophy [140].

The diagnostic usefulness of the miR-1 and miR-21, known in the pathogenesis of heart failure, and of galectin-3 protein was analysed in a group of patients with acute heart failure and coexistent asymptomatic type 2 diabetes mellitus [141]. A negative correlation was found between miR-1 expression and galectin-3 serum concentration and a positive correlation was observed between miR-21 expression and serum galectin-3 concentration in the studied group of patients. An application of a panel of determinations including miRNAs and biochemical biomarkers can lengthen the odds on an early identification of patients with acute heart failure among patients with asymptomatic type 2 diabetes mellitus [141].

The participation of the mentioned miR-1, miR-21 and serum galectin-3 concentration was also studied in patients with symptomatic heart failure and left ventricular hypertrophy and history-confirmed arterial hypertension [142]. It was shown that in patients with heart failure a reduction of miR-1 and miR-21 expression occurred, and these results were significantly correlated with the concentration of serum galectin-3, the important factor playing the key role in the fibrotic process. The changes of miR-1 and miR-21 expression and serum galectin-3 concentration were analysed also in patients with systolic heart failure with various degrees of intensity of left ventricular dilatation [143]. A relationship was observed between reduced miR-1 expression with increased serum galectin-3 concentration and the progression of unfavourable heart remodelling, assessed as left ventricle dilatation. An increased miR-21 expression was also found in patients with decompensated systolic heart failure. In the study, it has been demonstrated that both up-regulation of miR-1 and miR-21 as well as galectin-3 lead to myocardial hypertrophy and unfavourable heart remodelling.

In the study by Han et al. [144], the clinical importance was assessed of miR-214 and serum galectin-3 concentration in whole blood of patients with chronic heart failure. A statistically significant increase was noted of both miR-214 expression and galectin-3 serum concentration in the group with chronic heart failure. A positive correlation was demonstrated between miR-214 expression and serum galectin-3 concentration, what suggested the participation of both factors in the development of chronic heart failure. Galectin-3 exerts an effect promoting the myocardial fibrosis process, while miR-214 can regulate fibroblast proliferation [144].

Galectin-3 can be not only a clinically important biomarker of fibrotic process in cardiovascular diseases, but also an interesting therapeutic target that can possibly decelerate the progression of heart fibrosis. lncRNAs and miRNAs are involved in the regulation of signalling pathways in cardiac fibroblasts. Expression silencing or over-expression of lncRNAs or miRNAs in vivo can prevent the fibrotic process and improve the diastolic heart function. The interrelations in the lncRNA/miRNA/Gal-3 axis can constitute a susceptible target for therapeutic interventions, but further studies on this topic are required in order to increase the knowledge of their complex roles in the pathogenesis of cardiac fibrosis (Figure 3).

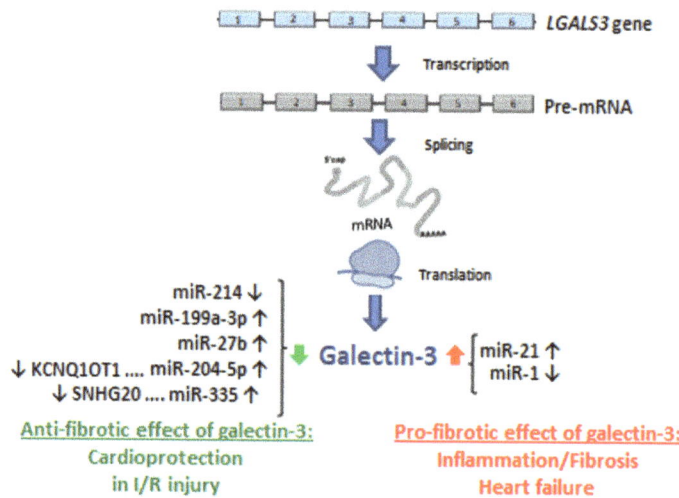

Figure 3. ncRNAs modulation of *LGALS3* gene expression in cardiovascular diseases. miR—microRNA; SNHG20—Small Nuclear RNA Host Gene 20; KCNQ1OT1—KCNQ1 Opposite Strand/Antisense Transcript 1. ↓—down-regulation; ↑—up-regulation; .... —interrelation between lncRNA and miRNA.

7. Conclusions and Perspectives

Based on the increasing literature data, galectin-3 emerges as a structurally unique and functionally extremely important galectin, expressed in various tissues and cell types and present not only inside but also outside cells, and also bound to cell membrane surfaces. The biological role of galectin-3 was initially ascribed to its carbohydrate-binding activity, but in the last decade a completely new spectrum of its functions was proved, not directly associated with the activity of that lectin. It has been found that galectin-3 participates in many pathological processes, in the first place in cardiovascular diseases but also in viral infections and many tumours.

Galectin-3 is an important factor in the pathophysiology of HF, mainly in view of its role in cardiac ventricular remodelling. Galectin-3 initially plays a protective role in the heart through its anti-apoptotic and anti-necrotic functions, while a prolonged expression of that protein leads to fibrosis and unfavourable remodelling of the damaged tissue. The sites of galectin-3 binding are mainly located in the extracellular matrix of the myocardium, fibroblasts and macrophages. Galectin-3 is released at the site of damage and activates the resting fibroblasts to become matrix-producing fibroblasts. The role of galectin-3 in fibroblast activation includes increased synthesis of cytoskeleton proteins, such as collagen type I, and inhibition of the activity of matrix metalloproteinases, what suggests that galectin-3 is involved in the initiation and development of the process of myocardial fibrosis. Usually, the expression of galectin-3 in a healthy heart is low, while its synthesis and release increase in fibrotic diseases, such as HF and AF. That creates wide possibilities of galectin-3 use in the diagnosis of cardiovascular diseases. Galectin-3 can provide additional information for the prognostication and stratification of HF risk. It seems, however, that the combination of biomarkers Galectin-3 and NT-proBNP or galectin-3 and cardiac troponins can provide a more precise clinical information than serum galectin-3 concentration alone. Apart from that use, galectin-3 seems to be a promising biomarker in cardiovascular diseases initiated and stimulated by inflammatory condition. It remains controversial whether this factor mediates unfavourable heart remodelling, or if it is merely a marker of heart failure. Some studies have provided evidence for a causal role of endogenous galectin-3 in the pathogenesis of myocardial fibrosis, hypertrophy and dysfunction in patients with heart failure. Based on these studies, it has been suggested that galectin-3 is not only a biomarker but also a mediator of the disease and it can be used as a therapeutic target [145].

Yu et al. [107] reported that the inhibition of galectin-3 was beneficial in mice that underwent transverse aortic construction (TAC) and in Ren2 hypertensive transgenic rats since it reduced fibrosis and improved function. Furthermore, in both murine and rat models of aldosterone-induced cardiac fibrosis, the loss of galectin-3 attenuated fibrotic changes and decreased cardiac dysfunction [108,109].

On the other hand, pharmacological and genetic inhibition of galectin-3 did not bring beneficial effects in a murine model of cardiomyopathy induced by transgenic activation of β_2-adrenoceptors [42]. Both pharmacological and genetic inhibition occurred before the development of cardiomyopathy, and the obtained results suggest that the loss of galectin-3 cannot prevent, let alone, reverse the dysfunction in this experimental model. The sympathetic nervous system is activated in the course of cardiovascular diseases, which leads to enhanced and sustained stimulation of cardiac β-adrenoreceptors [146]. The activation of β-adrenoreceptors modulates the role of galectin-3 in heart disease, both as a biomarker and the mediator of the disease, as well as increases the level of circulating galectin-3 and directly regulates galectin-3 expression in the heart [147]. This could be the reason for obtaining different results depending on the experimental model and for attempts to undermine the role of galectin-3 as both an important mediator in the pathogenesis of CVD and a therapeutic target. In our opinion, galectin-3 plays a significant role in the pathogenesis of fibrotic heart disease as a factor stimulating the development of inflammation and worsening the prognosis. Thus, it seems that galectin-3 may be recognized not only as a biomarker but also as a promising therapeutic target. However, this thesis should be evaluated in further experimental and clinical research.

The role of galectin-3 in physiological and pathophysiological processes has inspired the development of its inhibitors as not only new therapeutic methods, but also as experimental tools for the basic sciences. These inhibitors may be useful in studying of the role of galectin-3 both in vitro (cell and tissue cultures) and in animal models, therefore they can contribute to the extension of the knowledge and better understanding of the intra- and extracellular functions of galectin-3.

The molecules of non-coding RNA (lncRNA and miRNA) play an important role in the regulation of *LGALS3* expression in various pathological conditions of the myocardium. These molecules can silence *LGALS3* expression at post-transcription level and thus can have an influence on the disease development. Moreover, an experimental over-expression of lncRNA and miRNA can prevent or alleviate the process of myocardial fibrosis. The learning of the lncRNA/miRNA/Gal-3 interrelations is very important, since they can constitute a target for therapeutic interventions. To achieve that goal an extension is needed of our knowledge of their complex role in the pathogenesis of heart diseases.

Author Contributions: Conceptualization, G.S.; writing—original draft preparation, G.S., A.M.-J. and D.S.; writing—review and editing, G.S., A.M.-J. and D.S.; visualization, G.S.; supervision, G.S. and D.S. All authors have read and agreed to the published version of the manuscript.

Funding: This research received no external funding.

Institutional Review Board Statement: Not applicable.

Informed Consent Statement: Not applicable.

Data Availability Statement: Not applicable.

Conflicts of Interest: The authors declare no conflict of interest.

References

1. Barondes, S.H.; Cooper, D.N.; Gitt, M.A.; Leffler, H. Galectins. Structure and function of a large family of animal lectins. *J. Biol. Chem.* **1994**, *269*, 20807–20810. [CrossRef]
2. Cummings, R.D.; Liu, F.T.; Vasta, G.R. Galectins. In *Essentials of Glycobiology*, 3rd ed.; Varki, A., Cummings, R.D., Esko, J.D., Stanley, P., Hart, G.W., Aebi, M., Darvill, A.G., Kinoshita, T., Packer, N.H., Prestegard, J.H., et al., Eds.; Cold Spring Harbor Laboratory Press: Cold Spring Harbor, NY, USA, 2015–2017; pp. 469–480.
3. Kaminker, J.D.; Timoshenko, A.V. Expression, Regulation, and Functions of the Galectin-16 Gene in Human Cells and Tissues. *Biomolecules* **2021**, *11*, 1909. [CrossRef]
4. Timoshenko, A.V. Towards molecular mechanisms regulating the expression of galectins in cancer cells under microenvironmental stress conditions. *Cell. Mol. Life Sci.* **2015**, *72*, 4327–4340. [CrossRef]
5. Nio-Kobayashi, J. Tissue- and cell-specific localization of galectins, β-galactose-binding animal lectins, and their potential functions in health and disease. *Anat. Sci. Int.* **2017**, *92*, 25–36. [CrossRef]
6. Suthahar, N.; Meijers, W.C.; Sillje, H.H.W.; Ho, J.E.; Liu, F.T.; de Boer, R.A. Galectin-3 activation and inhibition in heart failure and cardiovascular disease: An update. *Theranostics* **2018**, *8*, 593–609. [CrossRef]
7. Karlsson, A.; Christenson, K.; Matlak, M.; Björstad, A.; Brown, K.L.; Telemo, E.; Salomonsson, E.; Leffler, H.; Bylund, J. Galectin-3 functions as an opsonin and enhances the macrophage clearance of apoptotic neutrophils. *Glycobiology* **2009**, *19*, 16–20. [CrossRef] [PubMed]
8. Funasaka, T.; Raz, A.; Nangia-Makker, P. Nuclear transport of galectin-3 and its therapeutic implications. *Semin. Cancer Biol.* **2014**, *27*, 30–38. [CrossRef]
9. Mazurek, N.; Conklin, J.; Byrd, J.C.; Raz, A.; Bresalier, R.S. Phosphorylation of the beta-galactoside-binding protein galectin-3 modulates binding to its ligands. *J. Biol. Chem.* **2000**, *275*, 36311–36315. [CrossRef] [PubMed]
10. Ahmad, N.; Gabius, H.J.; André, S.; Kaltner, H.; Sabesan, S.; Roy, R.; Liu, B.; Macaluso, F.; Brewer, C.F. Galectin-3 precipitates as a pentamer with synthetic multivalent carbohydrates and forms heterogeneous cross-linked complexes. *J. Biol. Chem.* **2004**, *279*, 10841–10847. [CrossRef] [PubMed]
11. Bambouskova, M.; Polakovicova, I.; Halova, I.; Goel, G.; Draberova, L.; Bugajev, V.; Doan, A.; Utekal, P.; Gardet, A.; Xavier, R.J.; et al. New Regulatory Roles of Galectin-3 in High-Affinity IgE Receptor Signaling. *Mol. Cell. Biol.* **2016**, *36*, 1366–1382. [CrossRef]
12. Mitchell, G.; Chen, C.; Portnoy, D.A. Strategies Used by Bacteria to Grow in Macrophages. *Microbiol. Spectr.* **2016**, *4*. [CrossRef] [PubMed]
13. Torina, A.; Villari, S.; Blanda, V.; Vullo, S.; La Manna, M.P.; Shekarkar Azgomi, M.; Di Liberto, D.; de la Fuente, J.; Sireci, G. Innate Immune Response to Tick-Borne Pathogens: Cellular and Molecular Mechanisms Induced in the Hosts. *Int. J. Mol. Sci.* **2020**, *21*, 5437. [CrossRef] [PubMed]
14. Torina, A.; Blanda, V.; Villari, S.; Piazza, A.; La Russa, F.; Grippi, F.; La Manna, M.P.; Di Liberto, D.; De La Fuente, J.; Sireci, G. Immune Response to Tick-Borne Hemoparasites: Host Adaptive Immune Response Mechanisms as Potential Targets for Therapies and Vaccines. *Int. J. Mol. Sci.* **2020**, *21*, 8813. [CrossRef] [PubMed]
15. Rabinovich, G.A.; Baum, L.G.; Tinari, N.; Paganelli, R.; Natoli, C.; Liu, F.T.; Iacobelli, S. Galectins and their ligands: Amplifiers, silencers or tuners of the inflammatory response? *Trends Immunol.* **2002**, *23*, 313–320. [CrossRef]
16. Brewer, C. Clusters, bundles, arrays and lattices: Novel mechanisms for lectin–saccharide-mediated cellular interactions. *Curr. Opin. Struct. Biol.* **2002**, *12*, 616–623. [CrossRef]
17. Funasaka, T.; Raz, A.; Nangia-Makker, P. Galectin-3 in angiogenesis and metastasis. *Glycobiology* **2014**, *24*, 886–891. [CrossRef]
18. Tazhitdinova, R.; Timoshenko, A.V. The emerging role of galectins and O-GlcNAc homeostasis in procesis of cellular differentiation. *Cells* **2020**, *9*, 1792. [CrossRef]
19. Blanda, V.; Bracale, U.M.; Di Taranto, M.D.; Fortunato, G. Galectin-3 in Cardiovascular Diseases. *Int. J. Mol. Sci.* **2020**, *21*, 9232. [CrossRef]
20. Djordjevic, A.; Zivkovic, M.; Stankovic, A.; Zivotic, I.; Koncar, I.; Davidovic, L.; Alavantic, D.; Djurić, T. Genetic Variants in the Vicinity of LGALS-3 Gene and LGALS-3 mRNA Expression in Advanced Carotid Atherosclerosis: An Exploratory Study. *J. Clin. Lab. Anal.* **2016**, *30*, 1150–1157. [CrossRef]
21. Iacobini, C.; Menini, S.; Ricci, C.; Scipioni, A.; Sansoni, V.; Cordone, S.; Taurino, M.; Serino, M.; Marano, G.; Federici, M.; et al. Accelerated lipid-induced atherogenesis in galectin-3-deficient mice: Role of lipoxidation via receptor-mediated mechanisms. *Arter. Thromb. Vasc. Biol.* **2009**, *29*, 831–836. [CrossRef]

22. Kadoglou, N.; Sfyroeras, G.; Spathis, A.; Gkekas, C.; Gastounioti, A.; Mantas, G.; Nikita, K.; Karakitsos, P.; Liapis, C.D. Galectin-3, Carotid Plaque Vulnerability, and Potential Effects of Statin Therapy. *Eur. J. Vasc. Endovasc. Surg.* **2015**, *49*, 4–9. [CrossRef] [PubMed]
23. Lee, Y.J.; Koh, Y.S.; Park, H.E.; Lee, H.J.; Hwang, B.H.; Kang, M.K.; Lee, S.Y.; Kim, P.J.; Ihm, S.H.; Seung, K.B.; et al. Spatial and Temporal Expression, and Statin Responsiveness of Galectin-1 and Galectin-3 in Murine Atherosclerosis. *Korean Circ. J.* **2013**, *43*, 223–230. [CrossRef]
24. Lu, Y.; Zhang, M.; Zhao, P.; Jia, M.; Liu, B.; Jia, Q.; Guo, J.; Dou, L.; Li, J. Modified citrus pectin inhibits galectin-3 function to reduce atherosclerotic lesions in apoE-deficient mice. *Mol. Med. Rep.* **2017**, *16*, 647–653. [CrossRef]
25. MacKinnon, A.C.; Liu, X.; Hadoke, P.W.; Miller, M.R.; Newby, D.E.; Sethi, T. Inhibition of galectin-3 reduces atherosclerosis in apolipoprotein E-deficient mice. *Glycobiology* **2013**, *23*, 654–663. [CrossRef]
26. Papaspyridonos, M.; McNeill, E.; de Bono, J.P.; Smith, A.; Burnand, K.G.; Channon, K.M.; Greaves, D.R. Galectin-3 is an amplifier of inflammation in atherosclerotic plaque progression through macrophage activation and monocyte chemoattraction. *Arterioscler. Thromb. Vasc. Biol.* **2008**, *28*, 433–440. [CrossRef] [PubMed]
27. Tabas, I.; García-Cardeña, G.; Owens, G.K. Recent insights into the cellular biology of atherosclerosis. *J. Cell Biol.* **2015**, *209*, 13–22. [CrossRef]
28. Davignon, J.; Ganz, P. Role of endothelial dysfunction in atherosclerosis. *Circulation* **2004**, *109*, III27–III32. [CrossRef]
29. Tian, L.; Chen, K.; Cao, J.; Han, Z.; Gao, L.; Wang, Y.; Fan, Y.; Wang, C. Galectin-3-induced oxidized low-density lipoprotein promotes the phenotypic transformation of vascular smooth muscle cells. *Mol. Med. Rep.* **2015**, *12*, 4995–5002. [CrossRef]
30. Ou, H.C.; Chou, W.C.; Hung, C.H.; Chu, P.M.; Hsieh, P.L.; Chan, S.H.; Tsai, K.L. Galectin-3 aggravates ox-LDL induced endothelial dysfunction through LOX-1 mediated signalling pathway. *Environ. Toxicol.* **2019**, *34*, 825–835. [CrossRef] [PubMed]
31. Agnello, L.; Bivona, G.; Lo Sasso, B.; Scazzone, C.; Bazan, V.; Bellia, C.; Ciaccio, M. Galectin-3 in acute coronary syndrome. *Clin. Biochem.* **2017**, *50*, 797–803. [CrossRef] [PubMed]
32. Madrigal-Matute, J.; Lindholt, J.S.; Fernandez-Garcia, C.E.; Benito-Martin, A.; Burillo, E.; Zalba, G.; Beloqui, O.; Llamas-Granda, P.; Ortiz, A.; Egido, J.; et al. Galectin-3, a biomarker linking oxidative stress and inflammation with the clinical outcomes of patients with atherothrombosis. *J. Am. Heart Assoc.* **2014**, *3*, e000785. [CrossRef] [PubMed]
33. Gucuk Ipek, E.; Akin Suljevic, S.; Kafes, H.; Basyigit, F.; Karalok, N.; Guray, Y.; Dinc Asarcikli, L.; Acar, B.; Demirel, H. Evaluation of galectin-3 levels in acute coronary syndrome. *Ann. Cardiol. Angeiol.* **2016**, *65*, 26–30. [CrossRef]
34. Arias, T.; Petrov, A.; Chen, J.; de Haas, H.; Pérez-Medina, C.; Strijkers, G.J.; Hajjar, R.J.; Fayad, Z.A.; Fuster, V.; Narula, J. Labeling galectin-3 for the assessment of myocardial infarction in rats. *EJNMMI Res.* **2014**, *4*, 75. [CrossRef]
35. González, G.E.; Cassaglia, P.; Noli Truant, S.; Fernández, M.M.; Wilensky, L.; Volberg, V.; Malchiodi, E.L.; Morales, C.; Gelpi, R.J. Galectin-3 is essential for early wound healing and ventricular remodeling after myocardial infarction in mice. *Int. J. Cardiol.* **2014**, *176*, 1423–1425. [CrossRef] [PubMed]
36. Maiolino, G.; Rossitto, G.; Pedon, L.; Cesari, M.; Frigo, A.C.; Azzolini, M.; Plebani, M.; Rossi, G.P. Galectin-3 predicts long-term cardiovascular death in high-risk patients with coronary artery disease. *Arterioscler. Thromb. Vasc. Biol.* **2015**, *35*, 725–732. [CrossRef] [PubMed]
37. Sharma, U.C.; Pokharel, S.; van Brakel, T.J.; van Berlo, J.H.; Cleutjens, J.P.; Schroen, B.; André, S.; Crijns, H.J.; Gabius, H.J.; Maessen, J.; et al. Galectin-3 marks activated macrophages in failure-prone hypertrophied hearts and contributes to cardiac dysfunction. *Circulation* **2004**, *110*, 3121–3128. [CrossRef] [PubMed]
38. Liu, Y.H.; D'Ambrosio, M.; Liao, T.D.; Peng, H.; Rhaleb, N.E.; Sharma, U.; André, S.; Gabius, H.J.; Carretero, O.A. N-acetyl-seryl-aspartyl-lysyl-proline prevents cardiac remodeling and dysfunction induced by galectin-3, a mammalian adhesion/growth-regulatory lectin. *Am. J. Physiol. Heart Circ. Physiol.* **2009**, *296*, H404–H412. [CrossRef]
39. Calvier, L.; Miana, M.; Reboul, P.; Cachofeiro, V.; Martinez-Martinez, E.; de Boer, R.A.; Poirier, F.; Lacolley, P.; Zannad, F.; Rossignol, P.; et al. Galectin-3 mediates aldosterone-induced vascular fibrosis. *Arterioscler. Thromb. Vasc. Biol.* **2013**, *33*, 67–75. [CrossRef]
40. Olivotto, I.; Maron, B.J.; Appelbaum, E.; Harrigan, C.J.; Salton, C.; Gibson, C.M.; Udelson, J.E.; O'Donnell, C.; Lesser, J.R.; Manning, W.J.; et al. Spectrum and clinical significance of systolic function and myocardial fibrosis assessed by cardiovascular magnetic resonance in hypertrophic cardiomyopathy. *Am. J. Cardiol.* **2010**, *106*, 261–267. [CrossRef]
41. Yakar Tülüce, S.; Tülüce, K.; Çil, Z.; Emren, S.V.; Akyıldız, Z.İ.; Ergene, O. Galectin-3 levels in patients with hypertrophic cardiomyopathy and its relationship with left ventricular mass index and function. *Anatol. J. Cardiol.* **2016**, *16*, 344–348. [CrossRef]
42. Nguyen, M.N.; Su, Y.; Kiriazis, H.; Yang, Y.; Gao, X.M.; McMullen, J.R.; Dart, A.M.; Du, X.J. Upregulated galectin-3 is not a critical disease mediator of cardiomyopathy induced by β2-adrenoceptor overexpression. *Am. J. Physiol. Heart Circ. Physiol.* **2018**, *314*, H1169–H1178. [CrossRef] [PubMed]
43. Frunza, O.; Russo, I.; Saxena, A.; Shinde, A.V.; Humeres, C.; Hanif, W.; Rai, V.; Su, Y.; Frangogiannis, N.G. Myocardial Galectin-3 Expression Is Associated with Remodeling of the Pressure-Overloaded Heart and May Delay the Hypertrophic Response without Affecting Survival, Dysfunction, and Cardiac Fibrosis. *Am. J. Pathol.* **2016**, *186*, 1114–1127. [CrossRef] [PubMed]
44. Magnani, J.W.; Rienstra, M.; Lin, H.; Sinner, M.F.; Lubitz, S.A.; McManus, D.D.; Dupuis, J.; Ellinor, P.T.; Benjamin, E.J. Atrial fibrillation: Current knowledge and future directions in epidemiology and genomics. *Circulation* **2011**, *124*, 1982–1993. [CrossRef] [PubMed]

45. Fujita, M.; Cheng, X.W.; Inden, Y.; Shimano, M.; Yoshida, N.; Inoue, A.; Yamamoto, T.; Takeshita, K.; Kyo, S.; Taguchi, N.; et al. Mechanisms with clinical implications for atrial fibrillation-associated remodeling: Cathepsin K expression, regulation, and therapeutic target and biomarker. *J. Am. Heart Assoc.* **2013**, *16*, e000503. [CrossRef]
46. Sygitowicz, G.; Maciejak-Jastrzębska, A.; Sitkiewicz, D. A review of the molecular mechanism underlying cardiac fibrosis and atrial fibrillation. *J. Clin. Med.* **2021**, *10*, 4430. [CrossRef]
47. Sonmez, O.; Ertem, F.U.; Vatankulu, M.A.; Erdogan, E.; Tasal, A.; Kucukbuzcu, S.; Goktekin, O. Novel fibro-inflammation markers in assessing left atrial remodeling in non-valvular atrial fibrillation. *Med. Sci. Monit.* **2014**, *20*, 463–470.
48. Gurses, K.M.; Yalcin, M.U.; Kocyigit, D.; Canpinar, H.; Evranos, B.; Yorgun, H.; Sahiner, M.L.; Kaya, E.B.; Ozer, N.; Tokgozoglu, L.; et al. Effects of persistent atrial fibrillation on serum galectin-3 levels. *Am. J. Cardiol.* **2015**, *115*, 647–651. [CrossRef]
49. Yalcin, M.U.; Gurses, K.M.; Kocyigit, D.; Canpinar, H.; Canpolat, U.; Evranos, B.; Yorgun, H.; Sahiner, M.L.; Kaya, E.B.; Hazirolan, T.; et al. The Association of Serum Galectin-3 Levels with Atrial Electrical and Structural Remodeling. *J. Cardiovasc. Electrophysiol.* **2015**, *26*, 635–640. [CrossRef]
50. Cha, J.H.; Wee, H.J.; Seo, J.H.; Ahn, B.J.; Park, J.H.; Yang, J.M.; Lee, S.W.; Kim, E.H.; Lee, O.H.; Heo, J.H.; et al. AKAP12 mediates barrier functions of fibrotic scars during CNS repair. *PLoS ONE* **2014**, *9*, e94695. [CrossRef]
51. Yao, Y.; Shen, D.; Chen, R.; Ying, C.; Wang, C.; Guo, J.; Zhang, G. Galectin-3 Predicts Left Ventricular Remodeling of Hypertension. *J. Clin. Hypertens.* **2016**, *18*, 506–511. [CrossRef]
52. Lin, Y.H.; Chou, C.H.; Wu, X.M.; Chang, Y.Y.; Hung, C.S.; Chen, Y.H.; Tzeng, Y.L.; Wu, V.C.; Ho, Y.L.; Hsieh, F.J.; et al. TAIPAI Study Group. Aldosterone induced galectin-3 secretion in vitro and in vivo: From cells to humans. *PLoS ONE* **2014**, *9*, e95254.
53. Liao, C.W.; Lin, Y.T.; Wu, X.M.; Chang, Y.Y.; Hung, C.S.; Wu, V.C.; Wu, K.D.; Lin, Y.H.; TAIPAI Study Group. The relation among aldosterone, galectin-3, and myocardial fibrosis: A prospective clinical pilot follow-up study. *J. Investig. Med.* **2016**, *64*, 1109–1113. [CrossRef]
54. Azibani, F.; Benard, L.; Schlossarek, S.; Merval, R.; Tournoux, F.; Fazal, L.; Polidano, E.; Launay, J.M.; Carrier, L.; Chatziantoniou, C.; et al. Aldosterone inhibits antifibrotic factors in mouse hypertensive heart. *Hypertension* **2012**, *59*, 1179–1187. [CrossRef]
55. Dong, H.; Wang, Z.H.; Zhang, N.; Liu, S.D.; Zhao, J.J.; Liu, S.Y. Serum Galectin-3 level, not Galectin-1, is associated with the clinical feature and outcome in patients with acute ischemic stroke. *Oncotarget* **2017**, *8*, 109752–109761. [CrossRef]
56. Wang, A.; Zhong, C.; Zhu, Z.; Xu, T.; Peng, Y.; Xu, T.; Peng, H.; Chen, C.S.; Wang, J.; Ju, Z.; et al. Serum Galectin-3 and Poor Outcomes among Patients with Acute Ischemic Stroke. *Stroke* **2018**, *49*, 211–214. [CrossRef] [PubMed]
57. Zeng, N.; Wang, A.; Xu, T.; Zhong, C.; Zheng, X.; Zhu, Z.; Peng, Y.; Peng, H.; Li, Q.; Ju, Z.; et al. Co-Effect of Serum Galectin-3 and High-Density Lipoprotein Cholesterol on the Prognosis of Acute Ischemic Stroke. *J. Stroke Cerebrovasc. Dis.* **2019**, *28*, 1879–1885. [CrossRef]
58. Burguillos, M.A.; Svensson, M.; Schulte, T.; Boza-Serrano, A.; Garcia-Quintanilla, A.; Kavanagh, E.; Santiago, M.; Viceconte, N.; Oliva-Martin, M.J.; Osman, A.M.; et al. Microglia-Secreted Galectin-3 Acts as a Toll-like Receptor 4 Ligand and Contributes to Microglial Activation. *Cell Rep.* **2015**, *10*, 1626–1638. [CrossRef] [PubMed]
59. Yeh, P.S.; Yang, C.M.; Lin, S.H.; Wang, W.M.; Chen, P.S.; Chao, T.H.; Lin, H.J.; Lin, K.C.; Chang, C.Y.; Cheng, T.J.; et al. Low levels of high-density lipoprotein cholesterol in patients with atherosclerotic stroke: A prospective cohort study. *Atherosclerosis* **2013**, *228*, 472–477. [CrossRef]
60. Winter, M.P.; Wiesbauer, F.; Alimohammadi, A.; Blessberger, H.; Pavo, N.; Schillinger, M.; Huber, K.; Wojta, J.; Lang, I.M.; Maurer, G.; et al. Soluble galectin-3 is associated with premature myocardial infarction. *Eur. J. Clin. Investig.* **2016**, *46*, 386–391. [CrossRef] [PubMed]
61. de Boer, R.A.; van Veldhuisen, D.J.; Gansevoort, R.T.; Muller Kobold, A.C.; van Gilst, W.H.; Hillege, H.L.; Bakker, S.J.; van der Harst, P. The fibrosis marker galectin-3 and outcome in the general population. *J. Intern. Med.* **2012**, *272*, 55–64. [CrossRef] [PubMed]
62. Lalancette-Hébert, M.; Swarup, V.; Beaulieu, J.M.; Bohacek, I.; Abdelhamid, E.; Weng, Y.C.; Sato, S.; Kriz, J. Galectin-3 is required for resident microglia activation and proliferation in response to ischemic injury. *J. Neurosci.* **2012**, *32*, 10383–10395. [CrossRef]
63. Kuśnierz-Cabala, B.; Maziarz, B.; Dumnicka, P.; Dembiński, M.; Kapusta, M.; Bociąga-Jasik, M.; Winiarski, M.; Garlicki, A.; Grodzicki, T.; Kukla, M. Diagnostic significance of serum galectin-3 in hospitalized patients with COVID-19—A preliminary study. *Biomolecules* **2021**, *11*, 1136. [CrossRef] [PubMed]
64. Ozturk, D.; Celik, O.; Satilmis, S.; Aslan, S.; Erturk, M.; Cakmak, H.A.; Kalkan, A.K.; Ozyilmaz, S.; Diker, V.; Gul, M. Association between serum galectin-3 levels and coronary atherosclerosis and plaque burden/structure in patients with type 2 diabetes mellitus. *Coron. Artery Dis.* **2015**, *26*, 396–401. [CrossRef]
65. Imran, T.F.; Shin, H.J.; Mathenge, N.; Wang, F.; Kim, B.; Joseph, J.; Gaziano, J.M.; Djoussé, L. Meta-Analysis of the Usefulness of Plasma Galectin-3 to Predict the Risk of Mortality in Patients With Heart Failure and in the General Population. *Am. J. Cardiol.* **2017**, *119*, 57–64. [CrossRef] [PubMed]
66. Zhuang, J.J.; Zhou, L.; Zheng, Y.H.; Ding, Y.S. The serum galectin-3 levels are associated with the severity and prognosis of ischemic stroke. *Aging* **2021**, *13*, 7454–7464. [CrossRef] [PubMed]
67. Jin, Q.H.; Lou, Y.F.; Li, T.L.; Chen, H.H.; Liu, Q.; He, X.J. Serum galectin-3: A risk factor for vascular complications in type 2 diabetes mellitus. *Chin. Med. J.* **2013**, *126*, 2109–2115.

68. Cao, Z.Q.; Yu, X.; Leng, P. Research progress on the role of gal-3 in cardio/cerebrovascular diseases. *Biomed. Pharmacother.* **2021**, *133*, 111066. [CrossRef]
69. Zhang, R.; Zhang, Y.; An, T.; Guo, X.; Yin, S.; Wang, Y.; Januzzi, J.L.; Cappola, T.P.; Zhang, J. Prognostic value of sST2 and galectin-3 for death relative to renal function in patients hospitalized for heart failure. *Biomark. Med.* **2015**, *9*, 433–441. [CrossRef]
70. Hara, A.; Niwa, M.; Kanayama, T.; Noguchi, K.; Niwa, A.; Matsuo, M.; Kuroda, T.; Hatano, Y.; Okada, H.; Tomita, H. Galectin-3: A potential prognostic and diagnostic marker for heart disease and detection of early stage pathology. *Biomolecules* **2020**, *10*, 1277. [CrossRef]
71. Stoltze Gaborit, F.; Bosselmann, H.; Kistorp, C.; Iversen, K.; Kumler, T.; Gustafsson, F.; Goetze, J.P.; Sölétormos, G.; Tønder, N.; Schou, M. Galectin 3: Association to neurohumoral activity, echocardiographic parameters and renal function in outpatients with heart failure. *BMC Cardiovasc. Disord.* **2016**, *16*, 117. [CrossRef]
72. Lisowska, A.; Knapp, M.; Tycińska, A.; Motybel, E.; Kamiński, K.; Święcki, P.; Musiał, W.J.; Dymicka-Piekarska, V. Predictive value of Galectin-3 for the occurrence of coronary artery disease and prognosis after myocardial infarction and its association with carotid IMT values in these patients: A mid-term prospective cohort study. *Atherosclerosis* **2016**, *246*, 309–317. [CrossRef]
73. Krintus, M.; Kozinski, M.; Fabiszak, T.; Kubica, J.; Panteghini, M.; Sypniewska, G. Establishing reference intervals for galectin-3 concentrations in serum requires careful consideration of its biological determinants. *Clin. Biochem.* **2017**, *50*, 599–604. [CrossRef]
74. Medvedeva, E.A.; Berezin, I.I.; Shchukin, Y.V. Galectin-3, Markers of Oxidative Stress and Renal Dysfunction in Patients with Chronic Heart Failure. *Kardiologia* **2017**, *57*, 46–50.
75. Mueller, T.; Egger, M.; Leitner, I.; Gabriel, C.; Haltmayer, M.; Dieplinger, B. Reference values of galectin-3 and cardiac troponins derived from a single cohort of healthy blood donors. *Clin. Chim. Acta* **2016**, *456*, 19–23. [CrossRef]
76. Shah, R.V.; Chen-Tournoux, A.A.; Picard, M.H.; van Kimmenade, R.R.; Januzzi, J.L. Galectin-3, cardiac structure and function, and long-term mortality in patients with acutely decompensated heart failure. *Eur. J. Heart Fail.* **2010**, *12*, 826–832. [CrossRef]
77. de Boer, R.A.; Voors, A.A.; Muntendam, P.; van Gilst, W.H.; van Veldhuisen, D.J. Galectin-3: A novel mediator of heart failure development and progression. *Eur. J. Heart Fail.* **2009**, *11*, 811–817. [CrossRef]
78. Felker, G.M.; Fiuzat, M.; Shaw, L.K.; Clare, R.; Whellan, D.J.; Bettari, L.; Shirolkar, S.C.; Donahue, M.; Kitzman, D.W.; Zannad, F.; et al. Galectin-3 in ambulatory patients with heart failure: Results from the HF-ACTION study. *Circ. Heart Fail.* **2012**, *5*, 72–78. [CrossRef]
79. Sygitowicz, G.; Tomaniak, M.; Błaszczyk, O.; Kołtowski, Ł.; Filipiak, K.J.; Sitkiewicz, D. Circulating microribonucleic acids: miR-1, miR-21 and miR-208a in patients with symptomatic heart failure: Preliminary results. *Arch. Cardiovasc. Dis.* **2015**, *108*, 634–642. [CrossRef] [PubMed]
80. Chen, K.; Jiang, R.J.; Wang, C.Q.; Yin, Z.F.; Fan, Y.Q.; Cao, J.T.; Han, Z.H.; Wang, Y.; Song, D.Q. Predictive value of plasma galectin-3 in patients with chronic heart failure. *Eur. Rev. Med. Pharmacol. Sci.* **2013**, *17*, 1005–1011. [PubMed]
81. Sygitowicz, G.; Tomaniak, M.; Filipiak, K.J.; Kołtowski, Ł.; Sitkiewicz, D. Galectin-3 in patients with acute heart failure—Preliminary report on first Polish experience. *Adv. Clin. Exp. Med.* **2016**, *25*, 617–623. [CrossRef] [PubMed]
82. Lok, D.J.; Van Der Meer, P.; de la Porte, P.W.; Lipsic, E.; Van Wijngaarden, J.; Hillege, H.L.; van Veldhuisen, D.J. Prognostic value of galectin-3, a novel marker of fibrosis, in patients with chronic heart failure: Data from the DEAL-HF study. *Clin. Res. Cardiol.* **2010**, *99*, 323–328. [CrossRef]
83. Tang, W.H.; Shrestha, K.; Shao, Z.; Borowski, A.G.; Troughton, R.W.; Thomas, J.D.; Klein, A.L. Usefulness of plasma galectin-3 levels in systolic heart failure to predict renal insufficiency and survival. *Am. J. Cardiol.* **2011**, *108*, 385–390. [CrossRef]
84. Fermann, G.J.; Lindsell, C.J.; Storrow, A.B.; Hart, K.; Sperling, M.; Roll, S.; Weintraub, N.L.; Miller, K.F.; Maron, D.J.; Naftilan, A.J.; et al. Galectin 3 complements BNP in risk stratification in acute heart failure. *Biomarkers* **2012**, *17*, 706–713. [CrossRef]
85. Szadkowska, I.; Wlazeł, R.N.; Migała, M.; Szadkowski, K.; Zielińska, M.; Paradowski, M.; Pawlicki, L. The association between galectin-3 and clinical parameters in patients with first acute myocardial infarction treated with primary percutaneous coronary angioplasty. *Cardiol. J.* **2013**, *20*, 577–582. [CrossRef]
86. Arkoumani, M.; Papadopoulou-Marketou, N.; Nicolaides, N.C.; Kanaka-Gantenbein, C.; Tentolouris, N.; Papassotiriou, I. The clinical impact of growth differentiation factor-15 in heart disease: A 2019 update. *Crit. Rev. Clin. Lab. Sci.* **2020**, *57*, 114–125. [CrossRef] [PubMed]
87. Merino-Merino, A.; Gonzalez-Bernal, J.; Fernandez-Zoppino, D.; Saez-Maleta, R.; Perez-Rivera, J.A. The role of galectin-3 and ST2 in cardiology: A short review. *Biomolecules* **2021**, *11*, 1167. [CrossRef] [PubMed]
88. Meijers, W.C.; van der Velde, A.R.; Kobold, A.C.M.; Dijck-Brouwer, J.; Wu, A.H.; Jaffe, A.S.; De Boer, R.A. Variability of biomarkers in patients with chronic heart failure and healthy controls. *Eur. J. Heart Fail.* **2017**, *19*, 357–365. [CrossRef] [PubMed]
89. Schindler, E.I.; Szymanski, J.J.; Hock, K.G.; Geltman, E.M.; Scott, M.G. Short- and Long-term Biologic Variability of Galectin-3 and Other Cardiac Biomarkers in Patients with Stable Heart Failure and Healthy Adults. *Clin. Chem.* **2016**, *62*, 360–366. [CrossRef] [PubMed]
90. Yancy, C.W.; Jessup, M.; Bozkurt, B.; Butler, J.; Casey, D.E., Jr.; Colvin, M.M.; Drazner, M.H.; Filippatos, G.S.; Fonarow, G.C.; Givertz, M.M.; et al. 2017 ACC/AHA/HFSA Focused Update of the 2013 ACCF/AHA Guideline for the Management of Heart Failure: A Report of the American College of Cardiology/American Heart Association Task Force on Clinical Practice Guidelines and the Heart Failure Society of America. *Circulation* **2017**, *136*, e137–e161.

91. Suthahar, N.; Meijers, W.C.; Silljé, H.H.W.; de Boer, R.A. From Inflammation to Fibrosis-Molecular and Cellular Mechanisms of Myocardial Tissue Remodelling and Perspectives on Differential Treatment Opportunities. *Curr. Heart Fail. Rep.* **2017**, *14*, 235–250. [CrossRef]
92. Tuegel, C.; Katz, R.; Alam, M.; Bhat, Z.; Bellovich, K.; de Boer, I.; Brosius, F.; Gadegbeku, C.; Gipson, D.; Hawkins, J.; et al. GDF-15, Galectin 3, Soluble ST2, and Risk of Mortality and Cardiovascular Events in CKD. *Am. J. Kidney Dis.* **2018**, *72*, 519–528. [CrossRef] [PubMed]
93. van der Velde, A.R.; Gullestad, L.; Ueland, T.; Aukrust, P.; Guo, Y.; Adourian, A.; Muntendam, P.; van Veldhuisen, D.J.; de Boer, R.A. Prognostic Value of Changes in Galectin-3 Levels Over Time in Patients With Heart Failure Data From CORONA and COACH. *Circ. Heart Fail.* **2013**, *6*, 219–226. [CrossRef] [PubMed]
94. Buszko, M.; Park, J.H.; Verthelyi, D.; Sen, R.; Young, H.A.; Rosenberg, A.S. The dynamic changes in cytokine responses in COVID-19: A snapshot of the current state of knowledge. *Nat. Immunol.* **2020**, *21*, 1146–1151. [CrossRef]
95. Kim, J.S.; Lee, J.Y.; Yang, J.W.; Lee, K.H.; Effenberger, M.; Szpirt, W.; Kronbichler, A.; Shin, J.I. Immunopathogenesis and treatment of cytokine storm in COVID-19. *Theranostics* **2020**, *11*, 316–329. [CrossRef] [PubMed]
96. Clementy, N.; Garcia, B.; André, C.; Bisson, A.; Benhenda, N.; Pierre, B.; Bernard, A.; Fauchier, L.; Piver, E.; Babuty, D. Galectin-3 level predicts response to ablation and outcomes in patients with persistent atrial fibrillation and systolic heart failure. *PLoS ONE* **2018**, *13*, e0201517. [CrossRef] [PubMed]
97. Clementy, N.; Piver, E.; Bisson, A.; Andre, C.; Bernard, A.; Pierre, B.; Fauchier, L.; Babuty, D. Galectin-3 in Atrial Fibrillation: Mechanisms and Therapeutic Implications. *Int. J. Mol. Sci.* **2018**, *19*, 976. [CrossRef]
98. Cui, Y.; Qi, X.; Huang, A.; Li, J.; Hou, W.; Liu, K. Differential and Predictive Value of Galectin-3 and Soluble Suppression of Tumorigenicity-2 (sST2) in Heart Failure with Preserved Ejection Fraction. *Med. Sci. Monit.* **2018**, *24*, 5139–5146. [CrossRef]
99. Alam, M.L.; Katz, R.; Bellovich, K.A.; Bhat, Z.Y.; Brosius, F.C.; de Boer, I.H.; Gadegbeku, C.A.; Gipson, D.S.; Hawkins, J.J.; Himmelfarb, J.; et al. Soluble ST2 and Galectin-3 and progression of CKD. *Kidney Int. Rep.* **2019**, *4*, 103–111. [CrossRef]
100. Tan, K.C.B.; Cheung, C.L.; Lee, A.C.H.; Lam, J.K.Y.; Wong, Y.; Shiu, S.W.M. Galectin-3 is independently associated with progression of nephropathy in type 2 diabetes mellitus. *Diabetologia* **2018**, *61*, 1212–1219. [CrossRef]
101. Noguchi, K.; Tomita, H.; Kanayama, T.; Niwa, A.; Hatano, Y.; Hoshi, M.; Sugie, S.; Okada, H.; Niwa, M.; Hara, A. Time-course analysis of cardiac and serum galectin-3 in viral myocarditis after an encephalomyocarditis virus inoculation. *PLoS ONE* **2019**, *14*, e0210971. [CrossRef]
102. de Oliveira, F.L.; Gatto, M.; Bassi, N.; Luisetto, R.; Ghirardello, A.; Punzi, L.; Doria, A. Galectin-3 in autoimmunity and autoimmune diseases. *Exp. Biol. Med.* **2015**, *240*, 1019–1028. [CrossRef]
103. Boza-Serrano, A.; Ruiz, R.; Sanchez-Varo, R.; García-Revilla, J.; Yang, Y.; Jimenez-Ferrer, I.; Paulus, A.; Wennström, M.; Vilalta, A.; Allendorf, D.; et al. Galectin-3, a novel endogenous TREM2 ligand, detrimentally regulates inflammatory response in Alzheimer's disease. *Acta Neuropathol.* **2019**, *138*, 251–273. [CrossRef] [PubMed]
104. Rahimian, R.; Béland, L.C.; Kriz, J. Galectin-3: Mediator of microglia responses in injured brain. *Drug Discov. Today* **2018**, *23*, 375–381. [CrossRef]
105. Siew, J.J.; Chen, H.M.; Chen, H.Y.; Chen, H.L.; Chen, C.M.; Soong, B.W.; Wu, Y.R.; Chang, C.P.; Chan, Y.C.; Lin, C.H.; et al. Galectin-3 is required for the microglia-mediated brain inflammation in a model of Huntington's disease. *Nat. Commun.* **2019**, *10*, 3473. [CrossRef]
106. Gonzalez, G.E.; Rhaleb, N.E.; D'Ambrosio, M.A.; Nakagawa, P.; Liao, T.D.; Peterson, E.L.; Leung, P.; Dai, X.G.; Janic, B.; Liu, Y.H.; et al. Cardiac-deleterious role of galectin-3 in chronic angiotensin II-induced hypertension. *Am. J. Physiol. Heart C* **2016**, *311*, H1287–H1296. [CrossRef]
107. Yu, L.; Ruifrok, W.P.; Meissner, M.; Bos, E.M.; van Goor, H.; Sanjabi, B.; van der Harst, P.; Pitt, B.; Goldstein, I.J.; Koerts, J.A.; et al. Genetic and pharmacological inhibition of galectin-3 prevents cardiac remodeling by interfering with myocardial fibrogenesis. *Circ. Heart Fail.* **2013**, *6*, 107–117. [CrossRef] [PubMed]
108. Calvier, L.; Martinez-Martinez, E.; Miana, M.; Cachofeiro, V.; Rousseau, E.; Sádaba, J.R.; Zannad, F.; Rossignol, P.; López-Andrés, N. The impact of galectin-3 inhibition on aldosterone-induced cardiac and renal injuries. *JACC Heart Fail.* **2015**, *3*, 59–67. [CrossRef] [PubMed]
109. Martinez-Martinez, E.; Calvier, L.; Fernandez-Celis, A.; Rousseau, E.; Jurado-Lopez, R.; Rossoni, L.V.; Jaisser, F.; Zannad, F.; Rossignol, P.; Cachofeiro, V.; et al. Galectin-3 blockade inhibits cardiac inflammation and fibrosis in experimental hyperaldosteronism and hypertension. *Hypertension* **2015**, *66*, 767–775. [CrossRef]
110. Mosleh, W.; Chaudhari, M.R.; Sonkawade, S.; Mahajan, S.; Khalil, C.; Frodey, K.; Shah, T.; Dahal, S.; Karki, R.; Katkar, R.; et al. The therapeutic potential of blocking Galectin-3 expression in acute myocardial infarction and mitigating inflammation of Infarct region: A clinical outcome-based translational study. *Biomark. Insights* **2018**, *13*, 1177271918771969. [CrossRef]
111. Al-Salam, S.; Hashmi, S. Myocardial ischemia reperfusion injury: Apoptotic, inflammatory and oxidative stress role of Galectin-3. *Cell. Physiol. Biochem.* **2018**, *50*, 1123–1139. [CrossRef]
112. Nguyen, M.N.; Ziemann, M.; Kiriazis, H.; Su, Y.; Thomas, Z.; Lu, Q.; Donner, D.G.; Zhao, W.B.; Rafehi, H.; Sadoshima, J.; et al. Galectin-3 deficiency ameliorates fibrosis and remodeling in dilated cardiomyopathy mice with enhanced Mst1 signaling. *Am. J. Physiol. Heart Circ. Physiol.* **2019**, *316*, H45–H60. [CrossRef] [PubMed]

113. Glinsky, V.V.; Kiriakova, G.; Glinskii, O.V.; Mossine, V.V.; Mawhinney, T.P.; Turk, J.R.; Glinskii, A.B.; Huxley, V.H.; Price, J.E.; Glinsky, G.V. Synthetic galectin-3 inhibitor increases metastatic cancer cell sensitivity to taxol-induced apoptosis in vitro and in vivo. *Neoplasia* **2009**, *11*, 901–909. [CrossRef]
114. Nangia-Makker, P.; Hogan, V.; Honjo, Y.; Baccarini, S.; Tait, L.; Bresalier, R.; Raz, A. Inhibition of human cancer cell growth and metastasis in nude mice by oral intake of modified citrus pectin. *J. Natl. Cancer Inst.* **2002**, *94*, 1854–1862. [CrossRef]
115. Frenay, A.R.; Yu, L.; van der Velde, A.R.; Vreeswijk-Baudoin, I.; López-Andrés, N.; van Goor, H.; Silljé, H.H.; Ruifrok, W.P.; de Boer, R.A. Pharmacological inhibition of galectin-3 protects against hypertensive nephropathy. *Am. J. Physiol. Renal Physiol.* **2015**, *308*, F500–F509. [CrossRef]
116. Oberg, C.T.; Leffler, H.; Nilsson, U.J. Inhibition of galectins with small molecules. *Chimia* **2011**, *65*, 18–23. [CrossRef] [PubMed]
117. Demotte, N.; Bigirimana, R.; Wieërs, G.; Stroobant, V.; Squifflet, J.L.; Carrasco, J.; Thielemans, K.; Baurain, J.F.; Van Der Smissen, P.; Courtoy, P.J.; et al. A short treatment with galactomannan GM-CT-01 corrects the functions of freshly isolated human tumor-infiltrating lymphocytes. *Clin. Cancer Res.* **2014**, *20*, 1823–1833. [CrossRef] [PubMed]
118. Slack, R.J.; Mills, R.; Mackinnon, A.C. The therapeutic potential of galectin-3 inhibition in fibrotic disease. *Int. J. Biochem. Cell Biol.* **2021**, *130*, 105881. [CrossRef]
119. Gao, X.; Zhi, Y.; Zhang, T.; Xue, H.; Wang, X.; Foday, A.D.; Tai, G.; Zhou, Y. Analysis of the neutral polysaccharide fraction of MCP and its inhibitory activity on galectin-3. *Glycoconj. J.* **2012**, *29*, 159–165. [CrossRef]
120. Xu, G.R.; Zhang, C.; Yang, H.X.; Sun, J.H.; Zhang, Y.; Yao, T.T.; Li, Y.; Ruan, L.; An, R.; Li, A.Y. Modified citrus pectin ameliorates myocardial fibrosis and inflammation via suppressing galectin-3 and TLR4/MyD88/NF-κB signaling pathway. *Biomed. Pharmacother.* **2020**, *126*, 110071. [CrossRef]
121. Hsieh, T.J.; Lin, H.Y.; Tu, Z.; Lin, T.C.; Wu, S.C.; Tseng, Y.Y.; Liu, F.T.; Hsu, S.T.; Lin, C.H. Dual thio-digalactoside-binding modes of human galectins as the structural basis for the design of potent and selective inhibitors. *Sci. Rep.* **2016**, *6*, 29457. [CrossRef]
122. van Hattum, H.; Branderhorst, H.M.; Moret, E.E.; Nilsson, U.J.; Leffler, H.; Pieters, R.J. Tuning the preference of thiodigalactoside- and lactosamine-based ligands to galectin-3 over galectin-1. *J. Med. Chem.* **2013**, *56*, 1350–1354. [CrossRef]
123. Duckworth, C.A.; Guimond, S.E.; Sindrewicz, P.; Hughes, A.J.; French, N.S.; Lian, L.Y.; Yates, E.A.; Pritchard, D.M.; Rhodes, J.M.; Turnbull, J.E.; et al. Chemically modified, non-anticoagulant heparin derivatives are potent galectin-3 binding inhibitors and inhibit circulating galectin-3-promoted metastasis. *Oncotarget* **2015**, *6*, 23671–23687. [CrossRef]
124. Michel, A.K.; Nangia-Makker, P.; Raz, A.; Cloninger, M.J. Lactose-functionalized dendrimers arbitrate the interaction of galectin-3/MUC1 mediated cancer cellular aggregation. *Chembiochem* **2014**, *15*, 2106–2112. [CrossRef] [PubMed]
125. Böcker, S.; Laaf, D.; Elling, L. Galectin Binding to Neo-Glycoproteins: LacDiNAc Conjugated BSA as Ligand for Human Galectin-3. *Biomolecules* **2015**, *5*, 1671–1696. [CrossRef]
126. Mirandola, L.; Yu, Y.; Chui, K.; Jenkins, M.R.; Cobos, E.; John, C.M.; Chiriva-Internati, M. Galectin-3C inhibits tumor growth and increases the anticancer activity of bortezomib in a murine model of human multiple myeloma. *PLoS ONE* **2011**, *6*, e21811. [CrossRef] [PubMed]
127. Mirandola, L.; Nguyen, D.D.; Rahman, R.L.; Grizzi, F.; Yuefei, Y.; Figueroa, J.A.; Jenkins, M.R.; Cobos, E.; Chiriva-Internati, M. Anti-galectin-3 therapy: A new chance for multiple myeloma and ovarian cancer? *Int. Rev. Immunol.* **2014**, *33*, 417–427. [CrossRef]
128. Sun, W.; Li, L.; Yang, Q.; Shan, W.; Zhang, Z.; Huang, Y. G3-C12 Peptide Reverses Galectin-3 from Foe to Friend for Active Targeting Cancer Treatment. *Mol. Pharm.* **2015**, *12*, 4124–4136. [CrossRef] [PubMed]
129. Ambros, V. The functions of animal microRNAs. *Nature* **2004**, *431*, 350–355. [CrossRef]
130. Bartel, D.P. MicroRNAs: Genomics, biogenesis, mechanism, and function. *Cell* **2004**, *116*, 281–297. [CrossRef]
131. Maciejak, A.; Kostarska-Srokosz, E.; Gierlak, W.; Dluzniewski, M.; Kuch, M.; Marchel, M.; Opolski, G.; Kiliszek, M.; Matlak, K.; Dobrzycki, S.; et al. Circulating miR-30a-5p as a prognostic biomarker of left ventricular dysfunction after acute myocardial infarction. *Sci. Rep.* **2018**, *8*, 9883. [CrossRef]
132. Wang, S.S.; Wu, L.J.; Li, J.J.; Xiao, H.B.; He, Y.; Yan, Y.X. A meta-analysis of dysregulated miRNAs in coronary heart disease. *Life Sci.* **2018**, *215*, 170–181. [CrossRef]
133. Maries, L.; Marian, C.; Sosdean, R.; Goanta, F.; Sirbu, I.O.; Anghel, A. MicroRNAs-The Heart of Post-Myocardial Infarction Remodeling. *Diagnostics* **2021**, *11*, 1675. [CrossRef]
134. Tanase, D.M.; Gosav, E.M.; Ouatu, A.; Badescu, M.C.; Dima, N.; Ganceanu-Rusu, A.R.; Popescu, D.; Floria, M.; Rezus, E.; Rezus, C. Current Knowledge of MicroRNAs (miRNAs) in Acute Coronary Syndrome (ACS): ST-Elevation Myocardial Infarction (STEMI). *Life* **2021**, *11*, 1057. [CrossRef]
135. Vegter, E.L.; Schmitter, D.; Hagemeijer, Y.; Ovchinnikova, E.S.; van der Harst, P.; Teerlink, J.R.; O'Connor, C.M.; Metra, M.; Davison, B.A.; Bloomfield, D.; et al. Use of biomarkers to establish potential role and function of circulating microRNAs in acute heart failure. *Int. J. Cardiol.* **2016**, *224*, 231–239. [CrossRef]
136. Song, X.W.; Li, Q.; Lin, L.; Wang, X.C.; Li, D.F.; Wang, G.K.; Ren, A.J.; Wang, Y.R.; Qin, Y.W.; Yuan, W.J.; et al. MicroRNAs are dynamically regulated in hypertrophic hearts, and miR-199a is essential for the maintenance of cell size in cardiomyocytes. *J. Cell. Physiol.* **2010**, *225*, 437–443. [CrossRef] [PubMed]
137. Vergaro, G.; Del Franco, A.; Giannoni, A.; Prontera, C.; Ripoli, A.; Barison, A.; Masci, P.G.; Aquaro, G.D.; Cohen Solal, A.; Padeletti, L.; et al. Galectin-3 and myocardial fibrosis in nonischemic dilated cardiomyopathy. *Int. J. Cardiol.* **2015**, *184*, 96–100. [CrossRef] [PubMed]

138. Zhang, M.; Cheng, K.; Chen, H.; Tu, J.; Shen, Y.; Pang, L.; Wu, W. MicroRNA-27 attenuates pressure overload-Induced cardiac hypertrophy and dysfunction by targeting galectin-3. *Arch. Biochem. Biophys.* **2020**, *689*, 108405. [CrossRef]
139. Rong, J.; Pan, H.; He, J.; Zhang, Y.; Hu, Y.; Wang, C.; Fu, Q.; Fan, W.; Zou, Q.; Zhang, L.; et al. Long non-coding RNA KCNQ1OT1/microRNA-204-5p/LGALS3 axis regulates myocardial ischemia/reperfusion injury in mice. *Cell. Signal.* **2020**, *66*, 109441. [CrossRef] [PubMed]
140. Li, M.; Qi, C.; Song, R.; Xiong, C.; Zhong, X.; Song, Z.; Ning, Z.; Song, X. Inhibition of Long Noncoding RNA SNHG20 Improves Angiotensin II-Induced Cardiac Fibrosis and Hypertrophy by Regulating the MicroRNA 335/Galectin-3 Axis. *Mol. Cell. Biol.* **2021**, *41*, e0058020. [CrossRef] [PubMed]
141. Al-Hayali, M.A.; Sozer, V.; Durmus, S.; Erdenen, F.; Altunoglu, E.; Gelisgen, R.; Atukeren, P.; Atak, P.G.; Uzun, H. Clinical Value of Circulating Microribonucleic Acids miR-1 and miR-21 in Evaluating the Diagnosis of Acute Heart Failure in Asymptomatic Type 2 Diabetic Patients. *Biomolecules* **2019**, *9*, 193. [CrossRef]
142. Tomaniak, M.; Sygitowicz, G.; Błaszczyk, O.; Kołtowski, Ł.; Puchta, D.; Malesa, K.; Kochanowski, J.; Sitkiewicz, D.; Filipiak, K.J. miR-1, miR-21, and galectin-3 in hypertensive patients with symptomatic heart failure and left ventricular hypertrophy. *Kardiol. Pol.* **2018**, *76*, 1009–1011. [CrossRef]
143. Tomaniak, M.; Sygitowicz, G.; Filipiak, K.J.; Błaszczyk, O.; Kołtowski, Ł.; Gąsecka, A.; Kochanowski, J.; Sitkiewicz, D. Dysregulations of miRNAs and galectin-3 may underlie left ventricular dilatation in patients with systolic heart failure. *Kardiol. Pol.* **2018**, *76*, 1012–1014. [CrossRef]
144. Han, R.; Li, K.; Li, L.; Zhang, L.; Zheng, H. Expression of microRNA-214 and galectin-3 in peripheral blood of patients with chronic heart failure and its clinical significance. *Exp. Ther. Med.* **2020**, *19*, 1322–1328. [CrossRef] [PubMed]
145. de Boer, R.A.; van der Velde, A.R.; Mueller, C.; van Veldhuisen, D.J.; Anker, S.D.; Peacock, W.F.; Adams, K.F.; Maisel, A. Galectin-3: A modifiable risk factor in heart failure. *Cardiovasc. Drugs Ther.* **2014**, *28*, 237–246. [CrossRef]
146. Lymperopoulos, A.; Rengo, G.; Koch, W.J. Adrenergic nervous system in heart failure: Pathophysiology and therapy. *Circ. Res.* **2013**, *113*, 739–753. [CrossRef] [PubMed]
147. Du, X.J.; Zhao, W.B.; Nguyen, M.N.; Lu, Q.; Kiriazis, H. β-adrenoreceptor activation affects galectin-3 as a biomarker and therapeutic target in heart disease. *Br. J. Pharmacol.* **2019**, *176*, 2449–2464. [CrossRef] [PubMed]

Review

The Complex Biological Effects of Pectin: Galectin-3 Targeting as Potential Human Health Improvement?

Lucas de Freitas Pedrosa [1], Avraham Raz [2] and João Paulo Fabi [1,3,4,*]

1. Department of Food Science and Experimental Nutrition, School of Pharmaceutical Sciences, University of São Paulo, São Paulo 05508000, SP, Brazil; lfpedrosa@usp.br
2. Department of Oncology and Pathology, School of Medicine, Karmanos Cancer Institute, Wayne State University, Detroit, MI 48201, USA; raza@karmanos.org
3. Food and Nutrition Research Center (NAPAN), University of São Paulo, São Paulo 05508080, SP, Brazil
4. Food Research Center (FoRC), CEPID-FAPESP (Research, Innovation and Dissemination Centers, São Paulo Research Foundation), São Paulo 05508080, SP, Brazil
* Correspondence: jpfabi@usp.br

Abstract: Galectin-3 is the only chimeric representative of the galectin family. Although galectin-3 has ubiquitous regulatory and physiological effects, there is a great number of pathological environments where galectin-3 cooperatively participates. Pectin is composed of different chemical structures, such as homogalacturonans, rhamnogalacturonans, and side chains. The study of pectin's major structural aspects is fundamental to predicting the impact of pectin on human health, especially regarding distinct molecular modulation. One of the explored pectin's biological activities is the possible galectin-3 protein regulation. The present review focuses on revealing the structure/function relationship of pectins, their fragments, and their biological effects. The discussion highlighted by this review shows different effects described within in vitro and in vivo experimental models, with interesting and sometimes contradictory results, especially regarding galectin-3 interaction. The review demonstrates that pectins are promissory food-derived molecules for different bioactive functions. However, galectin-3 inhibition by pectin had been stated in literature before, although it is not a fully understood, experimentally convincing, and commonly agreed issue. It is demonstrated that more studies focusing on structural analysis and its relation to the observed beneficial effects, as well as substantial propositions of cause and effect alongside robust data, are needed for different pectin molecules' interactions with galectin-3.

Keywords: galectin-3; pectin; structure and function; bioactive polysaccharides; galectin-3 inhibition

1. Introduction

Pectins are complex polysaccharides and versatile hydrocolloids, vastly available in plant cell walls and the middle lamella of higher plants. These polysaccharides are major components for maintaining rigidity and integrity of the plant tissues, positively impacting plant growth and health [1]. Every source of pectin has a variable amount of sub-structures, such as homogalacturonans (HG), rhamnogalacturonans (RG-I and II), and xylogalacturonan [2]. The HG region is primarily composed of a homopolymer of partially esterified 1-4-α-D-galactopyranuronic acid (Gal$p$A). RG-I regions are composed of repeated and intercalated α-D-Gal$p$A with α-L-rhamnopyranose (Rha$p$) [→2)-α-L-Rha$p$-(1,4)-α-D-Gal$p$A-(1→]. This region also has variable side chains of galactans (β-1,4-Gal$p$ residues with varying degree of polymerization), arabinans (α-1,5-L-Ara$f$ with 2- and 3-linked arabinose/arabinan branches), and arabinogalactans (Type I: β-1,4-D-galactans with O-3 linked L-arabinose or arabinan; Type II: β-1,3-linked D-galactans with β-O-6-linked galactans or arabinogalactan), attached to the Rha$p$ residues (side chains are linked at Rha$p$ O-4, as demonstrated in Figures 1 and 2) [2–4]. In diverse fruits, these structures undergo a series of chemical and enzymatic alterations during the ripening process, resulting in a wide variety of intramolecular changes inside the pectic chain, as illustrated before for the papaya pulp, a fruit that rapidly undergoes the ripening process [5–8].

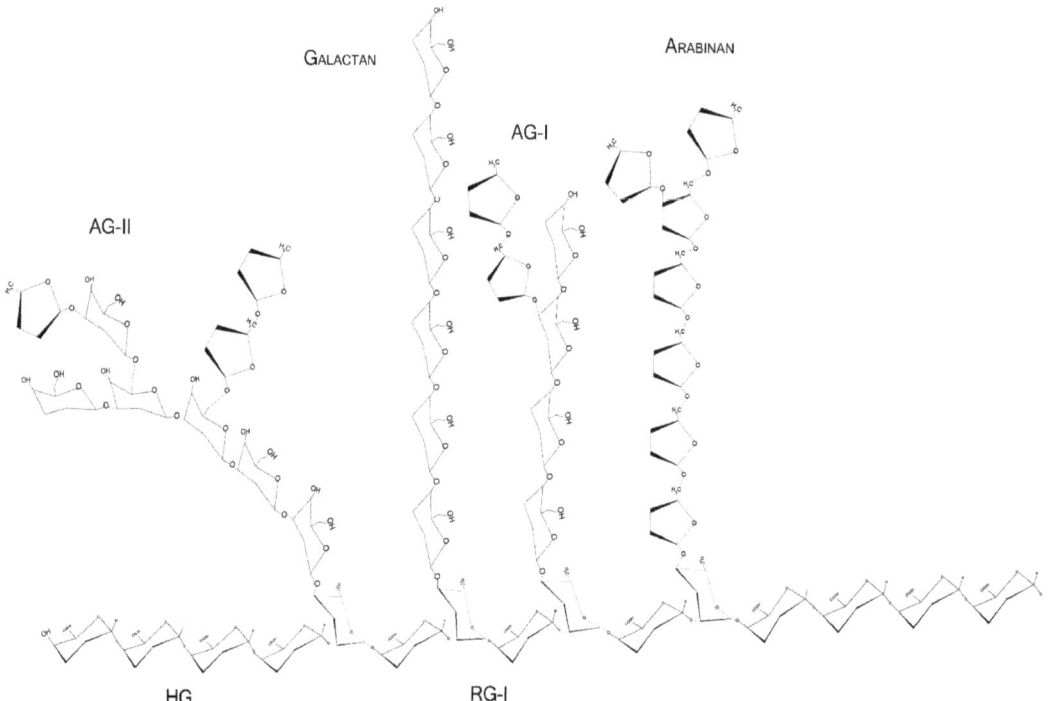

Figure 1. Schematic representation of major pectin components in chair conformation. HG—homogalacturonan, composed of linear α-1,4-D-galactopyranuronic acids; RG-I—intercalated α-D-galactopyranuronic acids and α-L-rhamnopyranose through α-1,4 and 1,2 glycosidic bindings; AG-I—β-1,4-D-galactopyranose with occasional O-3 α-L-arabinofuranose; AG-II—β-1,3-D-galactopyranose with O-6 α-L-arabinofuranose/arabinogalactans. Arabinans and galactans consist of linear α-1,5-L-arabinofuranoses and β-1,4-D-galactopyranoses, respectively.

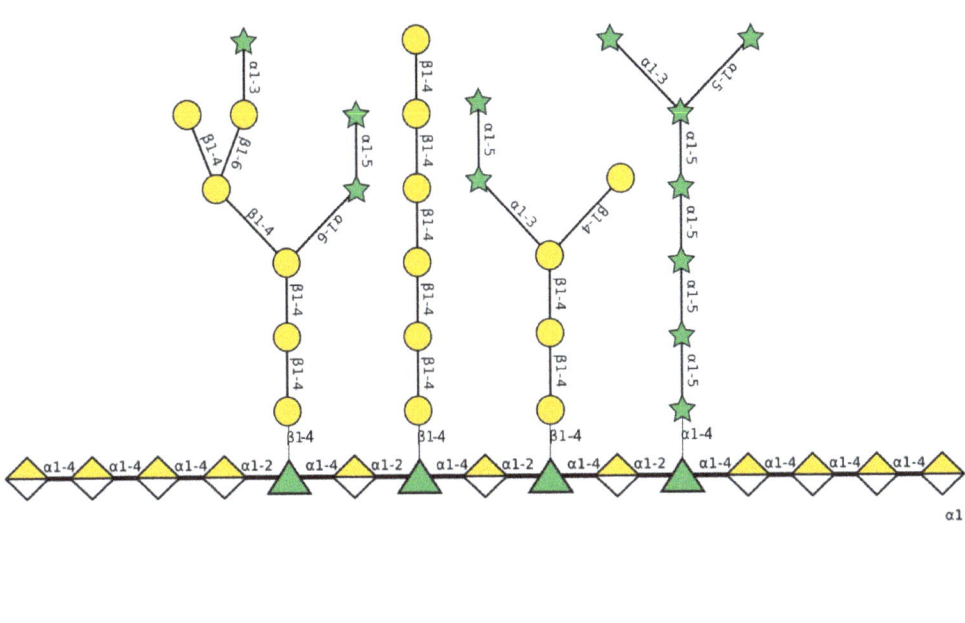

Figure 2. Schematic representation of major pectin components in Symbol Nomenclature for Glycans (SNFG-B) [9] model. Different glycosidic bounds common to pectins are illustrated.

As part of the dietary fiber group, pectins are not digested by the human tract, although the fermentation process by the human microbiota is greatly important for the maintenance of the colonic and systemic health through local signaling. The fermentation products and molecular fragments can improve metabolic syndrome and attenuate hypercholesterolemia, hypertriglyceridemia, and hyperglycemia, markers related to heart disease risks in mice, rats, and humans [10–14]. The pectic products and fragments after colonic fermentation also possess anti-oncogenic attributions in the colon for mitigating cancer-related risks in the colon and even in other types of neoplasia in humans [15].

Galectin-3 (Gal-3) is the chimeric representative member of the galectin protein family, located ubiquitously in the nucleus, cytoplasm, outer cell surface, and extracellular space in mammals. Gal-3 is classified as a β-galactoside binding protein, although—as it is discussed further ahead—its binding range cannot be restrained only to β-galactoside ligands. Gal-3 is composed of a flexible N-terminal domain, containing up to 150 amino acid residues with sequences rich in proline, tyrosine, glycine, and glutamine. This collagen-like region rich in proline, glycine, and tyrosine ends up in a C-terminal domain with a carbohydrate recognition domain (CRD) containing about 135 amino acids. The CRD region is responsible for the signature pattern of the galectins family [16–18]. Although the ubiquitous expression, the main biological source is derived from immunological and collagen-producing cells, where it helps to establish cell recognition and communication through protein–protein or glycan–protein interactions [19,20]. For example, one important and recently elucidated Gal-3 effect in physiological conditions is the recruitment of endosomal sorting complexes required for transport to damaged lysosomes (ESCRTs), so they can be effectively repaired [21]. However, alterations of Gal-3 expression are strongly related to tumor growth, cancer cell proliferation, cell-to-cell adhesion properties, fibrosis stimulation, T-lymphocytes apoptosis, macrophage differentiation into infiltrative forms that stabilize tumor environment, and other features [22–28]. Those attributions are shown in Figure 3, where the intestinal model was chosen, as it is simple to explain the interface between endogenous and exogenous stimuli that differently activate Gal-3 functions. Moreover, pectin is considered a bioactive compound when ingested as a food component or as a dietary supplement, a topic that is explained in the next sections. It can be noted in Figure 3 the different Gal-3 forms available at biological environments, such as the monomeric unit (intra and extracellular) recently secreted present at initial interactions with natural ligands and pentameric (chimeric appearance) after associations of other monomeric units through their N-terminal domain due to ligand stimulus [18,29].

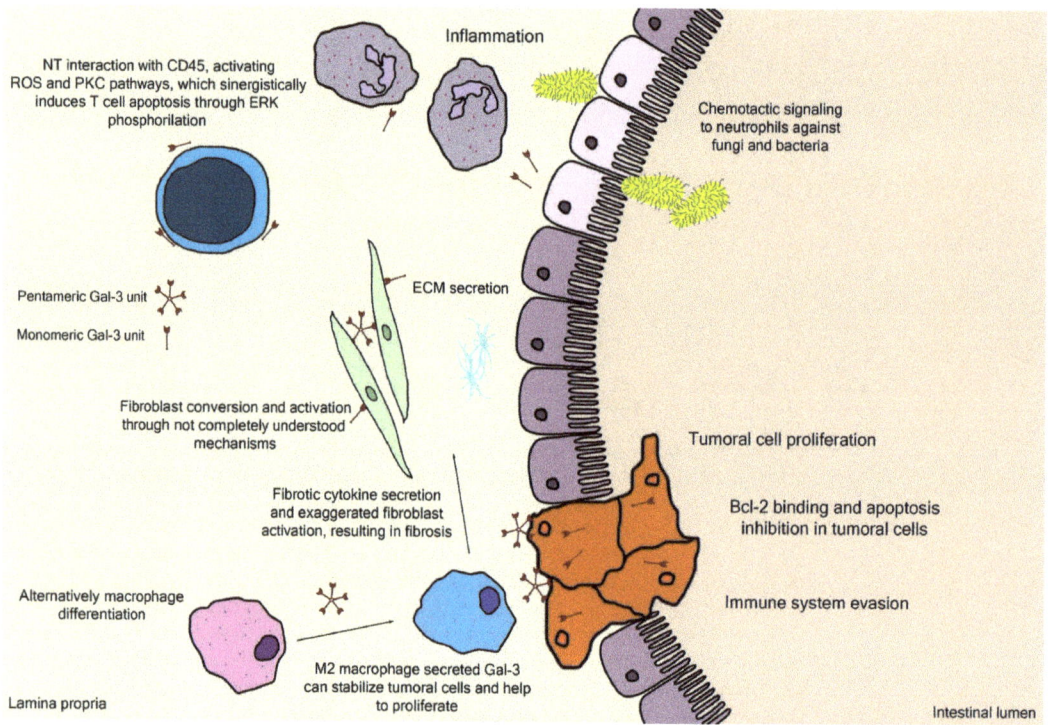

Figure 3. Schematic representation of (patho)physiological effects of galectin-3. As it is demonstrated, the physiology is separated by a thin line from the pathological scenario, such as the extracellular matrix (ECM) secretion stimuli or the chemotactic signaling for infiltrative immune cells. Bcl-2—B-cell lymphoma 2; ERK—extracellular signal-regulated kinases; NT—N-terminal tail/domain; PKC—protein kinase-C; ROS—reactive oxygen species.

While there are extensive high-quality data regarding Gal-3 binding sites modeled to small glycoconjugates and an important level of binding/inhibition by those [30,31], the present review looks forward to establishing ways described in the literature regarding pectin and its fragments and different molecular interactions with Gal-3 protein, as it is highly controversial due to structural variability and reproducibility of experimental conditions. This review also covers the Gal-3 independent pectin interactions. The gathering of these data is fundamental for a better understanding of the overall perspectives regarding this promising—but not yet fully understood—area to be explored. Likewise, all the structural parameters, such as distinct molecular sizes, a higher number of galactan/arabinan side chains or galacturonic acid contents, methyl esterification, and other properties important to pectin biological effects are also discussed by in vitro and in vivo perspectives [7,32,33].

2. Basic Pectin Molecular Aspects

2.1. Pectin Molecular Weight

Although having extremely variable structures, pectins are composed of large polysaccharide chains translated to solutions with high viscosity, a factor which would likely be an impairing factor for possible absorption through oral administration [34]. There are several approaches for extracting and, at the same time, lowering pectic molecular weight. The most commonly utilized method, especially in industrial extraction, is the chemical modification with mineral acids [35], such as hydrochloric and nitric acids, and organic acids [36–39], such as citric and malic acids. Enzymatic digestion is a more subtle and precise approach, using different cleavage agents, mostly arabinanases and galactanases, to remove excess side chains and polygalacturonases (exo- and endo-) to break down the larger HG backbones/pectic domain, aiming to obtain pre-planned molecular patterns [33,40]. Another category, thermic modifications, is less specific but far more practical and inexpensive. These modifications can be achieved by different methods, such as high temperature and pressure [41,42], ultrasound-assisted heating [43–45], and more underused but promising approaches, such as electromagnetic induction heating [46]. There are described extraction methods capable of conserving more of the native pectin structure, such as the use of low temperatures (40 °C) but with low extraction yields, and the decision which technique should be used depends on the final objective of the work [47], as it is discussed further that high molecular weight polysaccharides are not well suited for potential biological applications. Therefore, breaking down and/or manipulating the large polysaccharide chains into smaller portions while bringing better malleability to the molecule itself, improving viscosity and rheological properties, can set more biologically available binding sites, facilitating the potential interaction with proteins such as lectins or other cell-surface receptors.

2.2. Monosaccharides, Backbone, and Side Chains

The monomeric structure of polysaccharides is also extremely variable and inconsistent depending on the food source, extraction method, and modification strategy. These monomeric sugars are naturally presented as either pyranose (p, saccharides with chemical structure including six-member ring, composed of five carbon and one oxygen atoms) or furanose (f, saccharides with chemical structure including five-member ring, composed of four carbon and one oxygen atoms). This high variability represents a problem regarding standardization of recommended ratios between monosaccharides, such as the Gal$p$/Ara$f$, Ara$f$ + Gal$p$/Gal$p$A, associated with the proportions of molecular side chains (e.g., arabinans and galactans) or main structure sequences of α-D-Gal$p$A. For example, important works regarding molecular modeling of Gal-3 inhibitors that have standardized to low molecular weight molecules (<1000 Da) as higher protein-inhibitor ligands [48–50] did not consider polysaccharides. Galactan and galactosyl residues were the main focus as contributors to biological effects observed between pectin and Gal-3 interaction and inhibition since this protein has a preference for binding β-D-galactopyranoside [40,51]. However, while still important targets of interest, other monomeric compounds such as the Gal$p$A, Ara$f$, and their association have been described more consistently, as found for polysaccharides with Ara$f$ residues in higher quantities, mostly linked in α-(1→3,5)-L-arabinan side chains, demonstrating higher inhibition of Gal-3 through techniques such as Gal-3 hemagglutination assays (G3H) or binding through biolayer interferometry assays (BLI) and surface plasmon resonance assays (SPR) [3,33,52,53]. Therefore, this focus on specific monosaccharides that are not only galactosyl could help to understand the biological action of pectin and the possible Gal-3 interaction and inhibition.

2.3. Esterification Degree

Pectin can also be classified by the esterification degree throughout its molecule, where the common approach is the division in a lower degree of esterification (DE) (<50% DE—low-methylation or LMP) or higher DE (>50%—high-methylation or HMP), as well as how the esterification is distributed on pectin molecule (degree of blockiness). The determination of this parameter as a characterization step is crucial to establish the best application of the referred polysaccharide, such as the gelling property and potential where HMP achieves through hydrophobic interactions and hydrogen bonding, while LMP forms gels through the salt-bridge connecting adjacent or opposite carboxyl groups from divalent ions [54,55], emulsion property for protein complexation or drug delivering [56–58], in addition to the direct biologic relationship that is further discussed in the following topics. The RG-I backbone region's side chains are mostly a mixture of arabinans, galactans, and arabinogalactans attached to the rhamnose residue. RG-II is highly methyl-esterified and much more diverse in sugar composition side chains than RG-I. The main pectin backbone is composed of HG, also known as the "smooth" region. This main portion of pectic molecules can also exhibit varying degrees of methylation or acetylation (DM and DA, respectively) depending on the food source and extraction process [2]. Alkali treatment is the most common chemical method to achieve lower DM for pectic samples through direct saponification but may result in chemical waste or even slight alterations to the main pectin chain by β-elimination. Enzymatic treatment is less practical, requiring up to 25 h, depending on the expected DM. An alternative to this inconvenience is the high hydrostatic pressure-assisted enzymatic process, which is a promising operation to help attenuate this problem and facilitate the industrial application of this method [59].

2.4. Rheological Properties

When studying pectin rheology parameters, they are especially useful to determine potential applications of the characterized samples towards large-scale applications, both at food and non-food products, also helping at the definition of a better biologic destination [60]. The previously mentioned structural parameters directly impact the food product incorporation, such as viscoelasticity and thickness, which can be interpreted as gelling capacity, stabilizing potential in acidified milk beverages or fruit juices, emulsion capability with protein-rich solutions, and many others [61–64]. Therefore, the stratification generated from this type of analysis can lead to the best type of use for each pectin sample [65].

2.5. Food Source

The food source—exclusive from plants—together with the pectin extraction methods, and in some cases, the ripening parameters of fruits, are important factors in obtaining functional pectic molecules, as already mentioned. Usually, there is an additional ecologically sustainable status, as many of the possible sources are residues and byproducts from the juice and food processing industries. Some residues of apple [66], *Prunus domestica* and *Prunus mume* [67,68], jaboticaba [69], citrus [70,71], and papaya [7,8,72,73] have been explored for pectin extraction and biological activity studies.

3. Gal-3 Binding Sites and Pectin Interactions

Structurally, Gal-3 N-terminal tail (NT) transiently interacts with its CRD F-face and is linked to the glycine/proline-rich sequence, conferring its uniqueness in the galectin family, allowing self-oligomerization [74–76]. Gal-3 CRD is composed of 11 β-sheets, whereas five belong to the F-face and six belong to the conventional β-galactosides binding S-face, one opposed to the other, forming a β-sheet-sandwich (Figure 4A–C). In the S-face CRD, there is the NWGR conserved motif (Asparagine–Tryptophan–Glycine–Arginine), which is similarly encountered in the BH1 domain from B-cell Lymphoma-2 (Bcl-2) anti-apoptotic molecule, and suggested as the possible interaction that enables apoptosis evasion in tumoral cells (Figure 4B) [19,77,78]. Within the CRD, specifically the S-face, there are some subsites that can be named for easier comprehension of binding interactions. There are two conserved subsites (C and D), two non-conserved (A and B), and one not well-defined subsite (E) [30], where their interfaces are mainly constituted of hydrogen bonds. The CD-associated subsites are a target of natural ligands such as the N-acetyllactosamine (LacNAc) Gal-3 inhibitor or the β-D-galactopyranoside residue located in the Lactose molecule, where it remains tightly bonded mainly to C subsite amino acids (Figure 4D) [79].

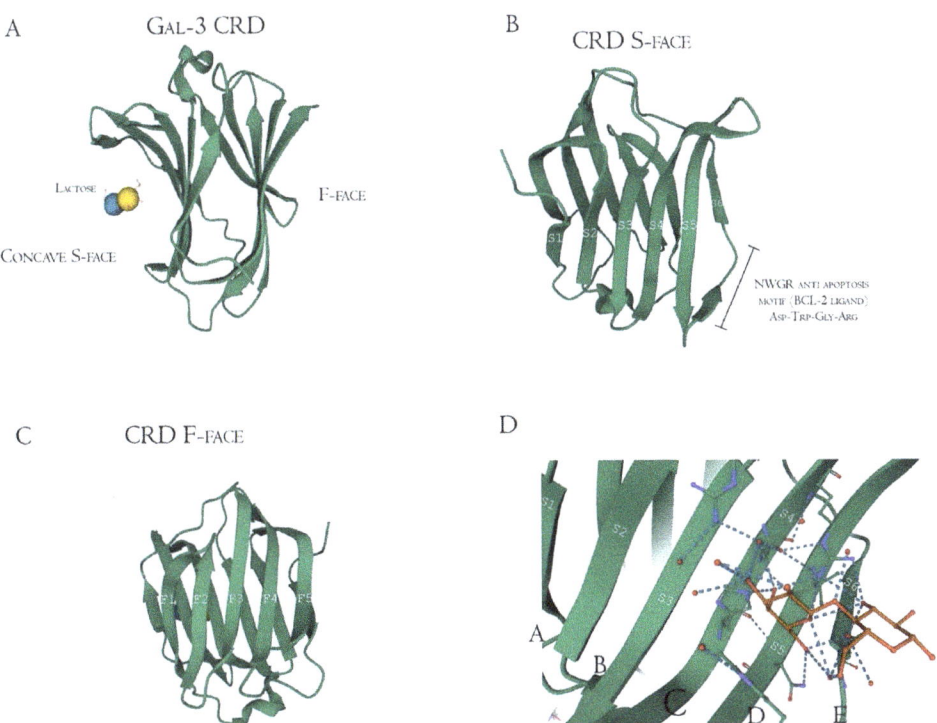

Figure 4. Galectin-3 crystalized tertiary structure X-ray diffraction, PDB ID 49RB [75,80,81]. (**A**) Gal-3 complete CRD, with the anti-parallel β-sheet sandwich; (**B**) CRD S-face β-sheets (S1–S6), which holds the ABCDE subsites. The NWGR motif is highlighted because of their biochemical importance; (**C**) CRD F-face β-sheets (F1–F5, also numbered as B-9, -8, -7, -2, and -11, respectively); (**D**) Schematic representation of β-D-galactopyranosyl-1,4- β-D glucopyranose (β-Lactose) binding at the canonical S-face. The hydrogens atoms are colored as red, the hydrogen bonds as blue dotted lines, and the lactose chain as orange. The binding is stronger at the C subsite (between S4 and S5) amino acids and the β-D-galactopyranosyl residue.

However, there is extensive literature regarding other synthetic and sugar-derived molecules (especially tetrasaccharides and thiosaccharides) that utilize the whole ABCD region (also mentioned as Gal-3 binding groove, as its morphological 3D structure conformation due to the interactions between the groups AB/CD and each one of them individually) for a better affinity performance and inhibition potential [30,49,74,82]. In addition to this, the interaction of pectic poly- and oligosaccharides with Gal-3 is suggested through N-tail epitope recognition by pectin side-chain residues (e.g., galactans). The F-face interacts with β-galactosides (but also other portions), disrupting the CRD F-face binding with NT. Additionally, the same pectic structure could show multiple interactions involving both the S face region (disturbance of amino acids residues from the canonical binding site, e.g., 154–176 sequence observed with Heteronuclear single quantum coherence spectroscopy

(HSQC spectra)) and F-face region (amino acids residues of β-sheets 7, 8, and 9 mostly, e.g., 210–225 sequence) (Figure 5) [76,83,84].

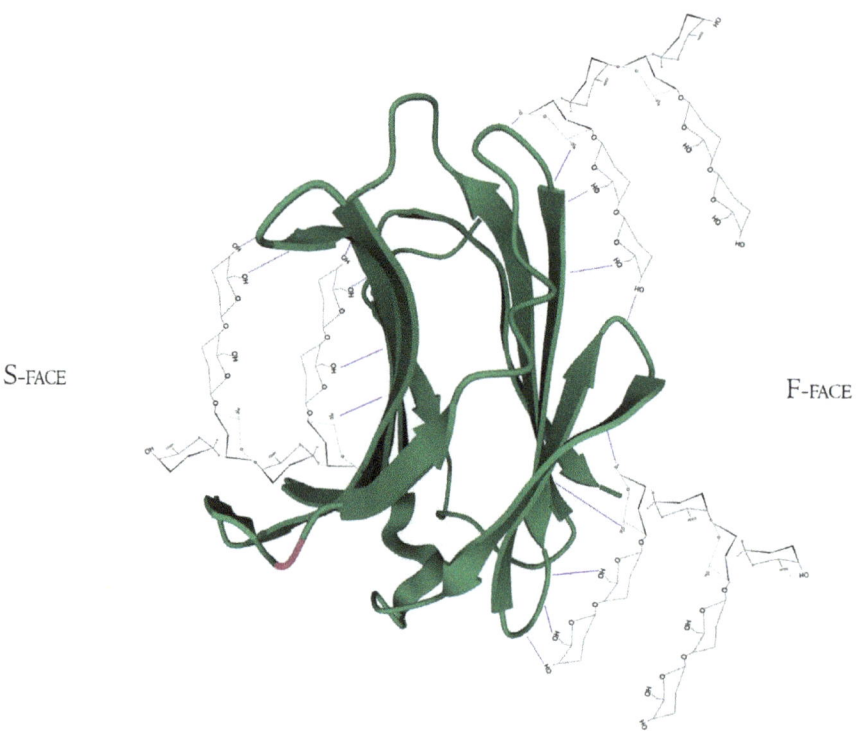

S-FACE F-FACE

Figure 5. Hypothetical interaction of RG-I fragments with both F and S-face of Galectin-3 simultaneously. Here, the main protagonists would be the β-D-galactopyranose and α-D-galactopyranuronic acid residues and would not act like specific pharmacological inhibitors, but maybe as potential Gal-3-ligand blockers through multivalency or allosteric occupation. Hydrogen bonds are represented by the blue dotted lines. PDB ID 49RB [75,80,81].

Other Gal-3 inhibitors that have been thoroughly and increasingly studied are pectin and its fragments. Some of the possible Gal-3 inhibition effects by pectin and fragments might include protecting pancreatic β-cells against oxidative and inflammatory stress [85]. Xu et al. [86] observed modified citrus pectin (MCP) downregulated pathways involved in myocardial fibrosis, but the authors did not study the Gal-3 inhibition by MCP in vivo, although Gal-3 was downregulated in the treated group [86]. Some other biological effects from Gal-3 inhibition are related to cancer proliferation control [7,32], with a strong positive relationship between overexpression of Gal-3 and carcinogenic processes, such as apoptosis evasion, higher cell invasion, and metastatic progression, which are key signatures in tumor and metastatic types of cells [87]. The structural relationship between the pectic chain and Gal-3 binding sites is also very important for the expected positive functional effects [33,88]. Extracellular Gal-3 biological functions are exerted mostly by interacting with glycoconjugates in cell surfaces, such as laminin and adhesin, signaling and activating specific pathways. Some studies regarding binding interactions between pectin and its fragments with Gal-3 are listed below (Table 1).

Table 1. Polysaccharide-Gal-3 binding articles summary.

Authors	Polysaccharide Residue	Analysis Method	Binding Evaluation
Wu et al., 2020 [33]	RG-I from citrus canning process water	Surface plasmon resonance	Smooth binding curve through SPR with decreased affinity with galactan side-chain removal
Zhang et al., 2016 [34]	MCP, RG-I-4, and p-galactan	Gal-3 hemagglutination, bio-layer interferometry, and surface plasmon resonance	RG-I-4 demonstrated higher Gal-3 avidity in comparison to the other two polysaccharides, with a K_D at sub-micromolar range (RG-I-4 and p-galactan), but no significant result when testing competitive assays with known S-face inhibitors such as lactose
Gao et al., 2013 [40]	Ginseng RG-I-4 domain	Gal-3 hemagglutination and surface plasmon resonance	RG-I-4 inhibited G3H and was bound specifically to CRD with high affinity with Ara residue location in the RG-I, changing the activity detected at the G3H assay
Gunning, Bongaerts, Morris et al., 2009 [51]	RG-I, PG, and galactans	Atomic force microscopy, fluorescence microscopy, nuclear magnetic resonance, and flow cytometry	Galactan binding to Gal-3 is lectin-saccharide highly specific, while RG-I has low specificity, and PG was not specific. The data suggest that the lesser "sterical crowding" of the galactans alongside its beta-1,4 linear chain could be the reason for the better performance observed
Shi et al., 2017 [52]	Ginseng RG-I-3A domain	Bio-layer interferometry, Gal-3 hemagglutination	Binding kinetics of RG-I-3A showed a high binding affinity with a K_D of 28 nM through and also presented notable G3H inhibition
Zhang et al., 2017 [83]	MCP-derived RG-I and HG portions	Gal-3 hemagglutination, bio-layer interferometry, ELISA, and nuclear magnetic resonance	Gal-3 bound to both portions separately but with a much more notable avidity when a combination of them (RG + HG) is performed, suggesting that this interaction exposes more binding sites at the lectin
Miller et al., 2015 [84]	Galactomannans (GM) and polymannan	Nuclear magnetic resonance	The primary binding surface of the GM's located mainly at F-face beta-sheets (7,8 and 9)
Zheng et al., 2020 [89]	MCP-derived HGs of varying molecular weights	Nuclear magnetic resonance heteronuclear single quantum coherence spectroscopy and crystallography	Higher molecular weight HGs demonstrated more perturbances at F-face resonances and involved more S-face beta-sheets at the binding footprint. A possible binding of Gal-3 to the non-terminal HG sites is suggested, and it is shown a different S-face binding pattern of HG's compared to lactose

Table 1. Cont.

Authors	Polysaccharide Residue	Analysis Method	Binding Evaluation
Miller et al., 2019 [90]	Galactan oligosaccharides of varying chain lengths	Nuclear magnetic resonance heteronuclear single quantum coherence spectroscopy	Binding affinity at the terminal non-reducing end of the galactans in the CRD S-face (beta-sheets 4, 5, and 6 chemical shifts mostly) increases with the increase in chain length
Zhao et al., 2017 [91]	Pumpkin RG-I-containing pectin	Surface plasmon resonance	Moderate binding affinity towards Gal-3 through SPR, with a fast association between protein and polysaccharide (K_A) and slow dissociation (K_D)
Miller et al., 2017 [92]	Ginseng RG-I-4 domain	Nuclear magnetic resonance heteronuclear single quantum coherence spectroscopy	Epitopes from RG-I-4 bind to three different labeled Gal-3 sites, two at the CRD and another one at NT. At lower concentrations, the F-face site is more activated, turning to S-face at higher ones

The Gal-3 binding range is considerably wide, including oligolactosamines, sulfated and sialylated glycans, and α-binding glycans, such as Fucα1-2, Galα1-3, and Galα1-4 added to the core of LacNAc molecule (as terminations) [89,90,93,94], which could indicate why non-β-Gal pectic fragments and other sources can still perform positively in some binding experiments. The β-1-4-galactan side chains inside the pectic RG-I domains have been highly attributed as the molecular factor involved in direct inhibition of Gal-3 (recombinant proteins and native proteins from cells), mainly discussed as a motif varying reaction, such as galactose residues in the middle or at terminal parts of those linear molecules. The longer galactan side chains in RG-I molecules identified in MCP and ginseng pectin were also associated with a stronger interaction with Gal-3 in vitro [40,95]. Similar positive binding affinity results with Gal-3, higher than observed with potato galactan, were obtained when isolating a 22.6 kDa RG-I polysaccharide from pumpkin [91]. Meanwhile, the RG-II enriched fractions extracted from *Panax ginseng* flower buds, with a high methylesterification and backbone substitution, as well as with lower content of galactose residues and low molecular weight, were associated with an absence of Gal-3 binding [3]. Highly esterified HG samples also did not inhibit Gal-3 [3]. However, other ginseng samples with an equivalent ratio between HG and RG-II regions had a positive binding affinity to Gal-3 in a similar way to the synergistic effect of HG and RG-I of citrus pectin [34,83]. In the latter case, the authors also suggested that the unesterified characteristic of pectin, alongside non-substituted GalA segments, were crucial structural elements for the observed biological effects [83]. Following this trend, commercial lemon pectin samples (from CP Kelco) with a low degree of methyl-esterification and low molecular weight were consistently and extremely potent in preventing negative outcomes in human islets with β-cell apoptosis (diabetes model) induced stress in vitro [85], in a dose–response manner. Gal-3 inflammatory stress induction was analyzed by evaluating oxygen consumption rate (OCR) in the presence and absence of a Gal-3 known inhibitor, α-lactose. After the previous incubation with α-lactose, the pectin sample effects that reversed OCR reduction induced by streptozocin and minimized inflammation induced by cytokine incubation were greatly compromised. The author's suggestions were based on the observed effects derived from pectin binding to Gal-3 [85]. However, specific data regarding the connection between the in vitro effects and Gal-3 inhibition by pectins were not proposed.

As already mentioned before in this topic, Gal-3 does not have only one direct binding site toward polysaccharides, known as the canonical S-face (sugar-binding) region inside the CRD. This can help to explain the different results in distinct binding experiments based upon competitive inhibition and chain length [88]. Detailed data suggest that the Gal-3 N-terminal binds to a 60 kDa RG-I-rich pectin portion of ginseng through galactans located in RG-I ramifications [92]. This same molecule with removed ramifications (Rha$p$ and Gal$p$A intercalated residues only) did not bind with Gal-3 [92]. Another molecule rich in galactose—but not derived from pectin—is a galactomannan isolated from *Cyamopsis tetragonoloba* guar gum flour (1-4-β-D-mannopyranose backbone with 1-6-α-D-galactose ramifications) also did not bind with Gal-3 [92]. However, experiments performed with potato galactans oligosaccharides derived from pectin had similar results as observed for the ginseng RG-I regarding Gal-3 binding [92], demonstrating the probable uniqueness structural features of pectin-derived molecules that can result in Gal-3 binding.

Gal-3 F-face has shown strong binding signals with pectin molecules, mainly at lower polysaccharide concentrations, opening more space to conventional S-face binding in higher concentrations. Like S-face, the F-face non-orthodox site is enriched by hydrophilic and charged amino acids residues. In addition to not having the Trp key-residue, other hydrophobic side chains could have similar functionality, as well as similar concave shape conformation [84,92] (Figure 4A–D). It is specified that resonance broadening was attributed as the primary capability of binding between Gal-3 and the RG-I-4 (a ginseng-derived rhamnogalacturonan), but also that broadening would be directly correlated to higher noise and lower sensitivity in analytical techniques [84,92], a factor which has to be taken into account. The presence of those non-conventional sites was suggested when lactose (CRD S-face inhibitor) did not interfere in RG-I binding to Gal-3 [40]. Furthermore, the binding of a determined ligand to the F-face of Gal-3 CRD could allosterically modify S-face residues. These shifts influence the conformation of the opposing face affecting the affinity between ligand (peptide, glycoconjugate, or polysaccharide) and the Gal-3, improving or attenuating the inhibitory/activity effects [76]. Moreover, Gal-3 N-terminal tail phosphorylation, although having little impact in CRD F-face, may be related to allosterically influencing carbohydrate-binding to the canonical CRD S-face site [76]. This structure-related information and respective identification method improvement contributed to more profoundly describing how the synergistic effects between pectin fragments work, as will be thoroughly discussed below.

Unusual non-galactoside poly-oligomeric ligands could act through multivalence interaction with Gal-3. The synthesis or identification of molecules that represent potential inhibitors of multimeric (chimeric) Gal-3 to cell-membrane glycoconjugates could be critical for minimizing known observed effects, such as the glycoclustering of receptors resulting in apoptosis induction of T cells [96]. Chelation, subsite binding, and reassociation of the binding site towards different monomeric structures within a molecule are also candidates for multivalent interactions with Gal-3 [97]. Although pectins and other natural ligands are harder to validate in comparison to synthetic compounds, parameters such as molecular flexibility and occupation of binding sites can be less manipulated/predicted [98]; the study of monosaccharide and consequently substructures ratios influencing Gal-3 binding is a promising area of investigation.

A recent in vitro study analyzed Gal-3 inhibition and MCF cell viability after specific enzymatic modification of citrus water-soluble fraction (WSF) rich in pectin. It was demonstrated that although β-1,4-galactan side chains in RG-I are still considered the most accountable molecular part for the observed effects, the partial removal of the side chains composed of other monosaccharide residues, such as α-1,5-arabinan, also contributed to inferior results both at cancer cell proliferation and direct gal-3 binding [33]. Other complex structures, such as the pectic acidic fractions extracted from *Camellia japonica* pollen, described as an RG-I-like polysaccharide, had their branched α-1-3,5 arabinan and type II arabinogalactans attributed to their strong Gal-3 inhibition effects [53]. It is suggested that, despite not having a significant relationship to the ratio of RG-I/HG, pectin-rich WSF bioactivity was dependent on cooperation between RG-I and HG regions [33]. In addition to this pectin fragment ratio, the monosaccharide residue composition ratio had an impactful performance. The Gal$p$ amount is necessary for the best Gal-3 inhibitory results since pectic fragments with lower Gal*p*/Ara*f* ratio resulted in lower Gal-3 inhibition even with similar molecular sizes than other studies (50 to 60 kDa) [52].

There are also in vitro data for anti-proliferative characteristics for sugar-beet pectin, in which both the RGI/HG backbone and the galactans/arabinan side chains exhibited those positive effects in HT-29 (human colorectal adenocarcinoma) cell populations [99]. The alkali treatment not only increased the RG-I/HG ratio (which translates to more neutral sugar side chains) but also enhanced the anti-proliferative effects. The removal of almost all side chains in the sample did not completely abolish the effects, denoting the importance of the cooperative effects between different pectic structures [99]. Similar results were obtained in our lab, in which papaya uronic fraction, a Gal*p*A fraction enriched with galactans, had the best in vitro results in inhibiting Gal-3 hemagglutination than neutral sugar-enriched fractions [7]. The difference of the samples was the degree of methylation, in which the former had low DM, and the latter had high DM.

Zhang and colleagues [83] indicated that a combination of RG and HG polysaccharides enhances the Gal-3 binding in vitro through HG interaction with RG, opening more Gal-3 binding epitopes in the RG molecule. At a particular RG/HG molar ratio, the interaction between polysaccharide and the F-face of Gal-3 could establish the new activated binding epitopes due to the higher prevalence of galactose residues, facilitating the S-face CRD interaction. This interaction between HG and RG could perform alterations in its conformation and increase or enhance the synergistic functional effects [83]. Each 130 kDa molecule of the isolated pectin fraction could bind up to 16 Gal-3 molecules. Overall, the combination of both structures showed biologically better activity than separated molecules [83]. Additional ginseng HG-rich fractions were responsible for inducing apoptotic process in vitro at higher doses and cell cycle arrest at lower doses in HT-29 cells, and these biological effects were increased after heat treatment of the polysaccharide fractions [100].

Another ginseng polysaccharide fraction had 91 kDa, an Ara*f*/Rha*p* ratio of 2:1, and a Gal*p*/Xyl*p* ratio of 1:1 and was characterized as xylo/rhamnogalacturonan I with arabinan/galactan side chains. This fraction was evaluated for in vitro Gal-3 inhibition, anticancer effect, and in vivo gut microbiota modulation [101]. The authors indicated that the xylans were mostly responsible for the microbiota's healthy recovery and protection. Meanwhile, arabinogalactan side chains can interact with Gal-3 down-regulating tyrosinase through its N-glycan binding site [101]. Overall, the polysaccharide was effective in restoring normal levels of the important interleukin for tumor rejection and T cell activation, such as IL-10. The polysaccharide also modulated IL-17, a mediating molecule over-produced by the tumor cell microenvironment, revealing a multi-targeted functionality of the ginseng polysaccharide [101]. Although studied for many years, this pectic polysaccharide immunomodulatory property is still a trending topic, especially when the structural differences influence the interaction between the pectin and immune receptors, such as the toll-like receptor family (TLR) and interleukins at the macrophage cell surface [73,102], and will be more profoundly discussed later.

After determining the direct inhibition potential by competing with Gal-3 ligands in ECM, it was demonstrated that MCP can also downregulate Gal-3 expression. The in vitro induction of cell cycle arrest at the G2/M phase, through Cyclin B1 decrease and cyclin-dependent kinase 1 (Cdc2) phosphorylation, disrupts the Gal-3 pathological function of maintaining cell cycle arrest at the late G1 phase that leads to evasion from apoptosis induction [103]. Gal-3 can also induce phosphorylation in the signal transducer and activator of transcription 3 (STAT-3) in ovarian cancer cell spheroids. Although demonstrating slight decreases in cell viability after treating with paclitaxel, a strong synergistic effect between this drug and MCP was observed. The IC50 values decreased when both compounds were together, alongside a 70% increase in caspase-3 activity and a 75% Cyclin D1 expression level decrease against the values obtained using only paclitaxel [104].

Regarding fibrosis induction, MCP alleviated liver fibrosis and stress-induced secretions, such as decreased malondialdehyde (MDA), TIMP metallopeptidase inhibitor 1 (TIMP-1), collagen-1 α-1 (Col1A1), and Gal-3 expressions, improving HSC apoptosis rate and the upregulation of glutathione and superoxide dismutase in vivo [105]. Renal fibrosis-related biomarkers in an adult male Wistar murine model were also attenuated by treatment with MCP in drinking water. Albuminuria, proinflammatory cytokines, such as small inducible cytokine A2, osteopontin, epithelial transforming growth factor-β1, and other epithelial to mesenchymal transition factors, were all controlled or restored to normal levels after treatment in normotensive experimental models of renal damage [106].

Gal-3 is up-regulated by aldosterone, mediating inflammatory and fibrotic response in vascular muscle cells in vitro and in vivo [107]. The mechanism was demonstrated as inducing Gal-3 secretion by macrophages via phosphatidylinositol 3-kinase inhibitor/AKT and nuclear factor κB transcription signaling pathways in vitro and in vivo [108]. Furthermore, in aldosterone-induced cardiac and renal injuries, MCP subcutaneous injection in mice—different from the major in vivo experiments when MCP was diluted in drinking water—downregulated Gal-3 at protein and mRNA levels while also minimizing cardiac adverse effects induced by aldosterone salt [109]. In addition, for the renal injuries, MCP inhibited Gal-3 at the tubular level but not in the glomeruli, which is highlighted by the authors since Gal-3 is not expressed at the glomerular level [109]. Nephrotoxicity is a major side-effect of cisplatin chemotherapy, which contributes to the number of acute kidney injury (AKI) hospitalizations. Mice that received 1% MCP in drinking water during the same period after cisplatin injection had better morphometric preservation and prognostic regarding renal fibrosis, such as lower serum creatinine levels, Gal-3, fibronectin, and collagen-1 expression, which could point to a protective effect of MCP in this treatment [110]. Specific mechanisms and routes of Gal-3 downregulation, however, were not successfully addressed by those studies. Therefore, those points should be considered for better knowledge of the possible link between treatment and the biological effect.

AKI-induced mice through the ischemia/reperfusion (IR) model led to Gal-3 expression, cardiac injury, and systemic inflammation. The authors indicate that the induction of inflammation alongside cardiac fibrosis was Gal-3 dependent, demonstrated through significant reduced deleterious outcomes in genetically Gal-3-KO mice that received orally MCP (100 mg/kg/day), which was also seen at WT-MCP-treated mice group [111], but the expression of other markers such as MCP-1 and ICAM-1 mRNA were also decreased by MCP treatment. Rats submitted to myocardial IR also had better prognostics when treated with MCP (in drinking water) one day before and eight days after the procedure, such as improved perfusion, serum brain natriuretic peptide normalization, IL-1b and C-reactive protein (CRP) reduction, lower Gal-3 expression levels at the ischemic tissue, and other parameters. The Gal-3 specific blockade suggested was measured by the authors through the expression of two proteins that are down-regulated by Gal-3, fumarase, and reticulocalbin-3, which were restored after MCP treatment [112]. Perindopril and MCP (in drinking water) were similarly effective as treatments for ischemic heart failures in rabbit models, lowering Collagen-I, III, and Gal-3 mRNA and protein expression, alongside slight reversion of histological remodeling (an important level of fibrosis was still maintained, visible by Masson staining of myocardial tissues). However, the exact mechanisms by which MCP or perindopril could exert their observed effects are still unclear and were not directly addressed [113]. Both the effects of Gal-3 and isoproterenol-induced left ventricular systolic dysfunction in the mice model with selective hyperaldosteronism, alongside myocardial fibrosis installation. Combined therapies with MCP in drinking water and canrenoate as an aldosterone blocker enhanced the anti-inflammatory and anti-fibrotic effects [114]. Rats in a pressure overload model that were treated with MCP in drinking water had lower Gal-3 mRNA expression and Gal-3 immunostaining-confirmed presence than control. Other fibrosis-related proteins such as α-smooth muscle actin (α-SMA), connective tissue growth factor (CTGF), transforming growth factor (TGF)β-1 and fibronectin, and also inflammation factors such as IL-6, IL-1β, and TNF-α were at lower levels than control [115]. The studies described herein regarding oral MCP supplementation, either isolated or in combination with other molecules: (i) did not only show effects influencing Gal-3, but also other pathways, receptors, and protein expressions; (ii) authors did not explain, or at least brought up for the discussion, if orally consumed MCP could reach the target organs, such as kidney, heart or liver. These factors should be considered whenever reading these interesting but overly biased results, denoting not-so exclusive observations that are also promising but opening many questions regarding the systemic distribution of pectin's pre- or post- colonic fermentation.

As mentioned before, there are physiological activation pathways that could involve Gal-3 action, important data to account for whenever testing new possible therapeutic agents that could selectively induce activation/apoptosis depending on the target [116]. This in vitro work has demonstrated that recombinant human Gal-3 initiates three distinct pathways, one for T cell activation (PIK3) and two hybrids (reactive oxygen species and protein kinase C—ROS and PKC, respectively). Furthermore, MCP and acidic fraction from ginseng roots inhibited T cell apoptosis in vitro by caspase-3 cleavage, and ginseng-derived fractions did not interfere in IL-2 secretion [116]. One of the Gal-3 connections established by the authors to the observed effects is proposed as upregulation of PI3K/Akt phosphorylation, in which the presence of an inhibitor for each molecule also inhibited IL-2 secretion. However, no similar effects regarding the effects of MCP in cardiac protection, fibrosis regulation, and normalized hypertension were observed in a recently published clinical trial from Lau et al. [117] in patients with high Gal-3 levels and established hypertension. The results raise many doubts and counterpoints regarding the MCP effects in in vitro and in vivo studies to be replicated in humans, which is discussed in the next chapter.

Recently, Gal-3 has been treated as a treatment target and prognostic marker for patients with severe acute respiratory syndrome-coronavirus 2 (SARS-CoV-2). Higher serum levels of Gal-3 were related to a tendency of severe acute respiratory distress syndrome (ARDS) development, and alongside IL-6 and CRP, Gal-3 was demonstrated to be the best predictive power for mortality outcome [118]. There are also positive correlations between Gal-3 and other inflammatory markers such as PTX-3, ferritin, and the marker of endothelial dysfunction, sFlt-1 [119], which could be utilized for intensive care unit (ICU) admission biomarker. A phase 2a study, the first clinical trial using an inhalator treatment targeting Gal-3-GB0139—associated with the standard of care procedure (dexamethasone)—identified lower Gal-3 serum levels, higher mean downward of CRP levels (although higher CRP was identified at first in the treatment group) and other inflammatory agents and better fibrosis marker levels than the control group. However, there was no statistical difference between patient mortality rates between groups [120]. This enhances and augments the discussion to further explore alternative treatments regarding Gal-3 inhibition.

4. Pectin and Gal-3 Controversies

As it has been widely studied and known, pectins are extensively fermented through the local intestinal microbiota, which could systematically impair "direct" mechanisms of action. In the above-discussed works showing systemic effects after oral ingestion of MCP, it is often ignored the thought process and viability on how those molecules could reach the systemic circulation. In counterpoint, a work that used antibody recognition of RG-I fragments from *Bupleurum falcatum* L.—demonstrating reactivity at mice bloodstream and liver—could indicate a partial small-intestine related absorption [2,121]. Similar studies based on β-glucan uptake mechanisms should be more explored for pectins in general, as both are non-digestible carbohydrates that suffer fermentation to an extensive degree [2]. Modified pectin from broccoli (*Brassica oleracea* L. Italica) was suggested to be absorbed after an increased number of activated macrophages and lymphocyte proliferation when administered through oral treatment in mice, but with no same effect in vitro [122]. One possible suggestion of pectin "absorption" is related to asialoglycoprotein receptors that could play a role in absorbing modified pectin fragments throughout the intestine, as they are notable galactoside-terminal glycoproteins transporters [123,124]; however, much more pieces of evidence need to be described to confirm this hypothesis. The oral consumption of MCP reduced liver metastasis on a mouse colon cancer model [125], while MCP-derived galactans ([-4-β-D-Gal$p$-1-]n) and arabinans ([-5-α-L-Ara$f$-1-]n) with a low degree of polymerization were absorbed through paracellular transport, and in lower rates through transcellular transport, similar to what was observed at in vitro culture of Caco-2 monolayer models [126]. Pectin-derived oligosaccharides with 1 kDa and rich in galactosyl residues were absorbed in BALB/c mice and human tumor cells while also changing membrane permeability in different human cancer cells, such as HepG2 and Colo 205 (hepatic and colon carcinoma, respectively) [127]. Microfold cells and gut-associated lymphoid tissue (GALT) are also proposed as explanations for the bloodstream presence of modified pectin fragments, where the former theoretically would serve as a facilitator to GALT macrophages to act internalizing these portions [122,128]. Once again, these assumptions of absorption models are still in germinative steps; therefore, they cannot be taken as unreasoning facts, but they can open new potential transcription elucidation of in vivo observed effects towards clinical significance in the future.

The above described highly selective effects, although not completely mechanistically understood, could be related to the mentioned diverse binding sites and different chemical conformations between those polysaccharides and Gal-3, but the range of varied mechanisms unrelated to Gal-3 specific inhibition cannot be excluded or ignored, as it could also play a role or even be the major protagonists regarding polysaccharides action.

In a related but slightly different context, it is highly important to clarify that a binding molecule is not the same as an inhibitor by itself. Additionally, this capability of pectins to exert both instances, binding to or inhibiting Gal-3, has been contested. As demonstrated by Stegmayr et al. [50], the use of a range of plant polysaccharides to study the capacity of interacting with representatives of the galectin family (including Gal-3) resulted in a very low or even absent binding and agglutination inhibition [50]. It is further discussed that the discrepancy observed in similar studies, such as the one from Gao et al. [95], could be due to a fine-tuning difference (concentration and temperature) or even multivalence interaction factors [50,95]. Furthermore, noteworthy, immobilized surface techniques may be prone to developing suitable conditions to the multivalent aspect, therefore potentially overestimating affinity results [88]. Nevertheless, the same authors also found a potential indirect effect. Re-incubating the JIMT-1 cells with the pectin samples led to Gal-3 accumulation around intracellular vesicles, feasibly inclining towards a "directional" change of location inside the cell, although more experiments regarding this property need to be performed to detect plausible applications of this observed scenario in vivo [50]. Other studies demonstrated biological effects in different in vitro cell cultures were totally or highly independent of Gal-3 inhibition; as demonstrated through Gal-3 hemagglutination (G3H) assay, cell lines with low expression of Gal-3 and/or usage of lactose (Gal-3 inhibitor) did not influence HG or RG-I activities, suggesting that cell migration inhibited by those polysaccharides did not rely upon human/mouse Gal-3 binding [8,129]. The main problem observed in literature is the overused statement of plant polysaccharides acting as specific pharmacological direct inhibitors of Gal-3, without specific data of inhibition shown, or even proposed study models to evaluate the integrity of those polysaccharides in reaching potential target organs. The facts were described in the letters written by Hakon Leffler, MD, Ph.D., and Anwen Shao, MD, Ph.D. [130,131], in response to pectin attribution of Gal-3 inhibition at blood–brain barrier disruption, relying on the lower expression detected by immunoblotting [132], but this is only one of the examples. Specific human Gal-3 inhibition of the S-face CRD region, mostly between C and D subsites, is achieved mainly by small glycoconjugates, such as lactose, N-acetyllactosamine, and many synthetized neo-glycoproteins [30,49,133–136].

To address those controversial perspectives, more studies regarding alternative mechanisms independent of Gal-3 interaction should be performed with polysaccharides while also exploring more structure-relation models. In the following chapters, alternative perspectives are further analyzed.

5. Pectin as Dietary Fiber: Some of the Gal-3 Independent Beneficial Effects to Human Health

Regarding pectin fermentation, it generates products that are essential for colonic and systemic health in general. One example is the inhibition of cholesterol intestinal absorption in an apoE$^{-/-}$ mice model through regulating mRNA levels of its transporters, resulting in controlled blood lipid levels in vivo [10]. This protection is also due to the physicochemical properties of the soluble fibers, where the bile acid excretion and cholesterol mobilization in the intestinal tract is compromised by the fiber viscosity [137]. Similar results were obtained in mildly hypercholesteremic humans, in which pectin with high DE and high molecular weight resulted in a cholesterol-lowering effect [12]. A rodent model of high-fat diets to induce non-alcoholic fatty liver disease also had positive results after receiving 8% citrus peel pectin. The diet had attenuated liver damage and lipid accumulation while also reducing some biomarkers such as carbohydrate-responsive element-binding protein (ChREBP) and reducing serum total triglyceride in vivo [11].

Regarding glucose metabolism, pectin (from apple and citrus) added to high fat/high sugar diets, even at low doses, were successful at ameliorating glucose serum levels [13,14], glucose tolerance, and insulin resistance biomarkers such as HOMA-IR and fasting insulin serum levels [13]. Those effects are suggested to be derived from a pectin capacity of lowering mucosal disaccharidase activities, specifically sucrase and maltase [14], and also through a potential pectin p-AKT upregulation, being beneficial to directing the insulin signaling [13].

Short-chain fatty acids (SCFA) are the most common by-product of pectin and other types of fibers fermentation, and there is extensive literature related to biological and health effects [138–141]. The SCFA help to maintain intestinal health through G-protein-coupled receptors (GPR) interaction, such as T regulatory cell homeostasis, epithelium integrity, and maintenance of an acute immune response and normal cytokine/chemokine expression of key modulators [142–145]. Different types of pectins and fragments can modulate microbiota and have different fermentation profiles. Sugar beet and soy pectin lowered *Akkermansia* relative abundance, while soy pectin showed high levels of propionate, butyrate, and branched SCFA concentrations in the colon of rats [146]. Supplementation of citrus pectin (CP) in piglets diets was also attributed to a higher relative abundance of Bacteroidetes members in colonic digesta and feces, and this pectin-enriched fraction also slowed the fermentation process, changing microbiota interaction [147]. In a dynamic digestion/fermentation simulator, CP could induce *Bifidobacterium* spp. growth, but not *Lactobacillus* spp. Both genera presence are considered beneficial to colonic health [148]. During in vitro fermentation, authors had similar results with high DM (70%) apple pectin [149], with in vivo observations also supporting the mentioned in vitro data [11]. The CP was capable of inducing growth of *Faecalibacterium prausnitzii*, a bacteria that has been pointed as a modulator in dysbiosis of Crohn's disease patients and a major agent of pectin utilization, with their lower levels correlated to inflammatory bowel disease [150–152]. Those data support the application of pectins in health investigations, even before going deeper into specific binding/modulation features, and they are summarized in Table 2 (for the diverse discussed biological effects of pectin).

Table 2. Summary of observed experimental effects in manuscripts studying pectin and its fragments.

Authors	Treatment	Study Type	Treatment Target	Observed Experimental Effects
Pedrosa, Lopes and Fabi, 2020 [7]	Papaya pectin acid and neutral fractions	In vitro	HCT 116, HT-29, and HCT-116 Gal-3$^{-/-}$	Gal-3-mediated agglutination inhibition, cell viability decrease in both WT and knockout cells (suggesting Gal-3 independent pathways)
Chen et al., 2018 [10]	SCFAs	In vivo	Male apoE$^{-/-}$ mice	Stimulation of Lxrα mediated genes expression related to intestinal cholesterol uptake and excretion; improved blood lipid profiles and anti-atherosclerotic property
Li, Zhang, and Yang 2018 [11]	CP	In vivo	Healthy male C57BL/6J mice	Pectin-supplemented high-fat diet mice had reduced lower liver damage, lipid accumulation, and total serum triglyceride
Brouns et al., 2012 [12]	Different DM and MW apple and citrus pectin (CP)	Human intervention	Mildly hyper-cholesterolemic men and women	Higher DM apple and citrus pectin lowered between 7 and 10% low-density lipoprotein cholesterol (LDL-C) compared to control
Liu et al., 2016 [13]	CP	In vivo	Male Sprague-Dawley rats with induced type 2 diabetes	Enhanced glucose tolerance, blood lipid levels, reduced insulin resistance, pAKT expression upregulation, and glycogen synthase kinase 3 β (GSK3β) downregulation
Fotschki et al., 2014 [14]	Apple fiber (low pectin)	In vivo	Male Wistar rats	Disaccharidase activity reduction, higher SCFA production, reduced serum glucose concentration
Prado et al., 2019 [32]	Chelate-soluble fraction of papaya pectin	In vitro	HCT 116 and HT-29 human colon cancer cells	Gal-3-mediated agglutination inhibition, similar to lactose control; pre-treatment with lactose suggests cell Gal-3 independent proliferation reduction for one of the fractions (3CSF)
Wu et al., 2020 [33]	CP fragments	In vitro	MCF-7 human breast cancer and A549 human lung carcinoma	Significant binding affinities to Gal-3; dose-responsive cell proliferation inhibition in vitro, not necessarily related to Gal-3

Table 2. Cont.

Authors	Treatment	Study Type	Treatment Target	Observed Experimental Effects
Gao et al., 2013 [40]	MCP, ginseng pectin fractions, potato galactans, and RG-I	In vitro	HT-29 human colon cancer cell line	RG I-4 from ginseng strongly inhibited Gal-3 mediated hemagglutination; better inhibition of cell adhesion and homotypic cell aggregation than lactose
Stegmayr et al., 2016 [50]	MCP	In vitro	JIMT-1 breast cancer cells	No Gal-3 inhibition was detected; however, MCP pre-incubation resulted in the accumulation of Gal-3 molecules around intracellular vesicles
Prado et al., 2020 [73]	Papaya pectins from different ripening periods	In vitro	THP-1 human monocytic cell	Different TLR's activation and inhibition depend on the ripening period
Hu et al., 2020 [85]	Lemon pectin	In vitro	Human pancreatic beta-cell	Unspecific and unspecified reduction of deleterious effects of inflammatory cytokines with very low (5%) degree of esterification pectin at cell culture
Xu et al., 2020 [86]	MCP	In vivo	Male Wistar rats	Down-regulation of Gal-3, TLR, and MyD88, decreased expression of IL-1β, IL-18, and TNF-α
Maxwell et al., 2016 [99]	Sugar beet and CP	In vitro	HT-29 human colon cancer cell line	Cell proliferation control and induction of apoptosis
Pynam and Dharmesh, 2019 [101]	Bael fruit pectin fragments	In vitro and in vivo	Healthy Swiss albino mice and B16F10 cell line	Microbiota protection, tyrosinase down-regulation, Gal-3 binding, downregulation of Gal-3 gene, IL10 and IL17 cytokines
Fang et al., 2018 [103]	MCP	In vitro	Human urinary bladder cancer (UBC) cells	Gal-3 down-regulation and inactivation of Akt signaling pathway, a decrease in Cyclin B1, G2/M phase arrest, Caspase-3 activation
Hossein et al., 2019 [104]	MCP	In vitro	SKOV-3 and SOC (serous ovarian cancer) cells	Synergistic effect of PTX and MCP increasing caspase-3 activity and decreasing cyclin D1 expression level
Abu-Elsaad and Elkashef, 2016 [105]	MCP	In vivo	Adult male Sprague-Dawley rats	Decreased liver fibrosis and necroinflammation, a decrease in MDA, TIMP-1, Col1A1, and Gal-3, increase in Caspase-3, gluthatione, and superoxide dismutase expression

Table 2. Cont.

Authors	Treatment	Study Type	Treatment Target	Observed Experimental Effects
Martinez-Martinez et al., 2016 [106]	MCP	In vivo	Adult male Wistar rats	Attenuation of renal fibrosis-related biomarkers, osteopontin, cytokine A2, albuminuria and TGF-β1
Calvier et al., 2015 [109]	MCP	In vivo	Adult male Wistar rats, C57BJ6 WT and Gal-3$^{-/-}$ mice	Reverted fibrosing markers and Gal-3 augmentation levels, similarly to spironolactone
Li et al., 2018 [110]	MCP	In vitro and in vivo	HEK293 cells and C57BL/6 male mice	Amelioration of renal interstitial fibrosis, lower collagen I and fibronectin in the kidney, reduced IL-1β mRNA levels, lower Gal-3 expression
Prud'homme et al., 2019 [111]	MCP	Cohort and in vivo	C57BL6/J and C57BL6/J Gal-3 KO male mice	Cardiac fibrosis induced by model prevented by MCP treatment, IL-1β level maintained, protected, treated mice against renal inflammation
Ibarrola et al., 2019 [112]	MCP	In vivo	Male Wistar rats	BNP serum level normalization, lower Gal-3 cardiac expression, reticulocalbin-3 and fumarase in the myocardium, IL-1β and CRP in serum
Li et al., 2019 [113]	MCP and perindopril	In vivo	New Zealand male rabbits	Gal-3, collagen I, and III downregulation
Vergaro et al., 2016 [114]	MCP	In vivo	Transgenic mice with aldosterone synthase gene overexpression	Reduced cardiac hypertrophy, fibrosis, Coll-1, and Coll-3 genes expression and also enhanced anti-inflammatory and anti-fibrotic effects when synergistically acting with Canrenoate
Ibarrola et al., 2017 [115]	MCP	In vivo	Male Wistar rats	Gal-3, mRNA expression normalized, collagen I, fibronectin, α-SMA, TGF-β1, and CTGF mRNA expression reduced compared to pressure overload group, vascular inflammatory markers expression was also controlled

Table 2. Cont.

Authors	Treatment	Study Type	Treatment Target	Observed Experimental Effects
Xue et al., 2019 [116]	Ginseng pectin fractions	In vitro and In vivo	Jurkat (human leukemia cells) and male IRC mice	MCP inhibited IL-2 expression, and the three pectin fractions utilized reversed cleaved caspase-3 formation alongside lactose. MCP and ginseng pectins inhibited ROS production in vitro. Reduced tumor weight and increased IL-2 secretion in vivo
Lau et al., 2021 [117]	MCP	Interventional trial	Participants with high Gal-3 levels and hypertension	MCP had no impact regarding attenuating of cardiac-related risk factors
Busato et al., 2020 [122]	Broccoli stalks pectin	In vitro and in vivo	Female albino swiss mice and peritoneal exsudate cells	Macrophage activation and higher phagocytic activity; IL-10 presence was higher at peritoneal fluid in vivo, but not at in vitro model
Liu et al., 2008 [125]	MCP	In vitro and in vivo	CT-26 cells and Balb/c female mice	MCP did not alter Gal-3 expression at metastatic liver cells, although it did inhibit tumor growth and metastatic rate
Courts, 2013 [126]	MCP	In vitro	Caco-2 monolayer	MCP fragments were absorbed through paracellular and to a lower degree by transcellular transports at in vitro culture
Huang et al., 2012 [127]	Enzyme-treated CP	In vitro and In vivo	HepG2, A549, Colo 205, and HEK293 cells, BALB/c mice	Altered membrane permeability (LDH release) in the cancer cell lines; low weight oligogalacturonide was absorbed by the mice and the tumor cells, enhancing Gal-3 release to the medium
Fan et al., 2018 [129]	Ginseng RG-I enriched pectins	In vitro	L-929 fibroblast cells	Modulation of cell migration and adhesion, independent of Gal-3
Nishikawa et al., 2018 [130]	Modified citrus pectin (MCP)	In vivo	Male C57BL/6 mice	Attenuated blood-brain barrier disruption Gal-3 upregulation, inactivation of ERK 1/2, STAT and MMP
Sivaprakasam et al., 2016 [143]	2% inulin, 2% pectin, and 1% cellulose	In vivo	Human colon cancer tissue and Ffar-2$^{-/-}$ C57BL/6J mice	Microbiota modulation, promotion of *Bifidobacterium* growth, and reduction of *Prevotellaceae*

Table 2. Cont.

Authors	Treatment	Study Type	Treatment Target	Observed Experimental Effects
Kim et al., 2013 [144]	SCFAs	In vivo	WT, GPR41$^{-/-}$ and GPR43$^{-/-}$ mice	Activation of intestinal epithelial cells to produce chemokines and cytokines, GPR's were essential in T effector cell activation and signaling pathways
Tian et al., 2016 [146]	Sugar beet, soy, low DM, and high DM citrus pectin	In vivo	Male Wistar rats	More stimulation of *Lactobacillus* and *Lachnospiraceae* growth in sugar beet pectin, higher production of SCFA's for low DM citrus and soy pectin
Tian et al., 2017 [147]	Low DM and high DM citrus pectin	In vivo	Piglets	The slower fermentation process, alteration of main fermentation region, and higher Bacteroidetes predominance
Ferreira-Lazarte et al., 2019 [148]	CP	In vitro	Dynamic gastric simulator with healthy volunteer fecal slurry donated	Growth stimulation of *Bifidobacterium* spp., *Bacteroides* spp., and *Faecalobacterium prausnitzii*, high increase in acetate and butyrate production
Chen et al., 2013 [149]	Apple pectin oligosaccharides	In vitro	Fecal batch culture fermentation	Increased numbers of *Lactobacillus* and *Bifidobacteria*, a higher concentration of acetic, lactic, and propionic acid decreased number of Clostridia and Bacteroides
Onumpai et al., 2011 [150]	Potato galactan, methylated citrus pectin, beet arabinan, *Arabidopsis thaliana* RG-I	In vitro	Fecal batch culture fermentation	Higher *Bifidobacterium* populations and higher SCFA's yield increased *Bacteroides-Prevotella* groups
Merheb, Abdel-Massih, and Karam, 2019 [153]	CP and MCP	In vivo	Female BALB/c mice	Upregulation of IL-17, IFN-γ, and TNF-α through IL-4 cytokine secretion in the spleen
Amorim et al., 2016 [154]	*Theobroma cacao* pod husk modified pectin	In vivo	Female albino Swiss mice	Promotion of macrophage differentiation, nitric oxide production, and upregulation of IL-12, TNF-α, and IL-10 secretion
Do Nascimento et al., 2017 [155]	Sweet pepper pectin	In vitro	THP-1 human monocytic cell	Modulation of TNF-α, IL-1β, and IL-10 production and secretion
Popov et al., 2011 [156]	Sweet pepper pectin	In vivo	Male BALB/c mice	Higher IL-10 production with lower TNF-α release

Table 2. Cont.

Authors	Treatment	Study Type	Treatment Target	Observed Experimental Effects
Ishisono et al., 2017 [157]	CP	In vivo	Male C57BL/6 mice	Suppression of IL-6 secretion from TLR activated macrophages and CD11c$^+$ cells
Vogt et al., 2016 [158]	Different DM lemon pectin	In vitro	T84 intestinal epithelial cells	NF-kB/AP-1 activation through TLR/MyD88 and protective effects in the intestinal barrier
Wang et al., 2018 [159]	*Hippophae rhamnoides* L. berries pectin	In vivo	Cyclophosphamide induced immunosuppressive mice	Macrophage activation, MyD88 increased expression and upregulated expression of TLR4
Park et al., 2013 [160]	RG-II from P. ginseng	In vivo and In vitro	C57BL6 WT, TCR KO, TLR KO mice, and BMDC cells	Facilitation of CD8$^+$ T cells, induced production of TNF-α, IL-12, IFN-γ, and IL-1β during dendritic cell maturation
Sahasrabudhe et al., 2018 [161]	Lemon pectins with different DM	In vitro and In vivo	HEK-Blue WT and mutated cell lines, female C57BL/6 mice	Inhibition of TLR2-1 heterodimer, prevention of ileitis in the mice model
Hu et al., 2021 [162]	Lemon pectins with different DM	In vivo	Sprague-Dawley male rats and C57BL/6 mice	Reduced peri-capsular fibrosis in vivo and decreased DAMP-induced TLR2 immune activation in vitro
Kolatsi-Jannou et al., 2011 [163]	MCP	In vivo	Male C57BL/6J mice	Reduced Gal-3 expression, reduced renal cell proliferation, apoptosis, fibrosis, and proinflammatory cytokine expression

6. Should Gal-3 Inhibition Be the Main Biological Effect Expected from Pectin?

Although the potential Gal-3 inhibition achieved by pectin could be a resourceful knowledge area, there is a great extent of literature showing parallel ways. Wild-type and Gal-3 knockout (Gal-3$^{-/-}$) HCT-116 (human) cells were exposed to different papaya pectin fractions in vitro. Specifically, the most acidic fraction (uronic fraction from fourth day after harvest), with a mean molecular weight of 128 kDa and high antibody reactivity to LM5 and LM16 (1,4-β-galactan and type-1-rhamnogalacturonan, respectively), kept a slightly lower efficacy at the Gal-3$^{-/-}$ cells, suggesting that even though the cancer cell viability decrease could be in part due to Gal-3 inhibition, it was not the only molecular modulated pathway [7]. Commercially available CP and MCP demonstrated to have a pro-inflammatory action independent of Gal-3 inhibition, upregulating cytokine secretion in the spleen of BALB/c mice, IL-17, IFN-γ, and TNF-α through IL-4 [153]. Modified pectin obtained from *Theobroma cacao* pod husks, highly composed of uronic acids, galactose, and rhamnose, with a low degree of methylation and amidation, was also attributed with a pro-inflammatory profile, similar to LPS stimulation in isolated macrophages from mice, upregulating secretion of IL-12 and TNF-α, although stimulating the anti-inflammatory IL-10 simultaneously. The pectins, however, did not enhance the phagocytic activity of the peritoneal macrophages [154]. Similar effects were detected in differentiated macrophage (THP-1) cell cultures that were in contact with native sweet pepper pectin. The polysaccharide was characterized as pectin composed of uronic acids, galactose, and arabinose, also confirmed through NMR with signals of methyl and acetyl groups linked to α-D-Gal$p$A with a high degree of methylation (85%) and low degree of acetylation (5%). There were also signals of (1→4)-linked-β-D-Gal$p$ units, which the authors attributed to type-1-arabinogalactans. The sweet pepper pectin induced in vitro TNF-α, IL-1β, and IL-10 secretion at the highest concentration used [155]. After modifying the native structure and removing its side chains by partial acid hydrolysis, the composition was uronic acids (91%) and rhamnose (9%). The respective signals identified in the native molecule related to galactan core residues had disappeared, with also a great reduction in DM (down to 17%). This modified sweet pepper pectin was still able to induce—at a lower rate than native pectin—the TNF-α and IL-10 secretion but at a higher rate the IL-1β [155]. Here, it is again highlighted the importance of the structure-dependence of pectic fractions with different potential targets. In earlier studies, sweet pepper pectin (1,4-α-D-galacturonan partially substituted with methyl and O-acetyl backbone) and low methoxyl CP performed similarly by lowering TNF-α and enhancing IL-10 secretion, which resulted in ameliorated survival rate in endotoxin-shock induced mice models [102,156]. Elsewhere, CP was also capable of reducing LPS-induced hypothermia and inflammatory cytokine gene expression, suppressing IL-6 production and TLR-4 signaling in vivo [157].

Toll-like receptors (TLR) have been thoroughly explored, especially the dichotomy of weighing between agonists or antagonists for cancer treatment, where the same TLR can exhibit anti or pro-tumor immune responses [164–166]. Lemon pectin ranging from 40 to 100 kDa and from 30 to 74% of DM, activated in vitro THP-1 phagocytic cells depending on *MyD88* in a TLR-mediated manner [158]. Additionally, the pectin DM and its structural backbone were correlated to NF-κB/AP-1 activation through TLR, where highly esterified polymers were strong activators, and their oligomers (produced after extensive hydrolytic processing) did not perform equally [158]. Similarly, ginseng polysaccharide extract composed of RG-II was a TLR-4 up-regulator and MyD88 activator in a structure-dependent manner in vivo [116]. A polysaccharide extracted from a European berry (*Hipphoides rhamnoides* L.) with 85% DM and consisting of repeating units of (1→4)-β-D-galactopyranosyluronic acid residues strongly stimulated TLR-4 [159,167]. The blockage of TLR4/MyD88 was used by authors to explore the interaction between the pectic structures and macrophages, where it inhibited the increase observed in nitric oxide and other cytokines induced by treatment with the polysaccharides [167]. Ginseng RG-II stimulated TLR-4, increasing dendritic cell maturation and activation through induced cytokine and mitogen-induced protein kinases (MAPKs) production. This cytotoxic T cell response inhibited tumor growth of EG7 lymphoma cells [160]. Interestingly, the ripening process for some sources of pectin, such as papaya, can be of utmost importance regarding main structural changes to effectively interact with diverse TLR receptors [73]. The authors demonstrated that although TLR-2 and -4 were activated by pectins isolated from unripe and ripe papayas, TLR-3, -5, and -9 were not activated by the pectins isolated from unripe fruits in two different time points (pectins with 580 and 610 kDa, higher galactose and glucose contents, alongside with a proportionally lower composition of GalA residues; the second unripe point had 15% DM). Specifically, TLR-3 and -9 could be inhibited by pectins isolated from unripe papayas due to high molecular weight structures [73], similar to what was previously observed with lemon pectins that inhibited the TLR-2 heterodimer formation with TLR-1, but not with TLR-6 [161]. Citrus pectin ranging from 18 to 69% DM (but without further structural details or suggestions) was able to inhibit TLR/MyD88 by oral administration in mice and decreased TLR-2 mediated immune response in rats when administered both in alginate microcapsules and directly at drinking water [162]. Administration of pectin samples and usage of checkpoint inhibitors could be powerful measures to assure immune responses in certain types of cancer [168]. Neohesperidin nanoliposomes incorporated with citrus pectin (65% DM) and chitosan (50 kDa, degree of deacetylation of 85%) had higher cellular uptake rates in comparison to chitosan or neohesperidin single treatments [169]. A nephropathy murine model with MCP added in drinking water found protective effects of the pectin sample independent of Gal-3 inhibition at early proliferation, but with Gal-3 downregulation later on [163]. An overall scheme of the possible biological effects of pectins that were discussed throughout the paper are depicted in Figure 6 using an intestine model for easier comprehension, and the different biological effects observed in in vitro, in vivo, and cohort/clinical studies are summarized as a table (Table 2).

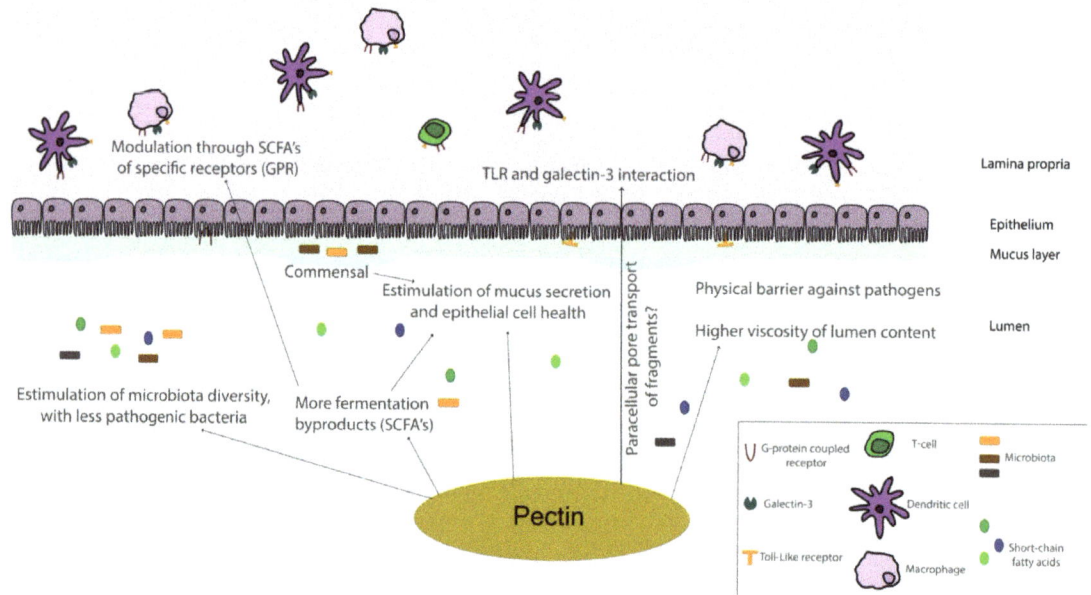

Figure 6. Schematic representation of the intestinal environment. Pectin molecules can interact in different ways with epithelial and immune components of intestinal tissues, regulating different responses directly and through fermentation by-products. SCFA—short-chain fatty acid; TLR—toll-like receptor.

Even though extensive human model studies are needed to confirm the positive outcomes of pectin ingestion regarding specific targets, the data obtained until now can help to demonstrate a variety of positive health-regulating effects of pectins. The pectin binding and inhibition of Gal-3 is one of them; however, this relationship is still highly controversial and contested. Human Gal-3 inhibition and the consequent beneficial effects on human health must be covered, from in vitro experiments to human clinical trials, passing through the bioavailability assays of pectin-derived fragments in the human body. The plausible wider range of usage for the polysaccharides is highlighted in this revision, enhancing the discussion of the inclusion of pectin molecules for synergistic effects with different molecules and drugs. Moreover, pectin could exert direct and indirect immunomodulatory effects, and their fragments could interact with different types of receptors. The fermentation products of pectins also help to sustain the intestinal and systemic environment, with all these possible beneficial effects of pectins being potentially achieved through supplementation by oral intake.

7. Conclusions

Targeted galectin-3 binding sites for therapeutic approaches are diverse. Pectin is an important food component classified as soluble dietary fiber. Its biological effects on human health go from colonic fermentation and microbiota modulation to potential direct interaction with intestinal cells and proteins, such as TLRs and Gal-3. Many studies can effectively suggest the multi-way interaction between pectin molecules and these ligands. The Gal-3 binding is suggested in the CRD motif, both through its F-face and S-face sites. Furthermore, the binding can occur in the N-terminal tail and even in a two-step interaction method, promoting subtle motif variations in the protein molecule and therefore enhancing an adequate interconnection with the pectin molecules. However, the biological effects of pectin transcend Gal-3 interaction and/or inhibition, which is far from being an established point, with several challenges to be overcome, and has undergone valid confrontations. All the literature and methodology improvements converge on the pectin diversity for enhancing human health. Moreover, pectin molecules exerting distinct regulatory/inhibitory effects, low side effects, and natural sources are interesting bioactive components to be added into dietary supplements. These pectin molecules could be used continuously to increase the natural intake of bioactive polysaccharides, e.g., in post-cardiac arrest and renal fibrosis pathologies, as well as to a great extent as an auxiliary factor in chemotherapy and possibly as immunomodulatory molecules. Extensive additional research is needed before confirming any of those promising illustrated scenarios.

Author Contributions: L.d.F.P.: Conceptualization, Data curation, and Writing (Original draft and Review and Editing). A.R.: Writing (Review and Editing). J.P.F.: Conceptualization, Supervision, and Writing (Review and Editing). All authors have read and agreed to the published version of the manuscript.

Funding: We thank São Paulo Research Foundation (FAPESP—Brazil; #2016/19657-8 and #2020/08063-5) and the Brazilian National Council for Scientific and Technological Development (CNPq; #142112/2019-4) for Lucas de Freitas Pedrosa scholarships. The study was financially supported by grants #2012/23970-2, #2013/07914-8, #2019/11816-8, and #2021/06419-0 from the São Paulo Research Foundation (FAPESP).

Conflicts of Interest: The authors declare no conflict of interest.

References

1. Gawkowska, D.; Cybulska, J.; Zdunek, A. Structure-related gelling of pectins and linking with other natural compounds: A review. *Polymers* **2018**, *10*, 762. [CrossRef]
2. Maxwell, E.G.; Belshaw, N.J.; Waldron, K.W.; Morris, V.J. Pectin—An emerging new bioactive food polysaccharide. *Trends Food Sci. Technol.* **2012**, *24*, 64–73. [CrossRef]
3. Cui, L.; Wang, J.; Huang, R.; Tan, Y.; Zhang, F.; Zhou, Y.; Sun, L. Analysis of pectin from *Panax ginseng* flower buds and their binding activities to galectin-3. *Int. J. Biol. Macromol.* **2019**, *128*, 459–467. [CrossRef] [PubMed]
4. Mohnen, D. Pectin structure and biosynthesis. *Curr. Opin. Plant Biol.* **2008**, *11*, 266–277. [CrossRef] [PubMed]
5. Fabi, J.P.; Seymour, G.B.; Graham, N.S.; Broadley, M.R.; May, S.T.; Lajolo, F.M.; Cordenunsi, B.R.; Oliveira do Nascimento, J.R. Analysis of ripening-related gene expression in papaya using an *Arabidopsis*-based microarray. *BMC Plant Biol.* **2012**, *12*, 242. [CrossRef] [PubMed]
6. Fabi, J.P.; Broetto, S.G.; da Silva, S.L.G.L.; Zhong, S.; Lajolo, F.M.; do Nascimento, J.R.O. Analysis of papaya cell wall-related genes during fruit ripening indicates a central role of polygalacturonases during pulp softening. *PLoS ONE* **2014**, *9*, e105685. [CrossRef] [PubMed]
7. De Freitas Pedrosa, L.; Lopes, R.G.; Fabi, J.P. The acid and neutral fractions of pectins isolated from ripe and overripe papayas differentially affect galectin-3 inhibition and colon cancer cell growth. *Int. J. Biol. Macromol.* **2020**, *164*, 2681–2690. [CrossRef] [PubMed]
8. Do Prado, S.B.R.; Ferreira, G.F.; Harazono, Y.; Shiga, T.M.; Raz, A.; Carpita, N.C.; Fabi, J.P. Ripening-induced chemical modifications of papaya pectin inhibit cancer cell proliferation. *Sci. Rep.* **2017**, *7*, 16564. [CrossRef] [PubMed]
9. Varki, A.; Cummings, R.D.; Aebi, M.; Packer, N.H.; Seeberger, P.H.; Esko, J.D.; Stanley, P.; Hart, G.; Darvill, A.; Kinoshita, T.; et al. Symbol nomenclature for graphical representations of glycans. *Glycobiology* **2015**, *25*, 1323–1324. [CrossRef]
10. Chen, Y.; Xu, C.; Huang, R.; Song, J.; Li, D.; Xia, M. Butyrate from pectin fermentation inhibits intestinal cholesterol absorption and attenuates atherosclerosis in apolipoprotein E-deficient mice. *J. Nutr. Biochem.* **2018**, *56*, 175–182. [CrossRef]
11. Li, W.; Zhang, K.; Yang, H. Pectin Alleviates High Fat (Lard) Diet-Induced Nonalcoholic Fatty Liver Disease in Mice: Possible Role of Short-Chain Fatty Acids and Gut Microbiota Regulated by Pectin. *J. Agric. Food Chem.* **2018**, *66*, 8015–8025. [CrossRef] [PubMed]
12. Brouns, F.; Theuwissen, E.; Adam, A.; Bell, M.; Berger, A.; Mensink, R.P. Cholesterol-lowering properties of different pectin types in mildly hyper-cholesterolemic men and women. *Eur. J. Clin. Nutr.* **2012**, *66*, 591–599. [CrossRef]
13. Liu, Y.; Dong, M.; Yang, Z.; Pan, S. Anti-diabetic effect of citrus pectin in diabetic rats and potential mechanism via PI3K/Akt signaling pathway. *Int. J. Biol. Macromol.* **2016**, *89*, 484–488. [CrossRef] [PubMed]

14. Fotschki, B.; Jurgoński, A.; Juśkiewicz, J.; Kołodziejczyk, K.; Sójka, M. Effects of dietary addition of a low-pectin apple fibre preparation on rats. *Pol. J. Food Nutr. Sci.* **2014**, *64*, 193–199. [CrossRef]
15. Kunzmann, A.T.; Coleman, H.G.; Huang, W.-Y.; Kitahara, C.M.; Cantwell, M.M.; Berndt, S.I. Dietary fiber intake and risk of colorectal cancer and incident and recurrent adenoma in the Prostate, Lung, Colorectal, and Ovarian Cancer Screening Trial. *Am. J. Clin. Nutr.* **2015**, *102*, 881–890. [CrossRef] [PubMed]
16. Song, L.; Tang, J.-W.; Owusu, L.; Sun, M.-Z.; Wu, J.; Zhang, J. Galectin-3 in cancer. *Clin. Chim. Acta* **2014**, *431*, 185–191. [CrossRef] [PubMed]
17. Sciacchitano, S.; Lavra, L.; Morgante, A.; Ulivieri, A.; Magi, F.; De Francesco, G.P.; Bellotti, C.; Salehi, L.B.; Ricci, A. Galectin-3: One molecule for an alphabet of diseases, from A to Z. *Int. J. Mol. Sci.* **2018**, *19*, 379. [CrossRef]
18. Fortuna-Costa, A.; Gomes, A.M.; Kozlowski, E.O.; Stelling, M.P.; Pavão, M.S.G. Extracellular galectin-3 in tumor progression and metastasis. *Front. Oncol.* **2014**, *4*. [CrossRef]
19. Suthahar, N.; Meijers, W.C.; Silljé, H.H.W.; Ho, J.E.; Liu, F.-T.; de Boer, R.A. Galectin-3 activation and inhibition in heart failure and cardiovascular disease: An update. *Theranostics* **2018**, *8*, 593–609. [CrossRef]
20. Filipová, M.; Bojarová, P.; Rodrigues Tavares, M.; Bumba, L.; Elling, L.; Chytil, P.; Gunár, K.; Křen, V.; Etrych, T.; Janoušková, O. Glycopolymers for Efficient Inhibition of Galectin-3: In Vitro Proof of Efficacy Using Suppression of T Lymphocyte Apoptosis and Tumor Cell Migration. *Biomacromolecules* **2020**, *21*, 3122–3133. [CrossRef]
21. Jia, J.; Claude-Taupin, A.; Gu, Y.; Choi, S.W.; Peters, R.; Bissa, B.; Mudd, M.H.; Allers, L.; Pallikkuth, S.; Lidke, K.A.; et al. Galectin-3 Coordinates a Cellular System for Lysosomal Repair and Removal. *Dev. Cell* **2020**, *52*, 69–87.e8. [CrossRef]
22. Maxwell, E.G.; Colquhoun, I.J.; Chau, H.K.; Hotchkiss, A.T.; Waldron, K.W.; Morris, V.J.; Belshaw, N.J. Rhamnogalacturonan i containing homogalacturonan inhibits colon cancer cell proliferation by decreasing ICAM1 expression. *Carbohydr. Polym.* **2015**, *132*, 546–553. [CrossRef] [PubMed]
23. Wu, K.-L.; Huang, E.-Y.; Jhu, E.-W.; Huang, Y.-H.; Su, W.-H.; Chuang, P.-C.; Yang, K.D. Overexpression of galectin-3 enhances migration of colon cancer cells related to activation of the K-Ras-Raf-Erk1/2 pathway. *J. Gastroenterol.* **2013**, *48*, 350–359. [CrossRef] [PubMed]
24. Song, S.; Ji, B.; Ramachandran, V.; Wang, H.; Hafley, M.; Logsdon, C.; Bresalier, R.S. Overexpressed galectin-3 in pancreatic cancer induces cell proliferation and invasion by binding ras and activating ras signaling. *PLoS ONE* **2012**, *7*, e42699. [CrossRef]
25. Margadant, C.; Van Den Bout, I.; Van Boxtel, A.L.; Thijssen, V.L.; Sonnenberg, A. Epigenetic regulation of galectin-3 expression by β1 integrins promotes cell adhesion and migration. *J. Biol. Chem.* **2012**, *287*, 44684–44693. [CrossRef] [PubMed]
26. Wang, W.; Guo, H.; Geng, J.; Zheng, X.; Wei, H.; Sun, R.; Tian, Z. Tumor-released galectin-3, a soluble inhibitory ligand of human NKp30, plays an important role in tumor escape from NK cell attack. *J. Biol. Chem.* **2014**, *289*, 33311–33319. [CrossRef] [PubMed]
27. Voss, J.J.L.P.; Ford, C.A.; Petrova, S.; Melville, L.; Paterson, M.; Pound, J.D.; Holland, P.; Giotti, B.; Freeman, T.C.; Gregory, C.D. Modulation of macrophage antitumor potential by apoptotic lymphoma cells. *Cell Death Differ.* **2017**, *24*, 971–983. [CrossRef] [PubMed]
28. Xue, H.; Liu, L.; Zhao, Z.; Zhang, Z.; Guan, Y.; Cheng, H.; Zhou, Y.; Tai, G. The N-terminal tail coordinates with carbohydrate recognition domain to mediate galectin-3 induced apoptosis in T cells. *Oncotarget* **2017**, *8*, 49824–49838. [CrossRef]
29. Freichel, T.; Heine, V.; Laaf, D.; Mackintosh, E.E.; Sarafova, S.; Elling, L.; Snyder, N.L.; Hartmann, L. Sequence-Defined Heteromultivalent Precision Glycomacromolecules Bearing Sulfonated/Sulfated Nonglycosidic Moieties Preferentially Bind Galectin-3 and Delay Wound Healing of a Galectin-3 Positive Tumor Cell Line in an In Vitro Wound Scratch Assay. *Macromol. Biosci.* **2020**, *20*, 2000163. [CrossRef]
30. Laaf, D.; Bojarová, P.; Elling, L.; Křen, V. Galectin—Carbohydrate Interactions in Biomedicine and Biotechnology. *Trends Biotechnol.* **2019**, *37*, 402–415. [CrossRef]
31. Rajput, V.K.; MacKinnon, A.; Mandal, S.; Collins, P.; Blanchard, H.; Leffler, H.; Sethi, T.; Schambye, H.; Mukhopadhyay, B.; Nilsson, U.J. A Selective Galactose-Coumarin-Derived Galectin-3 Inhibitor Demonstrates Involvement of Galectin-3-glycan Interactions in a Pulmonary Fibrosis Model. *J. Med. Chem.* **2016**, *59*, 8141–8147. [CrossRef] [PubMed]
32. Do Prado, S.B.R.; Santos, G.R.C.; Mourão, P.A.S.; Fabi, J.P. Chelate-soluble pectin fraction from papaya pulp interacts with galectin-3 and inhibits colon cancer cell proliferation. *Int. J. Biol. Macromol.* **2019**, *126*, 170–178. [CrossRef] [PubMed]
33. Wu, D.; Zheng, J.; Hu, W.; Zheng, X.; He, Q.; Linhardt, R.J.; Ye, X.; Chen, S. Structure-activity relationship of Citrus segment membrane RG-I pectin against Galectin-3: The galactan is not the only important factor. *Carbohydr. Polym.* **2020**, *245*, 116526. [CrossRef]
34. Zhang, T.; Lan, Y.; Zheng, Y.; Liu, F.; Zhao, D.; Mayo, K.H.; Zhou, Y.; Tai, G. Identification of the bioactive components from pH-modified citrus pectin and their inhibitory effects on galectin-3 function. *Food Hydrocoll.* **2016**, *58*, 113–119. [CrossRef]
35. Do Nascimento Oliveira, A.; de Almeida Paula, D.; de Oliveira, E.B.; Saraiva, S.H.; Stringheta, P.C.; Ramos, A.M. Optimization of pectin extraction from Ubá mango peel through surface response methodology. *Int. J. Biol. Macromol.* **2018**, *113*, 395–402. [CrossRef]
36. Chan, S.-Y.; Choo, W.-S. Effect of extraction conditions on the yield and chemical properties of pectin from cocoa husks. *Food Chem.* **2013**, *141*, 3752–3758. [CrossRef]
37. Pereira, P.H.F.; Oliveira, T.Í.S.; Rosa, M.F.; Cavalcante, F.L.; Moates, G.K.; Wellner, N.; Waldron, K.W.; Azeredo, H.M.C. Pectin extraction from pomegranate peels with citric acid. *Int. J. Biol. Macromol.* **2016**, *88*, 373–379. [CrossRef]

38. Oliveira, T.Í.S.; Rosa, M.F.; Cavalcante, F.L.; Pereira, P.H.F.; Moates, G.K.; Wellner, N.; Mazzetto, S.E.; Waldron, K.W.; Azeredo, H.M.C. Optimization of pectin extraction from banana peels with citric acid by using response surface methodology. *Food Chem.* **2016**, *198*, 113–118. [CrossRef]
39. Srivastava, P.; Malviya, R. Sources of pectin, extraction and its applications in pharmaceutical industry—An overview. *Indian J. Nat. Prod. Resour.* **2011**, *2*, 10–18.
40. Gao, X.; Zhi, Y.; Sun, L.; Peng, X.; Zhang, T.; Xue, H.; Tai, G.; Zhou, Y. The inhibitory effects of a rhamnogalacturonan I (RG-I) domain from ginseng Pectin on galectin-3 and its structure-activity relationship. *J. Biol. Chem.* **2013**, *288*, 33953–33965. [CrossRef]
41. Leclere, L.; Van Cutsem, P.; Michiels, C. Anti-cancer activities of pH- or heat-modified pectin. *Front. Pharmacol.* **2013**, *4*. [CrossRef] [PubMed]
42. Leclere, L.; Fransolet, M.; Cote, F.; Cambier, P.; Arnould, T.; Van Cutsem, P.; Michiels, C. Heat-modified citrus pectin induces apoptosis-like cell death and autophagy in HepG2 and A549 cancer cells. *PLoS ONE* **2015**, *10*, e0115831. [CrossRef]
43. Xu, Y.; Zhang, L.; Bailina, Y.; Ge, Z.; Ding, T.; Ye, X.; Liu, D. Effects of ultrasound and/or heating on the extraction of pectin from grapefruit peel. *J. Food Eng.* **2014**, *126*, 72–81. [CrossRef]
44. Wang, W.; Ma, X.; Xu, Y.; Cao, Y.; Jiang, Z.; Ding, T.; Ye, X.; Liu, D. Ultrasound-assisted heating extraction of pectin from grapefruit peel: Optimization and comparison with the conventional method. *Food Chem.* **2015**, *178*, 106–114. [CrossRef] [PubMed]
45. Wang, W.; Chen, W.; Zou, M.; Lv, R.; Wang, D.; Hou, F.; Feng, H.; Ma, X.; Zhong, J.; Ding, T.; et al. Applications of power ultrasound in oriented modification and degradation of pectin: A review. *J. Food Eng.* **2018**, *234*, 98–107. [CrossRef]
46. Zouambia, Y.; Ettoumi, K.Y.; Krea, M.; Moulai-Mostefa, N. A new approach for pectin extraction: Electromagnetic induction heating. *Arab. J. Chem.* **2017**, *10*, 480–487. [CrossRef]
47. Chen, J.; Cheng, H.; Zhi, Z.; Zhang, H.; Linhardt, R.J.; Zhang, F.; Chen, S.; Ye, X. Extraction temperature is a decisive factor for the properties of pectin. *Food Hydrocoll.* **2021**, *112*, 106160. [CrossRef]
48. Blanchard, H.; Bum-Erdene, K.; Bohari, M.H.; Yu, X. Galectin-1 inhibitors and their potential therapeutic applications: A patent review. *Expert Opin. Ther. Pat.* **2016**, *26*, 537–554. [CrossRef] [PubMed]
49. Laaf, D.; Bojarová, P.; Pelantová, H.; Křen, V.; Elling, L. Tailored Multivalent Neo-Glycoproteins: Synthesis, Evaluation, and Application of a Library of Galectin-3-Binding Glycan Ligands. *Bioconjugate Chem.* **2017**, *28*, 2832–2840. [CrossRef] [PubMed]
50. Stegmayr, J.; Lepur, A.; Kahl-Knutson, B.; Aguilar-Moncayo, M.; Klyosov, A.A.; Field, R.A.; Oredsson, S.; Nilsson, U.J.; Leffler, H. Low or No Inhibitory Potency of the Canonical Galectin Carbohydrate-binding Site by Pectins and Galactomannans. *J. Biol. Chem.* **2016**, *291*, 13318–13334. [CrossRef]
51. Gunning, A.P.; Bongaerts, R.J.M.; Morris, V.J. Recognition of galactan components of pectin by galectin-3. *FASEB J.* **2009**, *23*, 415–424. [CrossRef] [PubMed]
52. Shi, H.; Yu, L.; Shi, Y.; Lu, J.; Teng, H.; Zhou, Y.; Sun, L. Structural characterization of a rhamnogalacturonan I domain from ginseng and its inhibitory effect on galectin-3. *Molecules* **2017**, *22*, 1016. [CrossRef] [PubMed]
53. Zhou, L.; Ma, P.; Shuai, M.; Huang, J.; Sun, C.; Yao, X.; Chen, Z.; Min, X.; Zhang, T. Analysis of the water-soluble polysaccharides from *Camellia japonica* pollen and their inhibitory effects on galectin-3 function. *Int. J. Biol. Macromol.* **2020**, *159*, 455–460. [CrossRef] [PubMed]
54. Shao, P.; Wang, P.; Niu, B.; Kang, J. Environmental stress stability of pectin-stabilized resveratrol liposomes with different degree of esterification. *Int. J. Biol. Macromol.* **2018**, *119*, 53–59. [CrossRef] [PubMed]
55. Do Nascimento, G.E.; Simas-Tosin, F.F.; Iacomini, M.; Gorin, P.A.J.; Cordeiro, L.M.C. Rheological behavior of high methoxyl pectin from the pulp of tamarillo fruit (*Solanum betaceum*). *Carbohydr. Polym.* **2016**, *139*, 125–130. [CrossRef] [PubMed]
56. Schmidt, U.S.; Pietsch, V.L.; Rentschler, C.; Kurz, T.; Endreß, H.U.; Schuchmann, H.P. Influence of the degree of esterification on the emulsifying performance of conjugates formed between whey protein isolate and citrus pectin. *Food Hydrocoll.* **2016**, *56*, 1–8. [CrossRef]
57. Schmidt, U.S.; Schütz, L.; Schuchmann, H.P. Interfacial and emulsifying properties of citrus pectin: Interaction of pH, ionic strength and degree of esterification. *Food Hydrocoll.* **2017**, *62*, 288–298. [CrossRef]
58. Jacob, E.M.; Borah, A.; Jindal, A.; Pillai, S.C.; Yamamoto, Y.; Maekawa, T.; Kumar, D.N.S. Synthesis and characterization of citrus-derived pectin nanoparticles based on their degree of esterification. *J. Mater. Res.* **2020**, *35*, 1514–1522. [CrossRef]
59. Wan, L.; Chen, Q.; Huang, M.; Liu, F.; Pan, S. Physiochemical, rheological and emulsifying properties of low methoxyl pectin prepared by high hydrostatic pressure-assisted enzymatic, conventional enzymatic, and alkaline de-esterification: A comparison study. *Food Hydrocoll.* **2019**, *93*, 146–155. [CrossRef]
60. Dranca, F.; Oroian, M. Extraction, purification and characterization of pectin from alternative sources with potential technological applications. *Food Res. Int.* **2018**, *113*, 327–350. [CrossRef]
61. Begum, R.; Yusof, Y.A.; Aziz, M.G.; Uddin, M.B. Structural and functional properties of pectin extracted from jackfruit (*Artocarpus heterophyllus*) waste: Effects of drying. *Int. J. Food Prop.* **2017**, *20*, S190–S201. [CrossRef]
62. Karnik, D.; Wicker, L. Emulsion stability of sugar beet pectin fractions obtained by isopropanol fractionation. *Food Hydrocoll.* **2018**, *74*, 249–254. [CrossRef]
63. Juttulapa, M.; Piriyaprasarth, S.; Takeuchi, H.; Sriamornsak, P. Effect of high-pressure homogenization on stability of emulsions containing zein and pectin. *Asian J. Pharm. Sci.* **2017**, *12*, 21–27. [CrossRef] [PubMed]
64. Petkowicz, C.L.O.; Vriesmann, L.C.; Williams, P.A. Pectins from food waste: Extraction, characterization and properties of watermelon rind pectin. *Food Hydrocoll.* **2017**, *65*, 57–67. [CrossRef]

65. Picot-Allain, M.C.N.; Ramasawmy, B.; Emmambux, M.N. Extraction, Characterisation, and Application of Pectin from Tropical and Sub-Tropical Fruits: A Review. *Food Rev. Int.* **2020**, *38*, 282–312. [CrossRef]
66. Zhang, L.; Ye, X.; Ding, T.; Sun, X.; Xu, Y.; Liu, D. Ultrasound effects on the degradation kinetics, structure and rheological properties of apple pectin. *Ultrason. Sonochemistry* **2013**, *20*, 222–231. [CrossRef]
67. Basanta, M.F.; Ponce, N.M.A.; Rojas, A.M.; Stortz, C.A. Effect of extraction time and temperature on the characteristics of loosely bound pectins from Japanese plum. *Carbohydr. Polym.* **2012**, *89*, 230–235. [CrossRef]
68. Kosmala, M.; Milala, J.; Kołodziejczyk, K.; Markowski, J.; Zbrzeźniak, M.; Renard, C.M.G.C. Dietary fiber and cell wall polysaccharides from plum (*Prunus domestica* L.) fruit, juice and pomace: Comparison of composition and functional properties for three plum varieties. *Food Res. Int.* **2013**, *54*, 1787–1794. [CrossRef]
69. Moreno, L.; Nascimento, R.F.; Zielinski, A.A.F.; Wosiacki, G.; Canteri, M.H.G. Extraction and characterization of pectic substances in *Myrciaria cauliflora* (Jaboticaba sabará) fruit. *Rev. Strict. Sensu* **2016**, *1*, 1–11. [CrossRef]
70. Sayah, M.Y.; Chabir, R.; Benyahia, H.; Kandri, Y.R.; Chahdi, F.O.; Touzani, H.; Errachidi, F. Yield, esterification degree and molecular weight evaluation of pectins isolated from orange and grapefruit peels under different conditions. *PLoS ONE* **2016**, *11*, e0161751. [CrossRef]
71. Hao, M.; Yuan, X.; Cheng, H.; Xue, H.; Zhang, T.; Zhou, Y.; Tai, G. Comparative studies on the anti-tumor activities of high temperature- and pH-modified citrus pectins. *Food Funct.* **2013**, *4*, 960–971. [CrossRef] [PubMed]
72. Do Prado, S.B.R.; Melfi, P.R.; Castro-Alves, V.C.; Broetto, S.G.; Araújo, E.S.; Do Nascimento, J.R.O.; Fabi, J.P. Physiological degradation of pectin in papaya cell walls: Release of long chains galacturonans derived from insoluble fractions during postharvest fruit ripening. *Front. Plant Sci.* **2016**, *7*, 1–11. [CrossRef] [PubMed]
73. Prado, S.B.R.; Beukema, M.; Jermendi, E.; Schols, H.A.; de Vos, P.; Fabi, J.P. Pectin Interaction with Immune Receptors is Modulated by Ripening Process in Papayas. *Sci. Rep.* **2020**, *10*, 1–11. [CrossRef] [PubMed]
74. Flores-Ibarra, A.; Vértesy, S.; Medrano, F.J.; Gabius, H.J.; Romero, A. Crystallization of a human galectin-3 variant with two ordered segments in the shortened N-terminal tail. *Sci. Rep.* **2018**, *8*, 1–11. [CrossRef]
75. Su, J.; Zhang, T.; Wang, P.; Liu, F.; Tai, G.; Zhou, Y. The water network in galectin-3 ligand binding site guides inhibitor design. *Acta Biochim. Biophys. Sin.* **2015**, *47*, 192–198. [CrossRef]
76. Ippel, H.; Miller, M.C.; Vértesy, S.; Zheng, Y.; Cañada, F.J.; Suylen, D.; Umemoto, K.; Romanò, C.; Hackeng, T.; Tai, G.; et al. Intra- and intermolecular interactions of human galectin-3: Assessment by full-assignment-based NMR. *Glycobiology* **2016**, *26*, 888–903. [CrossRef]
77. Kim, S.-J.; Chun, K.-H. Non-classical role of Galectin-3 in cancer progression: Translocation to nucleus by carbohydrate-recognition independent manner. *BMB Rep.* **2020**, *53*, 173–180. [CrossRef]
78. Ruvolo, P.P. Galectin 3 as a guardian of the tumor microenvironment. *Biochim. Biophys. Acta—Mol. Cell Res.* **2016**, *1863*, 427–437. [CrossRef]
79. Chan, Y.-C.; Lin, H.-Y.; Tu, Z.; Kuo, Y.-H.; Hsu, S.-T.D.; Lin, C.-H. Dissecting the structure–Activity relationship of galectin—Ligand interactions. *Int. J. Mol. Sci.* **2018**, *19*, 392. [CrossRef]
80. Sehnal, D.; Bittrich, S.; Deshpande, M.; Svobodová, R.; Berka, K.; Bazgier, V.; Velankar, S.; Burley, S.K.; Koča, J.; Rose, A.S. Mol* Viewer: Modern web app for 3D visualization and analysis of large biomolecular structures. *Nucleic Acids Res.* **2021**, *49*, W431–W437. [CrossRef]
81. Berman, H.M.; Westbrook, J.; Feng, Z.; Gilliland, G.; Bhat, T.N.; Weissig, H.; Shindyalov, I.N.; Bourne, P.E. Protein Data Bank. *Nucleic Acids Res.* **2000**, *28*, 235–242. [CrossRef] [PubMed]
82. Johannes, L.; Jacob, R.; Leffler, H. Galectins at a glance. *J. Cell Sci.* **2018**, *131*, 1–9. [CrossRef] [PubMed]
83. Zhang, T.; Miller, M.C.; Zheng, Y.; Zhang, Z.; Xue, H.; Zhao, D.; Su, J.; Mayo, K.H.; Zhou, Y.; Tai, G. Macromolecular assemblies of complex polysaccharides with galectin-3 and their synergistic effects on function. *Biochem. J.* **2017**, *474*, 3849–3868. [CrossRef] [PubMed]
84. Miller, M.C.; Ippel, H.; Suylen, D.; Klyosov, A.A.; Traber, P.G.; Hackeng, T.; Mayo, K.H. Binding of polysaccharides to human galectin-3 at a noncanonical site in its carbohydrate recognition domain. *Glycobiology* **2015**, *26*, 88–99. [CrossRef]
85. Hu, S.; Kuwabara, R.; Beukema, M.; Ferrari, M.; de Haan, B.J.; Walvoort, M.T.C.; de Vos, P.; Smink, A.M. Low methyl-esterified pectin protects pancreatic β-cells against diabetes-induced oxidative and inflammatory stress via galectin-3. *Carbohydr. Polym.* **2020**, *249*, 116863. [CrossRef]
86. Xu, G.-R.; Zhang, C.; Yang, H.-X.; Sun, J.-H.; Zhang, Y.; Yao, T.-T.; Li, Y.; Ruan, L.; An, R.; Li, A.-Y. Modified citrus pectin ameliorates myocardial fibrosis and inflammation via suppressing galectin-3 and TLR4/MyD88/NF-κB signaling pathway. *Biomed. Pharmacother.* **2020**, *126*, 110071. [CrossRef]
87. Ilmer, M.; Mazurek, N.; Byrd, J.C.; Ramirez, K.; Hafley, M.; Alt, E.; Vykoukal, J.; Bresalier, R.S. Cell surface galectin-3 defines a subset of chemoresistant gastrointestinal tumor-initiating cancer cells with heightened stem cell characteristics. *Cell Death Dis.* **2016**, *7*, 1–9. [CrossRef]
88. Zhang, T.; Zheng, Y.; Zhao, D.; Yan, J.; Sun, C.; Zhou, Y.; Tai, G. Multiple approaches to assess pectin binding to galectin-3. *Int. J. Biol. Macromol.* **2016**, *91*, 994–1001. [CrossRef]
89. Zheng, Y.; Su, J.; Miller, M.C.; Geng, J.; Xu, X.; Zhang, T.; Mayzel, M.; Zhou, Y.; Mayo, K.H.; Tai, G. Topsy-turvy binding of negatively-charged homogalacturonan oligosaccharides to galectin-3. *Glycobiology* **2020**, *31*, 341–350. [CrossRef]

90. Miller, M.C.; Zheng, Y.; Zhou, Y.; Tai, G.; Mayo, K.H. Galectin-3 binds selectively to the terminal, non-reducing end of β(1→4)-galactans, with overall affinity increasing with chain length. *Glycobiology* **2019**, *29*, 74–84. [CrossRef]
91. Zhao, J.; Zhang, F.; Liu, X.; St. Ange, K.; Zhang, A.; Li, Q.; Linhardt, R.J. Isolation of a lectin binding rhamnogalacturonan-I containing pectic polysaccharide from pumpkin. *Carbohydr. Polym.* **2017**, *163*, 330–336. [CrossRef] [PubMed]
92. Miller, M.C.; Zheng, Y.; Yan, J.; Zhou, Y.; Tai, G.; Mayo, K.H. Novel polysaccharide binding to the N-terminal tail of galectin-3 is likely modulated by proline isomerization. *Glycobiology* **2017**, *27*, 1038–1051. [CrossRef] [PubMed]
93. Rapoport, E.M.; Bovin, N.V. Specificity of human galectins on cell surfaces. *Biochemistry* **2015**, *80*, 846–856. [CrossRef] [PubMed]
94. García Caballero, G.; Beckwith, D.; Shilova, N.V.; Gabba, A.; Kutzner, T.J.; Ludwig, A.-K.; Manning, J.C.; Kaltner, H.; Sinowatz, F.; Cudic, M.; et al. Influence of protein (human galectin-3) design on aspects of lectin activity. *Histochem. Cell Biol.* **2020**, *154*, 135–153. [CrossRef]
95. Gao, X.; Zhi, Y.; Zhang, T.; Xue, H.; Wang, X.; Foday, A.D.; Tai, G.; Zhou, Y. Analysis of the neutral polysaccharide fraction of MCP and its inhibitory activity on galectin-3. *Glycoconj. J.* **2012**, *29*, 159–165. [CrossRef]
96. Farhadi, S.A.; Liu, R.; Becker, M.W.; Phelps, E.A.; Hudalla, G.A. Physical tuning of galectin-3 signaling. *Proc. Natl. Acad. Sci. USA* **2021**, *118*, 1–10. [CrossRef]
97. Cecioni, S.; Imberty, A.; Vidal, S. Glycomimetics versus multivalent glycoconjugates for the design of high affinity lectin ligands. *Chem. Rev.* **2015**, *115*, 525–561. [CrossRef] [PubMed]
98. Hevey, R. Strategies for the development of glycomimetic drug candidates. *Pharmaceuticals* **2019**, *12*, 55. [CrossRef]
99. Maxwell, E.G.; Colquhoun, I.J.; Chau, H.K.; Hotchkiss, A.T.; Waldron, K.W.; Morris, V.J.; Belshaw, N.J. Modified sugar beet pectin induces apoptosis of colon cancer cells via an interaction with the neutral sugar side-chains. *Carbohydr. Polym.* **2016**, *136*, 923–929. [CrossRef] [PubMed]
100. Cheng, H.; Li, S.; Fan, Y.; Gao, X.; Hao, M.; Wang, J.; Zhang, X.; Tai, G.; Zhou, Y. Comparative studies of the antiproliferative effects of ginseng polysaccharides on HT-29 human colon cancer cells. *Med. Oncol.* **2011**, *28*, 175–181. [CrossRef]
101. Pynam, H.; Dharmesh, S.M. A xylorhamnoarabinogalactan I from Bael (*Aegle marmelos* L.) modulates UV/DMBA induced skin cancer via galectin-3 & gut microbiota. *J. Funct. Foods* **2019**, *60*, 103425. [CrossRef]
102. Popov, S.V.; Ovodov, Y.S. Polypotency of the immunomodulatory effect of pectins. *Biochemistry* **2013**, *78*, 823–835. [CrossRef] [PubMed]
103. Fang, T.; Liu, D.-D.; Ning, H.-M.; Liu, D.; Sun, J.-Y.; Huang, X.-J.; Dong, Y.; Geng, M.-Y.; Yun, S.-F.; Yan, J.; et al. Modified citrus pectin inhibited bladder tumor growth through downregulation of galectin-3. *Acta Pharmacol. Sin.* **2018**, *39*, 1885–1893. [CrossRef]
104. Hossein, G.; Halvaei, S.; Heidarian, Y.; Dehghani-Ghobadi, Z.; Hassani, M.; Hosseini, H.; Naderi, N.; Sheikh Hassani, S. Pectasol-C Modified Citrus Pectin targets Galectin-3-induced STAT3 activation and synergize paclitaxel cytotoxic effect on ovarian cancer spheroids. *Cancer Med.* **2019**, *8*, 4315–4329. [CrossRef] [PubMed]
105. Abu-Elsaad, N.M.; Elkashef, W.F. Modified citrus pectin stops progression of liver fibrosis by inhibiting galectin-3 and inducing apoptosis of stellate cells. *Can. J. Physiol. Pharmacol.* **2016**, *94*, 554–562. [CrossRef] [PubMed]
106. Martinez-Martinez, E.; Ibarrola, J.; Calvier, L.; Fernandez-Celis, A.; Leroy, C.; Cachofeiro, V.; Rossignol, P.; Lopez-Andres, N. Galectin-3 blockade reduces renal fibrosis in two normotensive experimental models of renal damage. *PLoS ONE* **2016**, *11*, e0166272. [CrossRef]
107. Calvier, L.; Miana, M.; Reboul, P.; Cachofeiro, V.; Martinez-Martinez, E.; De Boer, R.A.; Poirier, F.; Lacolley, P.; Zannad, F.; Rossignol, P.; et al. Galectin-3 mediates aldosterone-induced vascular fibrosis. *Arterioscler. Thromb. Vasc. Biol.* **2013**, *33*, 67–75. [CrossRef]
108. Lin, Y.-H.; Chou, C.-H.; Wu, X.-M.; Chang, Y.-Y.; Hung, C.-S.; Chen, Y.-H.; Tzeng, Y.-L.; Wu, V.-C.; Ho, Y.-L.; Hsieh, F.-J.; et al. Aldosterone induced galectin-3 secretion in vitro and in vivo: From cells to humans. *PLoS ONE* **2014**, *9*, e95254. [CrossRef] [PubMed]
109. Calvier, L.; Martinez-Martinez, E.; Miana, M.; Cachofeiro, V.; Rousseau, E.; Sádaba, J.R.; Zannad, F.; Rossignol, P.; López-Andrés, N. The impact of galectin-3 inhibition on aldosterone-induced cardiac and renal injuries. *JACC Heart Fail.* **2015**, *3*, 59–67. [CrossRef] [PubMed]
110. Li, H.-Y.; Yang, S.; Li, J.-C.; Feng, J.-X. Galectin 3 inhibition attenuates renal injury progression in cisplatin-induced nephrotoxicity. *Biosci. Rep.* **2018**, *38*. [CrossRef] [PubMed]
111. Prud'homme, M.; Coutrot, M.; Michel, T.; Boutin, L.; Genest, M.; Poirier, F.; Launay, J.M.; Kane, B.; Kinugasa, S.; Prakoura, N.; et al. Acute Kidney Injury Induces Remote Cardiac Damage and Dysfunction through the Galectin-3 Pathway. *JACC Basic Transl. Sci.* **2019**, *4*, 717–732. [CrossRef]
112. Ibarrola, J.; Matilla, L.; Martínez-Martínez, E.; Gueret, A.; Fernández-Celis, A.; Henry, J.P.; Nicol, L.; Jaisser, F.; Mulder, P.; Ouvrard-Pascaud, A.; et al. Myocardial Injury after Ischemia/Reperfusion Is Attenuated by Pharmacological Galectin-3 Inhibition. *Sci. Rep.* **2019**, *9*, 1–10. [CrossRef] [PubMed]
113. Li, S.; Li, S.; Hao, X.; Zhang, Y.; Deng, W. Perindopril and a Galectin-3 inhibitor improve ischemic heart failure in rabbits by reducing gal-3 expression and myocardial fibrosis. *Front. Physiol.* **2019**, *10*, 1–8. [CrossRef] [PubMed]
114. Vergaro, G.; Prud'Homme, M.; Fazal, L.; Merval, R.; Passino, C.; Emdin, M.; Samuel, J.L.; Cohen Solal, A.; Delcayre, C. Inhibition of Galectin-3 Pathway Prevents Isoproterenol-Induced Left Ventricular Dysfunction and Fibrosis in Mice. *Hypertension* **2016**, *67*, 606–612. [CrossRef]

115. Ibarrola, J.; Martínez-Martínez, E.; Sádaba, J.R.; Arrieta, V.; García-Peña, A.; Álvarez, V.; Fernández-Celis, A.; Gainza, A.; Rossignol, P.; Ramos, V.C.; et al. Beneficial effects of galectin-3 blockade in vascular and aortic valve alterations in an experimental pressure overload model. *Int. J. Mol. Sci.* **2017**, *18*, 1664. [CrossRef] [PubMed]
116. Xue, H.; Zhao, Z.; Lin, Z.; Geng, J.; Guan, Y.; Song, C.; Zhou, Y.; Tai, G. Selective effects of ginseng pectins on galectin-3-mediated T cell activation and apoptosis. *Carbohydr. Polym.* **2019**, *219*, 121–129. [CrossRef] [PubMed]
117. Lau, E.S.; Liu, E.; Paniagua, S.M.; Sarma, A.A.; Zampierollo, G.; López, B.; Díez, J.; Wang, T.J.; Ho, J.E. Galectin-3 Inhibition With Modified Citrus Pectin in Hypertension. *JACC Basic Transl. Sci.* **2021**, *6*, 12–21. [CrossRef]
118. Portacci, A.; Diaferia, F.; Santomasi, C.; Dragonieri, S.; Boniello, E.; Di Serio, F.; Carpagnano, G.E. Galectin-3 as prognostic biomarker in patients with COVID-19 acute respiratory failure. *Respir. Med.* **2021**, *187*, 106556. [CrossRef]
119. Kuśnierz-Cabala, B.; Maziarz, B.; Dumnicka, P.; Dembiński, M.; Kapusta, M.; Bociąga-Jasik, M.; Winiarski, M.; Garlicki, A.; Grodzicki, T.; Kukla, M. Diagnostic significance of serum galectin-3 in hospitalized patients with COVID-19—A preliminary study. *Biomolecules* **2021**, *11*, 1136. [CrossRef] [PubMed]
120. Gaughan, A.E.; Sethi, T.; Quinn, T.; Hirani, N.; Mills, A.; Annya, M.; Mackinnon, A.; Aslanis, V.; Li, F.; Connor, R.O.; et al. GB0139, an inhaled small molecule inhibitor of galectin-3, in COVID-19 pneumonitis: A randomised, controlled, open-label, phase 2a experimental medicine trial of safety, pharmacokinetics, and potential therapeutic value. *medRxiv* **2022**. [CrossRef]
121. Sakurai, M.H.; Matsumoto, T.; Kiyohara, H.; Yamada, H. Detection and tissue distribution of anti-ulcer peptic polysaccharides from *Bepleurum falcatum* by polyclonal antibody. *Planta Med.* **1996**, *62*, 341–346. [CrossRef] [PubMed]
122. Busato, B.; de Almeida Abreu, E.C.; de Oliveira Petkowicz, C.L.; Martinez, G.R.; Noleto, G.R. Pectin from *Brassica oleracea var. italica* triggers immunomodulating effects in vivo. *Int. J. Biol. Macromol.* **2020**, *161*, 431–440. [CrossRef] [PubMed]
123. Zhang, W.; Xu, P.; Zhang, H. Pectin in cancer therapy: A review. *Trends Food Sci. Technol.* **2015**, *44*, 258–271. [CrossRef]
124. Majee, S.B.; Avlani, D.; Ghosh, P.; Biswas, G.R. Therapeutic and pharmaceutical benefits of native and modified plant pectin. *J. Med. Plants Res.* **2018**, *12*, 1–6. [CrossRef]
125. Liu, H.-Y.; Huang, Z.-L.; Yang, G.-H.; Lu, W.-Q.; Yu, N.-R. Inhibitory effect of modified citrus pectin on liver metastases in a mouse colon cancer model. *World J. Gastroenterol.* **2008**, *14*, 7386–7391. [CrossRef]
126. Courts, F.L. Profiling of modified citrus pectin oligosaccharide transport across Caco-2 cell monolayers. *PharmaNutrition* **2013**, *1*, 22–31. [CrossRef]
127. Huang, P.-H.; Fu, L.-C.; Huang, C.-S.; Wang, Y.-T.; Wu, M.-C. The uptake of oligogalacturonide and its effect on growth inhibition, lactate dehydrogenase activity and galactin-3 release of human cancer cells. *Food Chem.* **2012**, *132*, 1987–1995. [CrossRef]
128. Mabbott, N.A.; Donaldson, D.S.; Ohno, H.; Williams, I.R.; Mahajan, A. Microfold (M) cells: Important immunosurveillance posts in the intestinal epithelium. *Mucosal Immunol.* **2013**, *6*, 666–677. [CrossRef]
129. Fan, Y.; Sun, L.; Yang, S.; He, C.; Tai, G.; Zhou, Y. The roles and mechanisms of homogalacturonan and rhamnogalacturonan I pectins on the inhibition of cell migration. *Int. J. Biol. Macromol.* **2018**, *106*, 207–217. [CrossRef]
130. Shao, A.; Wu, H.; Zhang, J. Letter by Shao et al Regarding Article, "Modified Citrus Pectin Prevents Blood-Brain Barrier Disruption in Mouse Subarachnoid Hemorrhage by Inhibiting Galectin-3". *J. Biol. Chem.* **2019**, *50*, e22. [CrossRef]
131. Leffler, H. Letter by Leffler Regarding Article, "Modified Citrus Pectin Prevents Blood-Brain Barrier Disruption in Mouse Subarachnoid Hemorrhage by Inhibiting Galectin-3". *J. Biol. Chem.* **2019**, *50*, e136. [CrossRef] [PubMed]
132. Nishikawa, H.; Liu, L.; Nakano, F.; Kawakita, F.; Kanamaru, H.; Nakatsuka, Y.; Okada, T.; Suzuki, H. Modified citrus pectin prevents blood-brain barrier disruption in mouse Subarachnoid hemorrhage by inhibiting Galectin-3. *Stroke* **2018**, *49*, 2743–2751. [CrossRef]
133. Hirani, N.; MacKinnon, A.C.; Nicol, L.; Ford, P.; Schambye, H.; Pedersen, A.; Nilsson, U.J.; Leffler, H.; Sethi, T.; Tantawi, S.; et al. Target inhibition of galectin-3 by inhaled TD139 in patients with idiopathic pulmonary fibrosis. *Eur. Respir. J.* **2021**, *57*, 1–13. [CrossRef]
134. Bumba, L.; Laaf, D.; Spiwok, V.; Elling, L.; Křen, V.; Bojarová, P. Poly-N-acetyllactosamine Neo-glycoproteins as nanomolar ligands of human galectin-3: Binding kinetics and modeling. *Int. J. Mol. Sci.* **2018**, *19*, 372. [CrossRef] [PubMed]
135. Laaf, D.; Steffens, H.; Pelantová, H.; Bojarová, P.; Křen, V.; Elling, L. Chemo-Enzymatic Synthesis of Branched N-Acetyllactosamine Glycan Oligomers for Galectin-3 Inhibition. *Adv. Synth. Catal.* **2017**, *359*, 4015–4024. [CrossRef]
136. Fischöder, T.; Laaf, D.; Dey, C.; Elling, L. Enzymatic synthesis of N-acetyllactosamine (LacNAc) type 1 oligomers and characterization as multivalent galectin ligands. *Molecules* **2017**, *22*, 1320. [CrossRef] [PubMed]
137. Mudgil, D.; Barak, S. Composition, properties and health benefits of indigestible carbohydrate polymers as dietary fiber: A review. *Int. J. Biol. Macromol.* **2013**, *61*, 1–6. [CrossRef] [PubMed]
138. Havenaar, R. Intestinal health functions of colonic microbial metabolites: A review. *Benef. Microbes* **2011**, *2*, 103–114. [CrossRef] [PubMed]
139. Ríos-Covián, D.; Ruas-Madiedo, P.; Margolles, A.; Gueimonde, M.; De los Reyes-Gavilán, C.G.; Salazar, N. Intestinal short chain fatty acids and their link with diet and human health. *Front. Microbiol.* **2016**, *7*, 1–9. [CrossRef] [PubMed]
140. Blaak, E.E.; Canfora, E.E.; Theis, S.; Frost, G.; Groen, A.K.; Mithieux, G.; Nauta, A.; Scott, K.; Stahl, B.; van Harsselaar, J.; et al. Short chain fatty acids in human gut and metabolic health. *Benef. Microbes* **2020**, *11*, 411–455. [CrossRef] [PubMed]
141. Chambers, E.S.; Preston, T.; Frost, G.; Morrison, D.J. Role of Gut Microbiota-Generated Short-Chain Fatty Acids in Metabolic and Cardiovascular Health. *Curr. Nutr. Rep.* **2018**, *7*, 198–206. [CrossRef] [PubMed]

142. Sivaprakasam, S.; Gurav, A.; Paschall, A.V.; Coe, G.L.; Chaudhary, K.; Cai, Y.; Kolhe, R.; Martin, P.; Browning, D.; Huang, L.; et al. An essential role of Ffar2 (Gpr43) in dietary fibre-mediated promotion of healthy composition of gut microbiota and suppression of intestinal carcinogenesis. *Oncogenesis* **2016**, *5*, e238. [CrossRef] [PubMed]
143. Sivaprakasam, S.; Prasad, P.D.; Singh, N. Benefits of short-chain fatty acids and their receptors in inflammation and carcinogenesis. *Pharmacol. Ther.* **2016**, *164*, 144–151. [CrossRef] [PubMed]
144. Kim, M.H.; Kang, S.G.; Park, J.H.; Yanagisawa, M.; Kim, C.H. Short-chain fatty acids activate GPR41 and GPR43 on intestinal epithelial cells to promote inflammatory responses in mice. *Gastroenterology* **2013**, *145*, 396–406. [CrossRef]
145. Beukema, M.; Faas, M.M.; de Vos, P. The effects of different dietary fiber pectin structures on the gastrointestinal immune barrier: Impact via gut microbiota and direct effects on immune cells. *Exp. Mol. Med.* **2020**, *52*, 1364–1376. [CrossRef]
146. Tian, L.; Scholte, J.; Borewicz, K.; van den Bogert, B.; Smidt, H.; Scheurink, A.J.W.; Gruppen, H.; Schols, H.A. Effects of pectin supplementation on the fermentation patterns of different structural carbohydrates in rats. *Mol. Nutr. Food Res.* **2016**, *60*, 2256–2266. [CrossRef]
147. Tian, L.; Bruggeman, G.; van den Berg, M.; Borewicz, K.; Scheurink, A.J.W.; Bruininx, E.; de Vos, P.; Smidt, H.; Schols, H.A.; Gruppen, H. Effects of pectin on fermentation characteristics, carbohydrate utilization, and microbial community composition in the gastrointestinal tract of weaning pigs. *Mol. Nutr. Food Res.* **2017**, *61*, 1–10. [CrossRef]
148. Ferreira-Lazarte, A.; Moreno, F.J.; Cueva, C.; Gil-Sánchez, I.; Villamiel, M. Behaviour of citrus pectin during its gastrointestinal digestion and fermentation in a dynamic simulator (simgi®). *Carbohydr. Polym.* **2019**, *207*, 382–390. [CrossRef]
149. Chen, J.; Liang, R.-H.; Liu, W.; Li, T.; Liu, C.-M.; Wu, S.-S.; Wang, Z.-J. Pectic-oligosaccharides prepared by dynamic high-pressure microfluidization and their in vitro fermentation properties. *Carbohydr. Polym.* **2013**, *91*, 175–182. [CrossRef]
150. Onumpai, C.; Kolida, S.; Bonnin, E.; Rastall, R.A. Microbial utilization and selectivity of pectin fractions with various structures. *Appl. Environ. Microbiol.* **2011**, *77*, 5747–5754. [CrossRef]
151. Lopez-Siles, M.; Khan, T.M.; Duncan, S.H.; Harmsen, H.J.M.; Garcia-Gil, L.J.; Flint, H.J. Cultured representatives of two major phylogroups of human colonic *Faecalibacterium prausnitzii* can utilize pectin, uronic acids, and host-derived substrates for growth. *Appl. Environ. Microbiol.* **2012**, *78*, 420–428. [CrossRef]
152. Scott, K.P.; Martin, J.C.; Duncan, S.H.; Flint, H.J. Prebiotic stimulation of human colonic butyrate-producing bacteria and bifidobacteria, in vitro. *FEMS Microbiol. Ecol.* **2014**, *87*, 30–40. [CrossRef]
153. Merheb, R.; Abdel-Massih, R.M.; Karam, M.C. Immunomodulatory effect of natural and modified Citrus pectin on cytokine levels in the spleen of BALB/c mice. *Int. J. Biol. Macromol.* **2019**, *121*, 1–5. [CrossRef]
154. Amorim, J.C.; Vriesmann, L.C.; Petkowicz, C.L.O.; Martinez, G.R.; Noleto, G.R. Modified pectin from *Theobroma cacao* induces potent pro-inflammatory activity in murine peritoneal macrophage. *Int. J. Biol. Macromol.* **2016**, *92*, 1040–1048. [CrossRef] [PubMed]
155. do Nascimento, G.E.; Winnischofer, S.M.B.; Ramirez, M.I.; Iacomini, M.; Cordeiro, L.M.C. The influence of sweet pepper pectin structural characteristics on cytokine secretion by THP-1 macrophages. *Food Res. Int.* **2017**, *102*, 588–594. [CrossRef] [PubMed]
156. Popov, S.V.; Ovodova, R.G.; Golovchenko, V.V.; Popova, G.Y.; Viatyasev, F.V.; Shashkov, A.S.; Ovodov, Y.S. Chemical composition and anti-inflammatory activity of a pectic polysaccharide isolated from sweet pepper using a simulated gastric medium. *Food Chem.* **2011**, *124*, 309–315. [CrossRef]
157. Ishisono, K.; Yabe, T.; Kitaguchi, K. Citrus pectin attenuates endotoxin shock via suppression of Toll-like receptor signaling in Peyer's patch myeloid cells. *J. Nutr. Biochem.* **2017**, *50*, 38–45. [CrossRef]
158. Vogt, L.M.; Sahasrabudhe, N.M.; Ramasamy, U.; Meyer, D.; Pullens, G.; Faas, M.M.; Venema, K.; Schols, H.A.; de Vos, P. The impact of lemon pectin characteristics on TLR activation and T84 intestinal epithelial cell barrier function. *J. Funct. Foods* **2016**, *22*, 398–407. [CrossRef]
159. Wang, H.; Bi, H.; Gao, T.; Zhao, B.; Ni, W.; Liu, J. A homogalacturonan from *Hippophae rhamnoides* L. Berries enhance immunomodulatory activity through TLR4/MyD88 pathway mediated activation of macrophages. *Int. J. Biol. Macromol.* **2018**, *107*, 1039–1045. [CrossRef] [PubMed]
160. Park, S.N.; Noh, K.T.; Jeong, Y.-I.; Jung, I.D.; Kang, H.K.; Cha, G.S.; Lee, S.J.; Seo, J.K.; Kang, D.H.; Hwang, T.-H.; et al. Rhamnogalacturonan II is a Toll-like receptor 4 agonist that inhibits tumor growth by activating dendritic cell-mediated CD8+ T cells. *Exp. Mol. Med.* **2013**, *45*, e8. [CrossRef] [PubMed]
161. Sahasrabudhe, N.M.; Beukema, M.; Tian, L.; Troost, B.; Scholte, J.; Bruininx, E.; Bruggeman, G.; van den Berg, M.; Scheurink, A.; Schols, H.A.; et al. Dietary fiber pectin directly blocks toll-like receptor 2—1 and prevents doxorubicin-induced ileitis. *Front. Immunol.* **2018**, *9*, 1–19. [CrossRef] [PubMed]
162. Hu, S.; Kuwabara, R.; Chica, C.E.N.; Smink, A.M.; Koster, T.; Medina, J.D.; de Haan, B.J.; Beukema, M.; Lakey, J.R.T.; García, A.J.; et al. Toll-like receptor 2-modulating pectin-polymers in alginate-based microcapsules attenuate immune responses and support islet-xenograft survival. *Biomaterials* **2021**, *266*, 120460. [CrossRef] [PubMed]
163. Kolatsi-Joannou, M.; Price, K.L.; Winyard, P.J.; Long, D.A. Modified citrus pectin reduces galectin-3 expression and disease severity in experimental acute kidney injury. *PLoS ONE* **2011**, *6*, e18683. [CrossRef]
164. Mai, C.W.; Kang, Y.B.; Pichika, M.R. Should a Toll-like receptor 4 (TLR-4) agonist or antagonist be designed to treat cancer? TLR-4: Its expression and effects in the ten most common cancers. *OncoTargets Ther.* **2013**, *6*, 1573–1587. [CrossRef]
165. Kaczanowska, S.; Joseph, A.M.; Davila, E. TLR agonists: Our best frenemy in cancer immunotherapy. *J. Leukoc. Biol.* **2013**, *93*, 847–863. [CrossRef]

166. Pradere, J.P.; Dapito, D.H.; Schwabe, R.F. The Yin and Yang of Toll-like receptors in cancer. *Oncogene* **2014**, *33*, 3485–3495. [CrossRef]
167. Wang, H.; Gao, T.; Du, Y.; Yang, H.; Wei, L.; Bi, H.; Ni, W. Anticancer and immunostimulating activities of a novel homogalacturonan from *Hippophae rhamnoides* L. berry. *Carbohydr. Polym.* **2015**, *131*, 288–296. [CrossRef] [PubMed]
168. Sato-Kaneko, F.; Yao, S.; Ahmadi, A.; Zhang, S.S.; Hosoya, T.; Kaneda, M.M.; Varner, J.A.; Pu, M.; Messer, K.S.; Guiducci, C.; et al. Combination immunotherapy with TLR agonists and checkpoint inhibitors suppresses head and neck cancer. *JCI Insight* **2017**, *2*, 1–18. [CrossRef]
169. Shishir, M.R.I.; Karim, N.; Gowd, V.; Xie, J.; Zheng, X.; Chen, W. Pectin-chitosan conjugated nanoliposome as a promising delivery system for neohesperidin: Characterization, release behavior, cellular uptake, and antioxidant property. *Food Hydrocoll.* **2019**, *95*, 432–444. [CrossRef]

Review

Towards a Better Understanding of the Relationships between Galectin-7, p53 and MMP-9 during Cancer Progression

Yves St-Pierre

INRS-Centre Armand-Frappier Santé Biotechnologie, Laval, QC H7V 1B7, Canada; yves.st-pierre@inrs.ca

Abstract: It has been almost 25 years since the discovery of galectin-7. This member of the galectin family has attracted interest from many working in the cancer field given its highly restricted expression profile in epithelial cells and the fact that cancers of epithelial origin (carcinoma) are among the most frequent and deadly cancer subtypes. Initially described as a p53-induced gene and associated with apoptosis, galectin-7 is now recognized as having a protumorigenic role in many cancer types. Several studies have indeed shown that galectin-7 is associated with aggressive behavior of cancer cells and induces expression of MMP-9, a member of the matrix metalloproteinases (MMP) family known to confer invasive behavior to cancer cells. It is therefore not surprising that many studies have examined its relationships with p53 and MMP-9. However, the relationships between galectin-7 and p53 and MMP-9 are not always clear. This is largely because p53 is often mutated in cancer cells and such mutations drastically change its functions and, consequently, its association with galectin-7. In this review, we discuss the functional relationships between galectin-7, p53 and MMP-9 and reconcile some apparently contradictory observations. A better understanding of these relationships will help to develop a working hypothesis and model that will provide the basis for further research in the hope of establishing a new paradigm for tackling the role of galectin-7 in cancer.

Keywords: galectin-7; p53; MMP-9; cancer; gain-of-function

1. Introduction: The Discovery of Galectin-7

The first reports on galectin-7 were published more than 25 years ago by two independent groups conducting systematic searches for differentially expressed proteins in keratinocytes. The first report was published by Prof. Julio Celis. Using two-dimensional gel electrophoresis, his group had been conducting systematic searches for differentially expressed proteins between normal and SV40-transformed human keratinocytes for several years [1]. The researchers noticed a protein, named IEF17, that was constitutively expressed in normal keratinocytes but not in transformed cells. They subsequently cloned the gene and showed that its sequence had strong homology to consensus sequences found in the carbohydrate-binding site of galectins. They further showed that the protein bound asialofetuin via a lactose-dependent interaction. Upon consultation with established galectin researchers who had just published a consensus on the criteria necessary for a protein to be recognized as a galectin the year before [2], they named this new protein galectin-7. During approximately the same period of time, a French group from L'Oréal was conducting a systematic search for epidermal-specific and stage-specific biomarkers using epidermal cDNA libraries [3]. Their attention was focused on the clone 1A12, whose sequence was identical to the sequence encoding the IEF7 protein identified by Madsen and colleagues that had just been submitted to the GenBank database. Magnaldo and colleagues constructed a GST-fusion protein with their cDNA clone, expressed it in an *E. coli* expression system and confirmed its affinity for lactose. Building on their initial discoveries, both the Danish and French groups actively pursued their research on galectin-7 and subsequently concluded that galectin-7 is a keratinocyte-specific marker found in all layers of the epidermis and other stratified epithelia of tissues, most notably in the tongue, cornea, esophagus, stomach, anus and Hassal's corpuscles of the thymus [4–8]. Subsequent studies showed, however, that galectin-7 is also present in other types of epithelia, including the myoepithelial cells of the mammary gland epithelium [9,10]. In most cases, galectin-7 was found to be expressed in both the cytosolic and nuclear compartments and released into the extracellular space by a nonclassical secretion mechanism, a pattern typical of members of the galectin family [11]. Although most of the literature has focused on the expression of galectin-7 in cells of epithelial origin, it is important to keep an open mind with regards to its expression in other cell types, as we now know that it is also expressed, albeit at a rarer frequency, in cells of other lineages. We and others have reported, for example, that *LGALS7* expression is found in specific subtypes of lymphoid/myeloid cells, including transformed human B and T lymphocytes [12,13] and normal CD8-positive dendritic cells [14,15]. Too often these data are lost in the sea of transcriptomic datasets and we still largely ignore the function of galectin-7 in these normal myeloid cells.

2. The First Stage of the Relationship between Galectin-7 and p53

In 1997, the Bert Vogelstein group published a seminal article in Nature on p53 [16]. Similar to many teams working on p53, the Vogelstein group was studying the molecular mechanisms involved in p53-induced cell death and p53-mediated regulation of the cell cycle. They were particularly interested in identifying transcripts that were under the control of p53. For this, they used an adenoviral vector to express wild-type p53 in the p53-defective human colon cancer cell line DLD-1. They identified 14 transcripts including a transcript encoding the *p21* gene and 13 other transcripts that they called p53-induced genes (PIGs) 1–13. Many of these transcripts were functionally linked to the generation of reactive oxygen species (ROS), allowing establishment of a clear and strong connection between p53 and oxidative stress. Other transcripts encoded multiple genes, including *LGALS7* (identified as PIG1). The authors paid very little attention to galectin-7 other than a comment that this gene was part of the galectin family and that some members (such as galectin-3) were capable of inducing superoxide production in neutrophils. In fact, there were at that time no more than a handful of papers that had been published on galectin-7 (cited above). Following this article by Polyak and colleagues, several groups confirmed that upregulation of galectin-7 was associated with apoptosis in a p53-dependent manner. For example, Bernerd et al. [17] showed that galectin-7 was upregulated in skin keratinocytes upon exposure to UVB irradiation, a known inducer of the *p53* gene and apoptosis in keratinocytes. They also found that induction of galectin-7 paralleled that of wild-type (wt) p53, further confirming that induction of galectin-7, at least in UVB-exposed skin keratinocytes, is dependent on wt p53. Consistent with this hypothesis, they found no detectable expression of galectin-7 in the SCC13 keratinocyte cell line, which does not express wt p53 but rather p53 mutated at position 179 [18]. This mutation at position 179 is known to destabilize the p53 protein by altering its ability to interact with a zinc molecule, causing loss of DNA-binding specificity [19]. Our group also reported that treatment of human MCF-7 breast cancer cells (which harbor wt alleles of p53) with doxorubicin, a well-known inducer of p53, induced the expression of galectin-7 at both the mRNA and protein levels and that this induction was abrogated following knockdown of wt p53 [20]. However, after all these years, the function of galectin-7 in this physiological context related to wt p53 is still unclear. Could GAL-7 be induced by p53 to promote DNA repair? This is certainly a real possibility considering that (1) DNA repair is a pathway activated by p53, (2) galectin-7 is expressed at high levels in the nucleus, (3) galectins have been shown to have nuclear functions and (4) other members of the galectin family, such as galectin-1 and galectin-3, are known to regulate DNA damage repair [21–23].

3. A Twisted Relationship

While initial studies in colon cancer cells indicated that p53-induced galectin-7 expression was associated with cell death, a number of reports published at that time brought the universality of this phenotype into question. These studies indicated that galectin-7 might instead be associated with cancer progression (Figure 1). For example, data obtained from a study using an experimental breast carcinoma model [24] and a study evaluating clinical specimens collected from patients with thyroid cancer [25] revealed that high galectin-7 expression levels were associated with cancer progression and aggressiveness. Conducting a comparative transcriptomic analysis between a low-tumorigenic lymphoma cell line (164T2) and its aggressive variants (S11 and S19) generated by serial in vivo passaging, our group further showed that the most prominent change that occurred in highly metastatic variants was the strong upregulation (160-fold) of galectin-7 [12]. This change was unexpected, as, at that time, galectin-7 was considered a keratinocyte marker expressed in all subtypes of stratified epidermis. Analysis of gene expression profiles stored in public datasets in the Gene Expression Omnibus (GEO) repository showed that such strong upregulation of galectin-7 in lymphoid cancer cells is not unusual, as it was also observed following in vivo passaging of the parental FL5.12 cell line, a murine prolymphocytic cell line [26]. Using our lymphoma model, we further provided evidence that such upregulation of galectin-7 in lymphoma cells is not an epiphenomenon. Transfection of an expression vector containing the galectin-7 gene into low-metastatic lymphoma cells increased their metastatic behavior [27]. Conversely, inhibition of galectin-7 in aggressive lymphoma variants decreased their invasive behavior in vivo [13]. These studies established for the first time that galectin-7 can promote metastasis, a conclusion that contradicted previous observations, indicating that galectin-7 leads to cell death. However, this should have come as no surprise given previous findings showing that other members of the galectin family are well known to have a dual role in controlling cell survival depending on the cellular context, differentiation state and/or cellular localization [28–30]. These reports showing that galectin-7 can promote cancer progression, however, challenged the existence of a universal p53–galectin-7 axis and questioned the relevance of continuing to define galectin-7 solely as a p53-induced gene, as originally described by Polyak et al. [16]. This conclusion is supported by data showing that many cancer cells with no detectable p53 transcriptional activity, such as breast cancer cells and HaCaT cells, which carry mutations in both alleles of p53 (R282W and H179Y) that render the protein transcriptionally inactive, express high levels of galectin-7 [17,20].

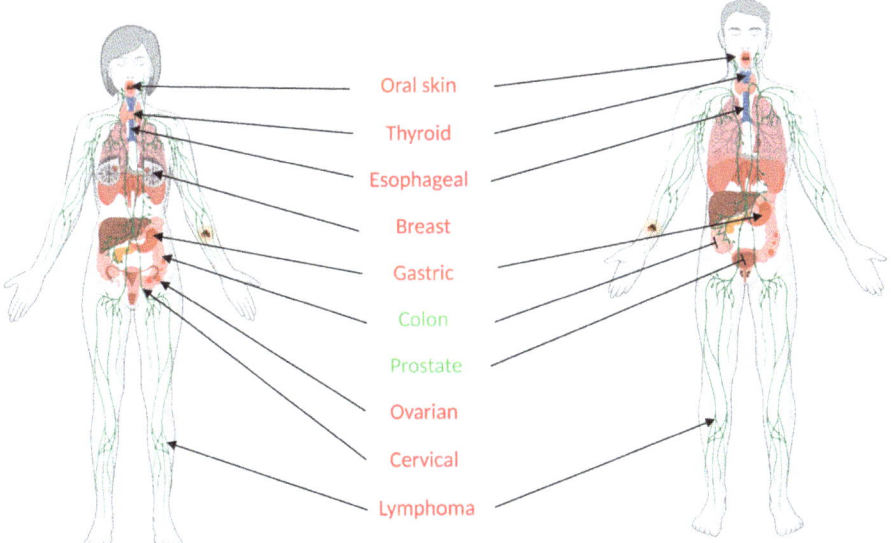

Figure 1. The dual role of galectin-7 in cancer. In most cancer types, overexpression of galectin-7 is associated with and/or promotes cancer progression. For some types of cancer, such as colon cancer and prostate cancer (in green), expression of galectin-7 is rather associated with an anti-tumor role.

4. De Novo Expression of Galectin-7 by Mutated p53

Although it is clear that the presence of wt p53 is not essential for a cell to express galectin-7, it is important to remember that p53 has protumorigenic functions when mutated [31,32]. Could it be that galectin-7 is induced by mutated forms of p53? Considering that p53 is mutated in approximately half of cancer cases and that galectin-7 is readily expressed in many cancer types, this is a real possibility. Our group decided to further examine this relationship between galectin-7 and p53 mutants. We studied the most frequent cancer-associated "hot spots" p53 mutations, including missense mutations within the region encoding the DNA-binding domain that involve amino acid residues in direct contact with DNA or amino acid residues that locally or globally affect the conformation of the p53 protein structure [33]. In all cases, these mutations not only eliminate the protective function normally mediated by wt p53 against stressors but also often confer cancer cells with new protumorigenic properties through a gain-of-function (GOF) mechanism. The GOF mechanism was first proposed by Arnold Levine [34]. Using a series of vectors encoding either mouse or human mutant p53 transfected into p53^{null} cells, new or additional phenotypes were conferred on these cells, suggesting that cancer cells with mutant p53 are likely to be more aggressive and have a poorer prognosis than cancer cells with no p53. This suggested that mutant p53 not only acts through a dominant-negative effect on the remaining wt p53 allele but can also trigger the selection of clonal cells with new properties. This could occur through at least two nonexclusive transcriptional and non-transcriptional mechanisms, thereby altering specific signaling pathways [35]. This is particularly well documented in the case of the NF-κB pathway, which is activated by mutant p53, causing chronic inflammatory conditions that favor the onset of transformation [36]. Using expression vectors encoding specific p53 mutants, we found that R175H induces de novo expression of galectin-7 in p53^{null} human breast (MDA-MB-453) and ovarian (SKOV-3) cancer cell lines [20,37]. This does not, however, mean that p53 mutants universally induce galectin-7 in every cancer cell line. The cellular context is likely to play a role. For example, another mutant, R273H, induces galectin-7 in breast cancer cells but not in ovarian cancer cells [20,37]. We also found that treatment of MDA-MB-231 (R280K) and MCF-7 (harboring wt p53) breast cancer cells with doxorubicin, a widely used chemotherapeutic agent, induced galectin-7 in both cell lines. In fact, we found that both can bind to the *LGALS7* promoter in breast cancer cells [20]. It is thus logical to suggest that both wild-type and mutant p53 can induce galectin-7 in breast cancer cells, but given the intrinsic resistance of these cells to cell death, the balance is shifted towards the protumorigenic role of galectin-7 (Figure 2). Interestingly, we found that R273H-induced expression of galectin-7 was inhibited by caffeic acid phenethyl ester (CAPE) and quercetin, two inhibitors of NF-κB [20,38,39]. It is important to note, however, that constitutive NF-κB activity is not sufficient to induce galectin-7. For example, p53-null MDA-MB-453 cells exhibit high levels of NF-κB activity but do not express detectable levels of galectin-7.

Given the multiple roles of galectins in controlling the tumor microenvironment, it is reasonable to hypothesize that the production and release of galectin-7 by cancer cells is one of multiple mechanisms used by GOF mutants of p53 to facilitate a protumorigenic microenvironment [40]. Such a role for galectin-7 in the tumor microenvironment would include its ability to induce local immunosuppression, as we know now that galectin-7, like many galectins, induces apoptosis in activated immune cells [37,41,42]. This role would also include the regulation of the expression of protumorigenic genes in cancer cells. In this context, a number of studies have established a new functional link between galectin-7 and MMP-9, a member of the zinc metalloproteinases family involved in the degradation of the extracellular matrix and several steps of the metastatic process [43]. Such an association between galectin-7 and MMP-9 has been observed in many cancer types. It was first reported in an experimental lymphoma mouse model, in which the expression of galectin-7 by lymphoma cells was shown to increase the aggressive behavior of lymphoma cells by inducing MMP-9 [27]. We and others have since shown that galectin-7 can induce MMP-9 in breast cancer, ovarian cancer, head and neck cancer, cervical and oral squamous cell carcinomas and, more recently, gastric cancer cells [44–49]. In some cases, this association between galectin-7 and MMP-9 was found to be specific, as no such association was found between MMP-9 and galectin-1 or galectin-3 [44,45]. Today, when studying the role of galectin-7 in tumor progression, most notably during metastasis, it is difficult to ignore the relationship of galectin-7 with MMP-9.

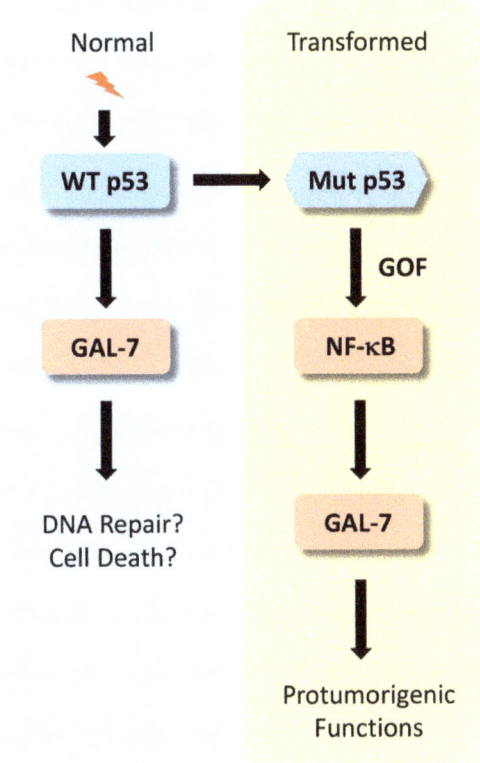

Figure 2. Reconciling the relationship between galectin-7 with wt and mutant p53. In normal cells harboring wt p53 alleles, galectin-7 is induced by de novo expression induced by wt p53 following stress signals. This pathway is also active in some cancer types, such as colon cancer [16]. Expression of nuclear galectin-7 may then regulate the cell cycle and/or DNA repair. In cancer cells with intrinsic resistance to cell death, galectin-7 can be induced by a gain-of-function (GOF) mechanism via mutant p53, shifting the balance towards the pro-tumorigenic role of galectin-7.5. The galectin-7–MMP-9 relationship in cancer.

5. How Does Galectin-7 Induce MMP-9?

Increased expression of MMP-9 induced by galectin-7, which occurs at both the mRNA and protein levels, has in many cases been documented by adding recombinant human galectin-7 to cancer cells. The use of this approach is largely due to the fact that galectin-7 is relatively easy to produce and purify by standard lactosyl-Sepharose affinity chromatography [41,50]. We can propose a model that integrates the multiple findings on the mechanisms by which galectin-7 can induce MMP-9 (Figure 3). Binding of recombinant galectin-7 to specific cell surface receptors can trigger a number of signaling pathways, including the p38 MAPK, ERK1/2 and JNK pathways [41,47]. The use of pharmacological inhibitors, such as SB203580, an inhibitor of the p38 pathway, and inhibitors of the ERK1/2 and JNK pathways, has shown that both pathways are likely important for galectin-7-

mediated induction of MMP-9 [44,47]. This is not very surprising, as these pathways are commonly used by growth factors to induce MMP-9 expression in cancer cells [43]. The ability of galectin-7 to augment MMP-9 expression in cancer cells has also been established using cancer cells transfected with eukaryotic expression vectors containing the gene encoding galectin-7 (i.e., *LGALS7*). Given the propensity of galectins to be released into the extracellular microenvironment, the ability of galectin-7 to induce MMP-9 expression is again likely to occur via extracellular binding of galectin-7 to cell-surface receptors. This hypothesis is supported by data showing that the increased MMP-9 expression induced by galectin-7 transfectants is inhibited by the addition of lactose to the culture medium [27]. However, another possible pathway that may contribute to the increased expression of MMP-9 induced by galectin-7 involves galectin-7 re-entry by endocytosis via clathrin-coated pits in a manner similar to that reported for other galectins [51]. This intracellular pool of galectin-7 is targeted by a relatively large diversity of binding partners depending on the cell type [41]. This has been well documented in the case of galectin-7/Bcl-2 interactions [52]. Galectin-7 is also targeted by Tidf1, which represses the oncogenic role of galectin-7 by promoting its degradation [49]. Whether de novo expression of Tidf1 represses galectin-7-induced MMP-9 expression is an interesting possibility that has not, to our knowledge, been studied.

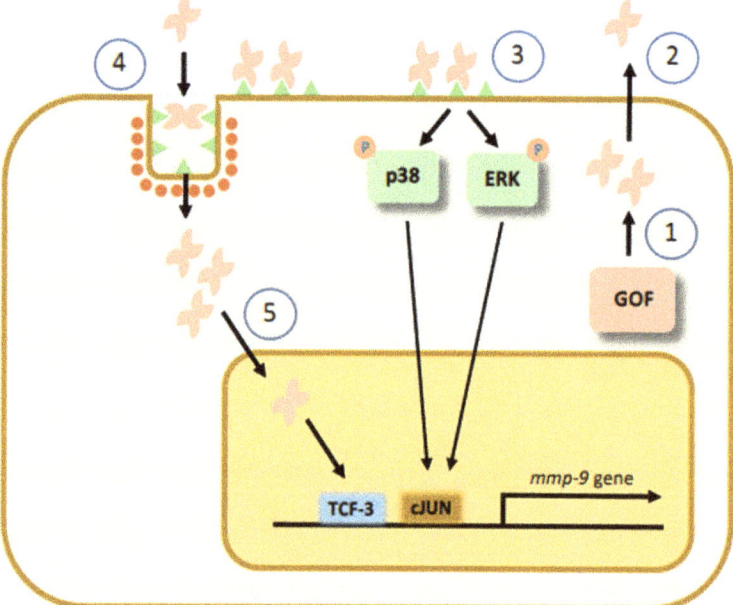

Figure 3. A working model for the control of MMP-9 gene expression in cancer cells via galectin-7. De novo expression of galectin-7 in cancer cells by the GOF pathway leads to accumulation of intracellular galectin-7 (1). Intracellular galectin-7 translocates directly to the nucleus to trigger MMP-9 gene expression (5) or be released by cancer cells in the extracellular tumor microenvironment (2). Once outside cancer cells, extracellular galectin-7 can bind and activate glycoreceptor-mediated signaling pathways that lead to MMP-9 gene expression (3). Such a mechanism can be activated in adjacent cancer cells or in the same cell following an autocrine mechanism (4).

The main question that remains is what cell-surface glycoreceptors bind galectin-7 and trigger intracellular signaling that ultimately leads to MMP-9 expression? Given the promiscuity of galectins for multiple glycosylated cell-surface receptors, it is likely that the signals implicated in the upregulation of MMP-9 involve multiple receptors that vary according to the cancer subtype. Accordingly, a logical hypothesis is that extracellular galectin-7 induces MMP-9 via stabilization and/or reorganization of cell-surface glycoreceptors to lower the signaling threshold necessary for triggering intracellular signaling [53]. This has been well described for galectin-1 and galectin-3 [54–56]. We must also be open to the possibility that such a signaling cascade could in fact be initiated by carbohydrate-independent interactions. This has been well documented in the case of galectin-1, which binds to unglycosylated membrane pre-B cell receptor, which regulates glycan-dependent formation of lattice [57]. A recent study indeed showed that galectin-7 could bind to nonglycosylated cell-surface receptors such as E-cadherin [58]. The binding of galectin-7 to E-cadherin, which normally undergoes endocytosis, stabilizes the receptor at the membrane and restrains its endocytosis. Interestingly, E-cadherin is a well-known substrate of MMPs, including MMP-9 [59–63]. E-cadherin being one of these critical receptors that induces MMP-9 in cancer cells is a real possibility. In such a case, the cleavage of E-cadherin by MMP-9 would ensure strict autoregulation of the galectin-7–MMP-9 axis. Future investigations will be necessary to test this hypothesis.

6. Conclusions

The objective of this review was to shed light on the relationships between galectin-7, p53 and MMP-9. It is clear that we should avoid referring to galectin-7 simply as a p53-induced gene for two reasons. First, galectin-7 is not always associated with apoptosis and is often, if not most of the time, associated with tumor progression. Second, there is no direct relationship between galectin-7 expression and p53 transcriptional activity. Many $p53^{null}$ cells express high levels of galectin-7, while transcriptionally inactive mutant p53 induces galectin-7. However, it may be too early to question the relationship between wt p53 and galectin-7. Indeed, it is important to note that wt p53 can behave like a mutant form if its structural integrity is not properly maintained by a chaperone that prevents misfolding into a mutant-like form [64]. The reverse is also true, as mutant p53 can be reverted into a conformationally wt-like form of p53 [65]. In contrast, galectin-7 seems to have a strong and long-lasting relationship with MMP-9. Clearly, both can be induced by similar mechanisms, including the GOF activity of mutant p53. This relationship among galectin-7, MMP-9 and mutant p53 could be more common than we think, considering the increasing number of studies showing that galectin-7 promotes cancer progression. Future studies should also pay attention to another member of the p53 family, i.e., p63, which has an expression profile close to that of galectin-7 (stratified and glandular epithelia) [66]. Moreover, it has been reported that galectin-7 could regulate keratinocyte proliferation and differentiation through c-Jun N-terminal kinase (JNK1)-miR-203-p63 pathway [67].

Funding: This research received no external funding.

Conflicts of Interest: The author declares no conflict of interest.

References

1. Madsen, P.; Rasmussen, H.H.; Flint, T.; Gromov, P.; Kruse, T.A.; Honoré, B.; Vorum, H.; Celis, J.E. Cloning, expression, and chromosome mapping of human galectin-7. *J. Biol. Chem.* **1995**, *270*, 5823–5829. [CrossRef]
2. Barondes, S.H.; Castronovo, V.; Cooper, D.N.; Cummings, R.D.; Drickamer, K.; Feizi, T.; Gitt, M.A.; Hirabayashi, J.; Hughes, C.; Kasai, K.I. Galectins: A family of animal beta-galactoside-binding lectins. *Cell* **1994**, *76*, 597–598. [CrossRef]
3. Magnaldo, T.; Bernerd, F.; Darmon, M. Galectin-7, a human 14-kDa S-lectin, specifically expressed in keratinocytes and sensitive to retinoic acid. *Dev. Biol.* **1995**, *168*, 259–271. [CrossRef] [PubMed]
4. Østergaard, M.; Rasmussen, H.H.; Nielsen, H.V.; Vorum, H.; Ørntoft, T.F.; Wolf, H.; Celis, J.E. Proteome profiling of bladder squamous cell carcinomas: Identification of markers that define their degree of differentiation. *Cancer Res.* **1997**, *57*, 4111–4117. [PubMed]
5. Magnaldo, T.; Fowlis, D.; Darmon, M. Galectin-7, a marker of all types of stratified epithelia. *Differentiation* **1998**, *63*, 159–168. [CrossRef] [PubMed]
6. Timmons, P.M.; Colnot, C.É.; Cail, I.S.; Poirier, F.R.; Magnaldo, T.H. Expression of galectin-7 during epithelial development coincides with the onset of stratification. *Int. J. Dev. Biol.* **2002**, *43*, 229–235.
7. Chung, C.H.; Bernard, P.S.; Perou, C.M. Molecular portraits and the family tree of cancer. *Nat. Genet.* **2002**, *32*, 533–540. [CrossRef]

8. Sato, M.; Nishi, N.; Shoji, H.; Kumagai, M.; Imaizumi, T.; Hata, Y.; Hirashima, M.; Suzuki, S.; Nakamura, T. Quantification of galectin-7 and its localization in adult mouse tissues. *J. Biochem.* **2002**, *131*, 255–260. [CrossRef]
9. Jones, C.; Mackay, A.; Grigoriadis, A.; Cossu, A.; Reis-Filho, J.S.; Fulford, L.; Dexter, T.; Davies, S.; Bulmer, K.; Ford, E.; et al. Expression profiling of purified normal human luminal and myoepithelial breast cells: Identification of novel prognostic markers for breast cancer. *Cancer Res.* **2004**, *64*, 3037–3045. [CrossRef]
10. Demers, M.; Rose, A.A.; Grosset, A.A.; Biron-Pain, K.; Gaboury, L.; Siegel, P.M.; St-Pierre, Y. Overexpression of galectin-7, a myoepithelial cell marker, enhances spontaneous metastasis of breast cancer cells. *Am. J. Pathol.* **2010**, *176*, 3023–3031. [CrossRef]
11. Johannes, L.; Jacob, R.; Leffler, H. Galectins at a glance. *J. Cell Sci.* **2018**, *131*. [CrossRef] [PubMed]
12. Moisan, S.; Demers, M.; Mercier, J.; Magnaldo, T.; Potworowski, E.F.; St-Pierre, Y. Upregulation of galectin-7 in murine lymphoma cells is associated with progression toward an aggressive phenotype. *Leukemia* **2003**, *17*, 751–759. [CrossRef] [PubMed]
13. Demers, M.; Biron-Pain, K.; Hébert, J.; Lamarre, A.; Magnaldo, T.; St-Pierre, Y. Galectin-7 in lymphoma: Elevated expression in human lymphoid malignancies and decreased lymphoma dissemination by antisense strategies in experimental model. *Cancer Res.* **2007**, *67*, 2824–2829. [CrossRef]
14. Becker, A.M.; Dao, K.H.; Han, B.K.; Kornu, R.; Lakhanpal, S.; Mobley, A.B.; Li, Q.Z.; Lian, Y.; Wu, T.; Reimold, A.M.; et al. SLE peripheral blood B cell, T cell and myeloid cell transcriptomes display unique profiles and each subset contributes to the interferon signature. *PLoS ONE* **2013**, *8*, e67003. [CrossRef]
15. Ding, Y.; Guo, Z.; Liu, Y.; Li, X.; Zhang, Q.; Xu, X.; Gu, Y.; Zhang, Y.; Zhao, D.; Cao, X. The lectin Siglec-G inhibits dendritic cell cross-presentation by impairing MHC class I–peptide complex formation. *Nat. Immunol.* **2016**, *17*, 1167–1175. [CrossRef] [PubMed]
16. Polyak, K.; Xia, Y.; Zweier, J.L.; Kinzler, K.W.; Vogelstein, B. A model for p53-induced apoptosis. *Nature* **1997**, *389*, 300–305. [CrossRef] [PubMed]
17. Bernerd, F.; Sarasin, A.; Magnaldo, T. Galectin-7 overexpression is associated with the apoptotic process in UVB-induced sunburn keratinocytes. *Proc. Natl. Acad. Sci. USA* **1999**, *96*, 11329–11334. [CrossRef]
18. John, L.S.; Sauter, E.R.; Herlyn, M.; Litwin, S.; Adler-Storthz, K. Endogenous p53 gene status predicts the response of human squamous cell carcinomas to wild-type p53. *Cancer Gene Ther.* **2000**, *7*, 749–756. [CrossRef]
19. Joerger, A.C.; Fersht, A.R. Structural biology of the tumor suppressor p53. *Annu. Rev. Biochem.* **2008**, *77*, 557–582. [CrossRef]
20. Campion, C.G.; Labrie, M.; Lavoie, G.; St-Pierre, Y. Expression of galectin-7 is induced in breast cancer cells by mutant p53. *PLoS ONE* **2013**, *8*, e72468. [CrossRef]
21. Huang, E.Y.; Chen, Y.F.; Chen, Y.M.; Lin, I.H.; Wang, C.C.; Su, W.H.; Chuang, P.C.; Yang, K.D. A novel radioresistant mechanism of galectin-1 mediated by H-Ras-dependent pathways in cervical cancer cells. *Cell Death Dis.* **2012**, *3*, e251. [CrossRef] [PubMed]
22. Carvalho, R.S.; Fernandes, V.C.; Nepomuceno, T.C.; Rodrigues, D.C.; Woods, N.T.; Suarez-Kurtz, G.; Chammas, R.; Monteiro, A.N.; Carvalho, M.A. Characterization of LGALS3 (galectin-3) as a player in DNA damage response. *Cancer Biol. Ther.* **2014**, *15*, 840–850. [CrossRef]
23. Boutas, I.; Potiris, A.; Brenner, W.; Lebrecht, A.; Hasenburg, A.; Kalantaridou, S.; Schmidt, M. The expression of galectin-3 in breast cancer and its association with chemoresistance: A systematic review of the literature. *Arch. Gynecol. Obstet.* **2019**, *300*, 1113–1120. [CrossRef]
24. Lu, J.; Pei, H.; Kaeck, M.; Thompson, H.J. Gene expression changes associated with chemically induced rat mammary carcinogenesis. *Mol. Carcinog.* **1997**, *20*, 204–215. [CrossRef]
25. Rorive, S.; Eddafali, B.; Fernandez, S.; Decaestecker, C.; André, S.; Kaltner, H.; Kuwabara, I.; Liu, F.T.; Gabius, H.J.; Kiss, R.; et al. Changes in galectin-7 and cytokeratin-19 expression during the progression of malignancy in thyroid tumors: Diagnostic and biological implications. *Mod. Pathol.* **2002**, *15*, 1294–1301. [CrossRef]
26. Fan, G.; Simmons, M.J.; Ge, S.; Dutta-Simmons, J.; Kucharczak, J.; Ron, Y.; Weissmann, D.; Chen, C.C.; Mukherjee, C.; White, E.; et al. Defective ubiquitin-mediated degradation of antiapoptotic Bfl-1 predisposes to lymphoma. *Blood* **2010**, *115*, 3559–3569. [CrossRef]
27. Demers, M.; Magnaldo, T.; St-Pierre, Y. A novel function for galectin-7: Promoting tumorigenesis by up-regulating MMP-9 gene expression. *Cancer Res.* **2005**, *65*, 5205–5210. [CrossRef] [PubMed]
28. Calific, S.; Castronovo, V.; Bracke, M.; van den Brûle, F. Dual activities of galectin-3 in human prostate cancer: Tumor suppression of nuclear galectin-3 vs tumor promotion of cytoplasmic galectin-3. *Oncogene* **2004**, *23*, 7527–7536. [CrossRef]
29. Vladoiu, M.C.; Labrie, M.; St-Pierre, Y. Intracellular galectins in cancer cells: Potential new targets for therapy. *Int. J. Oncol.* **2014**, *44*, 1001–1014. [CrossRef] [PubMed]
30. Chou, F.C.; Chen, H.Y.; Kuo, C.C.; Sytwu, H.K. Role of galectins in tumors and in clinical immunotherapy. *Int. J. Mol. Sci.* **2018**, *19*, 430. [CrossRef]
31. Brosh, R.; Rotter, V. When mutants gain new powers: News from the mutant p53 field. *Nat. Rev. Cancer* **2009**, *9*, 701–713. [CrossRef]
32. Muller, P.A.; Vousden, K.H. Mutant p53 in cancer: New functions and therapeutic opportunities. *Cancer Cell* **2014**, *25*, 304–317. [CrossRef]
33. Olivier, M.; Hollstein, M.; Hainaut, P. TP53 mutations in human cancers: Origins, consequences, and clinical use. *Cold Spring Harb. Perspect. Biol.* **2010**, *2*, a001008. [CrossRef] [PubMed]

34. Dittmer, D.; Pati, S.; Zambetti, G.; Chu, S.; Teresky, A.K.; Moore, M.; Finlay, C.; Levine, A.J. Gain of function mutations in p53. *Nat. Genet.* **1993**, *4*, 42–46. [CrossRef] [PubMed]
35. Freed-Pastor, W.A.; Prives, C. Mutant p53: One name, many proteins. *Genes Dev.* **2012**, *26*, 1268–1286. [CrossRef] [PubMed]
36. Cooks, T.; Pateras, I.S.; Tarcic, O.; Solomon, H.; Schetter, A.J.; Wilder, S.; Lozano, G.; Pikarsky, E.; Forshew, T.; Rozenfeld, N.; et al. Mutant p53 prolongs NF-κB activation and promotes chronic inflammation and inflammation-associated colorectal cancer. *Cancer Cell* **2013**, *23*, 634–646. [CrossRef] [PubMed]
37. Labrie, M.; Vladoiu, M.C.; Grosset, A.A.; Gaboury, L.; St-Pierre, Y. Expression and functions of galectin-7 in ovarian cancer. *Oncotarget* **2014**, *5*, 7705. [CrossRef]
38. Natarajan, K.; Singh, S.; Burke, T.R.; Grunberger, D.; Aggarwal, B.B. Caffeic acid phenethyl ester is a potent and specific inhibitor of activation of nuclear transcription factor NF-kappa B. *Proc. Natl. Acad. Sci. USA* **1996**, *93*, 9090–9095. [CrossRef]
39. Min, Y.D.; Choi, C.H.; Bark, H.; Son, H.Y.; Park, H.H.; Lee, S.; Park, J.W.; Park, E.K.; Shin, H.I.; Kim, S.H. Quercetin inhibits expression of inflammatory cytokines through attenuation of NF-κB and p38 MAPK in HMC-1 human mast cell line. *Inflamm. Res.* **2007**, *56*, 210–215. [CrossRef]
40. Amelio, I.; Melino, G. Context is everything: Extrinsic signalling and gain-of-function p53 mutants. *Cell Death Discov.* **2020**, *6*, 16. [CrossRef]
41. Vladoiu, M.C.; Labrie, M.; Létourneau, M.; Egesborg, P.; Gagné, D.; Billard, É.; Grosset, A.A.; Doucet, N.; Chatenet, D.; St-Pierre, Y. Design of a peptidic inhibitor that targets the dimer interface of a prototypic galectin. *Oncotarget* **2015**, *6*, 40970. [CrossRef] [PubMed]
42. López de Los Santos, Y.; Bernard, D.N.; Egesborg, P.; Létourneau, M.; Lafortune, C.; Cuneo, M.J.; Urvoas, A.; Chatenet, D.; Mahy, J.P.; St-Pierre, Y.; et al. Binding of a Soluble meso-Tetraarylporphyrin to Human Galectin-7 Induces Oligomerization and Modulates Its Pro-Apoptotic Activity. *Biochemistry* **2020**, *59*, 4591–4600. [CrossRef] [PubMed]
43. St-Pierre, Y.; Couillard, J.; Van Themsche, C. Regulation of MMP-9 gene expression for the development of novel molecular targets against cancer and inflammatory diseases. *Expert Opin. Ther. Targets* **2004**, *8*, 473–489. [CrossRef] [PubMed]
44. Park, J.E.; Chang, W.Y.; Cho, M. Induction of matrix metalloproteinase-9 by galectin-7 through p38 MAPK signaling in HeLa human cervical epithelial adenocarcinoma cells. *Oncol. Rep.* **2009**, *22*, 1373–1379. [PubMed]
45. Saussez, S.; Cludts, S.; Capouillez, A.; Mortuaire, G.; Smetana, K.; Kaltner, H.; André, S.; Leroy, X.; Gabius, H.J.; Decaestecker, C. Identification of matrix metalloproteinase-9 as an independent prognostic marker in laryngeal and hypopharyngeal cancer with opposite correlations to adhesion/growth-regulatory galectins-1 and -7. *Int. J. Oncol.* **2009**, *34*, 433–439. [CrossRef] [PubMed]
46. Zhu, H.; Wu, T.C.; Chen, W.Q.; Zhou, L.J.; Wu, Y.; Zeng, L.; Pei, H.P. Roles of galectin-7 and S100A9 in cervical squamous carcinoma: Clinicopathological and in vitro evidence. *Int. J. Cancer* **2013**, *132*, 1051–1059. [CrossRef]
47. Guo, J.P.; Li, X.G. Galectin-7 promotes the invasiveness of human oral squamous cell carcinoma cells via activation of ERK and JNK signaling. *Oncol. Lett.* **2017**, *13*, 1919–1924. [CrossRef]
48. Chen, Y.S.; Chang, C.W.; Tsay, Y.G.; Huang, L.Y.; Wu, Y.C.; Cheng, L.H.; Yang, C.C.; Wu, C.H.; Teo, W.H.; Hung, K.F.; et al. HSP40 co-chaperone protein Tid1 suppresses metastasis of head and neck cancer by inhibiting Galectin-7-TCF3-MMP9 axis signaling. *Theranostics* **2018**, *8*, 3841–3855. [CrossRef]
49. Wang, S.F.; Huang, K.H.; Tseng, W.C.; Lo, J.F.; Li, A.F.Y.; Fang, W.L.; Chen, C.F.; Yeh, T.S.; Chang, Y.L.; Chou, Y.C.; et al. DNAJA3/Tid1 Is Required for Mitochondrial DNA Maintenance and Regulates Migration and Invasion of Human Gastric Cancer Cells. *Cancers* **2020**, *12*, 3463. [CrossRef]
50. Wu, S.C.; Paul, A.; Ho, A.; Patel, K.R.; Allen, J.W.L.; Verkerke, H.; Arthur, C.M.; Stowell, S.R. Generation and Use of Recombinant Galectins. *Curr. Protoc.* **2021**, *1*, e63. [CrossRef]
51. Bibens-Laulan, N.; St-Pierre, Y. Intracellular galectin-7 expression in cancer cells results from an autocrine transcriptional mechanism and endocytosis of extracellular galectin-7. *PLoS ONE* **2017**, *12*, e0187194. [CrossRef] [PubMed]
52. Villeneuve, C.; Baricault, L.; Canelle, L.; Barboule, N.; Racca, C.; Monsarrat, B.; Magnaldo, T.; Larminat, F. Mitochondrial proteomic approach reveals galectin-7 as a novel BCL-2 binding protein in human cells. *Mol. Biol. Cell.* **2011**, *22*, 999–1013. [CrossRef]
53. Brewer, C.F.; Miceli, M.C.; Baum, L.G. Clusters, bundles, arrays and lattices: Novel mechanisms for lectin–saccharide-mediated cellular interactions. *Curr. Opin. Struct. Biol.* **2002**, *12*, 616–623. [CrossRef]
54. Demetriou, M.; Granovsky, M.; Quaggin, S.; Dennis, J.W. Negative regulation of T-cell activation and autoimmunity by Mgat5 N-glycosylation. *Nature* **2001**, *409*, 733–739. [CrossRef]
55. He, J.; Baum, L.G. Presentation of galectin-1 by extracellular matrix triggers T cell death. *J. Biol. Chem.* **2004**, *279*, 4705–4712. [CrossRef]
56. Garner, O.B.; Baum, L.G. Galectin–glycan lattices regulate cell-surface glycoprotein organization and signalling. *Biochem. Soc. Trans.* **2008**, *36*, 1472–1477. [CrossRef]
57. Bonzi, J.; Bornet, O.; Betzi, S.; Kasper, B.T.; Mahal, L.K.; Mancini, S.J.; Schiff, C.; Sebban-Kreuzer, C.; Guerlesquin, F.; Elantak, L. Pre-B cell receptor binding to galectin-1 modifies galectin-1/carbohydrate affinity to modulate specific galectin-1/glycan lattice interactions. *Nat. Commun.* **2015**, *6*, 6194. [CrossRef] [PubMed]
58. Advedissian, T.; Proux-Gillardeaux, V.; Nkosi, R.; Peyret, G.; Nguyen, T.; Poirier, F. E-cadherin dynamics is regulated by galectin-7 at epithelial cell surface. *Sci. Rep.* **2017**, *7*, 17086. [CrossRef] [PubMed]

59. Lochter, A.; Galosy, S.; Muschler, J.; Freedman, N.; Werb, Z.; Bissell, M.J. Matrix metalloproteinase stromelysin-1 triggers a cascade of molecular alterations that leads to stable epithelial-to-mesenchymal conversion and a premalignant phenotype in mammary epithelial cells. *J. Cell. Biol.* **1997**, *139*, 1861–1872. [CrossRef]
60. Noë, V.; Fingleton, B.; Jacobs, K.; Crawford, H.C.; Vermeulen, S.; Steelant, W.; Bruyneel, E.; Matrisian, L.M.; Mareel, M. Release of an invasion promoter E-cadherin fragment by matrilysin and stromelysin-1. *J. Cell Sci.* **2001**, *114*, 111–118. [CrossRef]
61. Zheng, G.; Lyons, J.G.; Tan, T.K.; Wang, Y.; Hsu, T.T.; Min, D.; Succar, L.; Rangan, G.K.; Hu, M.; Henderson, B.R.; et al. Disruption of E-cadherin by matrix metalloproteinase directly mediates epithelial-mesenchymal transition downstream of transforming growth factor-β1 in renal tubular epithelial cells. *Am. J. Pathol.* **2009**, *175*, 580–591. [CrossRef]
62. Lynch, C.C.; Vargo-Gogola, T.; Matrisian, L.M.; Fingleton, B. Cleavage of E-cadherin by matrix metalloproteinase-7 promotes cellular proliferation in nontransformed cell lines via activation of RhoA. *J. Oncol.* **2010**, *2010*, 530745. [CrossRef] [PubMed]
63. Liu, Y.; Burkhalter, R.; Symowicz, J.; Chaffin, K.; Ellerbroek, S.; Stack, M.S. Lysophosphatidic Acid disrupts junctional integrity and epithelial cohesion in ovarian cancer cells. *J. Oncol.* **2012**, *2012*, 501492. [CrossRef] [PubMed]
64. Trinidad, A.G.; Muller, P.A.; Cuellar, J.; Klejnot, M.; Nobis, M.; Valpuesta, J.M.; Vousden, K.H. Interaction of p53 with the CCT complex promotes protein folding and wild-type p53 activity. *Mol. Cell.* **2013**, *50*, 805–817. [CrossRef]
65. Rivlin, N.; Katz, S.; Doody, M.; Sheffer, M.; Horesh, S.; Molchadsky, A.; Koifman, G.; Shetzer, Y.; Goldfinger, N.; Rotter, V.; et al. Rescue of embryonic stem cells from cellular transformation by proteomic stabilization of mutant p53 and conversion into WT conformation. *Proc. Natl. Acad. Sci. USA* **2014**, *111*, 7006–7011. [CrossRef] [PubMed]
66. Advedissian, T.; Deshayes, F.; Viguier, M. Galectin-7 in epithelial homeostasis and carcinomas. *Int. J. Mol. Sci.* **2017**, *18*, 2760. [CrossRef] [PubMed]
67. Chen, H.L.; Chiang, P.C.; Lo, C.H.; Lo, Y.H.; Hsu, D.K.; Chen, H.Y.; Liu, F.T. Galectin-7 regulates keratinocyte proliferation and differentiation through JNK-miR-203-p63 signaling. *J. Investig. Dermatol.* **2016**, *136*, 182–191. [CrossRef]

Review

Functions and Inhibition of Galectin-7, an Emerging Target in Cellular Pathophysiology

Nishant V. Sewgobind, Sanne Albers and Roland J. Pieters *

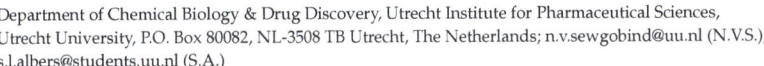

Department of Chemical Biology & Drug Discovery, Utrecht Institute for Pharmaceutical Sciences, Utrecht University, P.O. Box 80082, NL-3508 TB Utrecht, The Netherlands; n.v.sewgobind@uu.nl (N.V.S.); s.l.albers@students.uu.nl (S.A.)
* Correspondence: r.j.pieters@uu.nl; Tel.: +31-620-293-387

Abstract: Galectin-7 is a soluble unglycosylated lectin that is able to bind specifically to β-galactosides. It has been described to be involved in apoptosis, proliferation and differentiation, but also in cell adhesion and migration. Several disorders and diseases are discussed by covering the aforementioned biological processes. Structural features of galectin-7 are discussed as well as targeting the protein intracellularly or extracellularly. The exact molecular mechanisms that lie behind many biological processes involving galectin-7 are not known. It is therefore useful to come up with chemical probes or tools in order to obtain knowledge of the physiological processes. The objective of this review is to summarize the roles and functions of galectin-7 in the human body, providing reasons why it is necessary to design inhibitors for galectin-7, to give the reader structural insights and describe its current inhibitors.

Keywords: galectin-7; epithelial tissues; apoptosis; targeting; inhibitors

Citation: Sewgobind, N.V.; Albers, S.; Pieters, R.J. Functions and Inhibition of Galectin-7, an Emerging Target in Cellular Pathophysiology. *Biomolecules* **2021**, *11*, 1720. https://doi.org/10.3390/biom11111720

Academic Editor: Alexander V. Timoshenko

Received: 13 September 2021
Accepted: 13 November 2021
Published: 18 November 2021

Publisher's Note: MDPI stays neutral with regard to jurisdictional claims in published maps and institutional affiliations.

Copyright: © 2021 by the authors. Licensee MDPI, Basel, Switzerland. This article is an open access article distributed under the terms and conditions of the Creative Commons Attribution (CC BY) license (https://creativecommons.org/licenses/by/4.0/).

1. Introduction: Galectin-7

Galectin-7 belongs to a family of lectins that bind specifically to β-galactosides, i.e., the galectins. To date, 16 different members of galectins have been described in mammals, and 12 members have been characterized in humans. Although galectins share primary structural resemblance in their carbohydrate-recognition domains (CRDs), they are subdivided into three groups based on the molecular architecture. Prototype galectins contain a single CRD and form homodimers (human galectin-1, -2, -7, -10, -13, -14, and -16). Tandem-repeat galectins (human galectin-4, -8, -9, and -12) contain two CRDs that are connected by a short peptide linker that can range from 5 up to 70 amino acids. Finally, there are chimera-type galectins (only member: human galectin-3) when a single CRD is connected to an amino-terminal polypeptide non-lectin domain through which it can form oligomers [1–3].

Galectin-7 was first reported by Celis in 1995 while searching for keratinocyte proteins that may play a role in the maintenance of the normal phenotype and various skin diseases. One of these proteins corresponded to IEF17 in the keratinocyte database and had a shared identity with the galectin family. It contained all the amino acids that are central to the β-galactoside binding. For this reason, the protein was named galectin-7 after consultation with researchers in the field [4]. The findings of the Celis group were supported by Magnaldo and colleagues [5]. Both groups concluded that galectin-7 is a keratinocyte-specific marker often found in all layers of the epidermis and other stratified epithelia of tissues; in the tongue, cornea, esophagus, stomach, anus, Hassal's corpuscles of the thymus and even in myoepithelial cells of the mammary gland epithelium [6].

Galectin-7 is synthesized in the cytoplasm, and it accumulates in the cytosol or nucleus before secretion to the outer plasma membrane or extracellular matrix. Like all other galectins, the secretion or export of galectin-7 from the cytoplasm occurs via an undefined nonclassical secretory mechanism [1,7].

The X-ray crystal structure of human galectin-7 in its native form is described by the Celis and Acharya groups as a dimer. It has a significant amino acid sequence identity to the known prototype of galectin-1, -2 and -10 [8]. Although it was reported as a monomer in solution [8,9], the observed molecular weight as determined by ultracentrifugation and sedimentation experiments strongly suggests that it is a dimer in solution [10]. Nesmelova and co-workers confirmed these findings and reported ^{1}H, ^{13}C, and ^{15}N chemical shift assignments for the human galectin-7 dimer as determined by heteronuclear, triple resonance NMR spectroscopy in solution [11].

Acting intra- or extracellularly, galectin-7 participates in diverse processes, such as controlling apoptosis, cell migration and cell adhesion. In addition, it also plays a crucial role in the re-epithelialization process of corneal or epidermal wounds and in several human diseases/disorders, such as cancer [12]. Because of its diverse roles in human cellular pathology and the fact that the precise modes of action of galectin-7 are not well understood in many cases, there is a need for strong inhibitors that target galectin-7 specifically in order to provide insights into the biological mechanisms and as a string point for therapeutic intervention. This review intends to present an up-to-date overview on galectin-7 and its various roles in the human body from a chemical as well as a biological point of view. We aim to do this by covering the following subjects: biological importance of galectin-7, targeting galectin-7, and structural features. We will refer to the current synthetic inhibitors of galectin-7.

2. Galectin-7, a Convergence of Pathology with Physiology

Being mainly expressed in stratified epithelia, galectin-7 is described in epithelial tissues as being involved in apoptotic responses, proliferation and differentiation, but also in cell adhesion and migration [13]. In the following section, we will examine its involvement by elaborating on several biological processes and disorders which are linked to (the functions of) galectin-7.

2.1. Role in Epidermal Homeostasis of Skin, Corneal and Periodontal Tissue

Bernerd et al. showed that UVB irradiation of skin keratinocytes, reconstructed in vitro and of human skin *ex vivo*, lead to sunburn/apoptotic skin keratinocytes. These sunburn/apoptotic keratinocytes express higher levels of galectin-7 than other keratinocytes, suggesting that galectin-7 is strongly associated with UVB-induced apoptosis in the epidermis [14].

The previously obtained result by Bernerd et al. was confirmed and revealed that the expression of galectin-7 is induced by UVB irradiation and also *cis*-UCA (*cis*-urocanic acid). The latter is an epidermal chromophore that undergoes *trans* to *cis* isomerization after UVB irradiation. Notably, *cis*-UCA is a potent inhibitor of cutaneous acquired immunity. It was concluded that galectin-7 induces apoptosis and demonstrated that it is highly expressed in the epidermis of patients with actinic keratosis, compared with normal skin [15].

Gendronneau et al. found evidence for the role of galectin-7 in the process of skin wound healing. They generated galectin-7–deficient mice that were viable and exhibited no phenotypical abnormalities in skin structure, organization, differentiation or expression of epidermal markers. However, the epidermal response to UVB radiation as well as mechanical injury in vivo proved to be disturbed. Sunburn cells occurred earlier, the apoptotic response was less acute, and it lasted longer, compared with wt (wild-type) tissue. It was concluded that galectin-7 modulates keratinocyte apoptosis and proliferation as well as migration [16].

In addition, the same group studied the role of galectin-7 overexpression in basal keratinocytes of skin repair after environmental stress. The epidermal response to a scratch on the surface was delayed (timing of wound closure). The re-epithelialization of cells located at each edge of the wound depends on cellular interactions, notably through adherens junctions. It was proposed that the overexpression of galectin-7 causes the loosening of adherens junctions and hence, the delay in wound closure. When the transgenic mice (with keratinocytes overexpressing galectin-7) were exposed to UVB radiation, more keratinocyte apoptosis was induced. The effects on the maintenance of epidermal homeostasis of deficient and overexpressed galectin-7 were proven to be very similar [17]. Advedissian and co-workers continued the study of the involvement of galectin-7 in cell migration and found that there is an interaction with a key component of adherens junctions, E-cadherin. They showed an interaction between galectin-7 and E-cadherin at the plasma membrane, which causes intercellular adhesion [18].

Mechanistic evidence was provided for the aforementioned findings of Gendronneau et al. The galectin-7 knockdown results in reduced differentiation and increased proliferation of keratinocytes. Moreover, it was shown that galectin-7 positively regulates microRNA (miR)-203 expression, which in turn is used for regulating keratinocyte differentiation and proliferation. To determine how galectin-7 regulates keratinocyte proliferation and differentiation through miR-203, the expression of a known miR-203 target, p63 (an essential transcription factor involved in skin development), in galectin-7 knockdown cells was examined. Knocking down either galectin-7 or miR-203 in keratinocytes increased the expression of p63. The rescue of miR-203 expression in a galectin-7 knockdown model reduced p63 expression. Further extensive research showed that increased galectin-7 expression upregulates c-Jun N-terminal kinase 1 (JNK1) by a direct interaction, which is required for miR-203 expression. Finally, they established that galectin-7 has an intracellular function in keratinocytes through the JNK1-miR-203-p63 pathway [19]. More recently, it was found that the expression of galectin-7 is reduced by cytokines in the skin lesions of patients with psoriasis. This results in the hyperproliferation of keratinocytes and skin inflammation [20].

Systemic sclerosis (SSc) is a multisystem connective tissue disorder characterized by vascular injury, fibrosis of the skin, various internal organs following autoimmune inflammation and tissue injury [21]. Saigusa and co-workers investigated the potential contribution of galectin-7 to the development of clinical manifestations in SSc, using clinical samples from patients and cultured keratinocytes. Galectin-7 proved to be remarkably downregulated in the basal and suprabasal layers of the lesional epidermis of involved skin in contrast to the abundant expression throughout the epidermis of normal control skin. In addition, SSc patients with diffuse pigmentation and those with esophageal dysfunction had significantly decreased serum galectin-7 levels as compared to those without each symptom. Suppression of the galectin-7 level is believed to be stimulated by autocrine endothelin signaling stimulation in SSc keratinocytes [22].

Patients who suffer from diabetes mellitus [23] have a high risk of impaired wound healing that sometimes may lead to infection and amputation. As cell migration is an important process involved in proper wound healing, Huang and co-workers demonstrated that a high glucose environment reduced galectin-7 expression in keratinocytes, due to enhanced O-GlcNAc (O-linked N-acetyl-D-glucosamine) glycosylation of certain regulators of galectin-7 expression. This dysregulation of galectin-7 causes a significant reduction of keratinocyte migration and thus, improper wound healing [24]. The context of this dysregulation can be associated with O-GlcNAc-mediated processes controlling cellular differentiation [25]. A more detailed review regarding the re-epithelialization of skin wounds is reported [26].

In their search for novel, galectin-based therapeutic strategies for the treatment of non-healing corneal tissue epithelial defects, Cao et al. demonstrated via Western blot analysis that healing corneas contained increased levels of galectin-7 throughout the epithelium, compared with normal corneas after injury. Furthermore, it was reported that exogenous galectin-7 stimulated the rate of corneal epithelial wound closure. Inhibition of this stimulatory effect of galectin-7 occurred by a competing lactose but not by non-binding sucrose. It was suggested that the CRD of the lectin is directly involved in wound closure [27].

Er: YAG (erbium-doped yttrium–aluminum–garnet) laser therapy is used for periodontal treatment by removing soft and hard tissues as well as calculus, with minimal heat-related side effects. The bactericidal effect makes the therapy even more useful. Er:

YAG laser irradiation promotes faster adhesion and growth of human gingival fibroblasts (HGFs) and periodontal ligament fibroblasts (PDL fibroblasts). The cell proliferation of HGFs is reported to be stimulated, and this might be caused due to an increase in the protein expression of galectin-7 in the HGFs. Er: YAG laser irradiation causes a direct effect of promoting proliferation, migration, and invasion of PDL fibroblasts through the upregulation of galectin-7, yet its signaling pathway needs to be verified [28].

2.2. Roles in Cancer

Approximately 85% of cancers occur in epithelial cells: the carcinomas [29]. Like many galectins, galectin-7 displays opposite effects in terms of tumor progression from one histological type to another. It may contribute to the growth and/or development of certain tumor types, while acting negatively on the development of other tumor types [12]. Galectin-7 does not only have a role in carcinomas [13], but also in lymphomas and melanomas by contributing either to neoplastic transformation and tumor progression through the regulation of cell growth, cell cycle, angiogenesis, apoptosis and cell migration. In addition, galectin-7 may have a protective effect on cancer, depending on the tissue type [30]. Hanahan and Weinberg defined hallmarks of most cancers which describe the biological capabilities essential for carcinogenesis [31]. There are a number of papers published regarding the subject of cancer (development) and the roles of (targeting) galectins, and even galectin-7 in particular [13,30,32–50]. Nevertheless, our goal for this section is to provide the reader with a brief overview of the presence and roles of galectin-7 in most cancers/cancer types by covering mostly recent publications.

Analysis of the expression of galectin-7 in benign and malignant thyroid cancers showed a downregulation of galectin-7 in adenomas, compared to carcinomas [51]. It was shown that galectin-7 is constitutively expressed in aggressive (metastatic) lymphoma cells at both mRNA and protein levels. Highly metastatic variants of the lymphoma cell line showed strong upregulation of galectin-7 in the spleen, the thymus and kidneys, due to the methylation of the galectin-7 gene (*LGALS7*) [52,53]. Methylation of the *LGALS7* gene, leading to the silencing of galectin-7 during gastric cancer tumorigenesis, was also suggested by Kim and colleagues. They revealed significantly lower expression levels of galectin-7 in malignant tissues of gastric cancer patients, compared with matched normal tissues. The overexpression of galectin-7 in AGS gastric adenocarcinoma cells suppressed cell proliferation, migration, and invasion, whereas the removal of galectin-7 in KATO III gastric carcinoma cells reversed these properties [54].

To determine its critical role in lymphoma progression, Demers and co-workers hypothesized two years later that the promalignant activity of galectin-7 in thymic lymphoma is related to its capacity to induce MMP-9 (matrix metalloproteinase-9, a metastatic gene) expression. Their hypothesis was based on the evidence that galectin-7 transfectants have higher levels of MMP-9 expression, while the addition of lactose completely inhibits the expression of MMP-9. Furthermore, murine or human recombinant galectin-7 induces the expression of MMP-9 in both mouse and human lymphoma cells [55]. In continuation, the same group found evidence that galectin-7 is expressed in human lymphoid malignancies and proposed that it is a critical tumor-modulating gene that controls the dissemination of lymphoma cells via MMP-9 [56]. The reader is also referred to the review by St-Pierre regarding the relationships between galectin-7, p53 and MMP-9 during cancer progression [6].

Galectin-7 was reported to be highly expressed in ESCC (esophageal squamous cell carcinoma) during a study that was designed to isolate and identify ESCC biomarkers, using proteomic tools. The level of galectin-7 expression was related to the degree of ESCC differentiation [57].

Galectin-7 is also believed to increase the invasive behavior of breast cancer cells; the ability to metastasize to the lungs and bones increased in mouse models. It is believed that breast cancer cells overexpressing galectin-7 are related to the ability of galectin-7 to protect against apoptosis [58]. An important mediator of galectin-7 gene activation in breast cancer cells, CCAAT/enhancer-binding protein beta or C/EBPβ, was suggested to contribute by the same group in 2014 [59]. Grosset et al. generated a mutant form of galectin-7 in which arginine 74 was mutated to obtain galectin-7^{R74S}, a CRD-defective mutant form of galectin-7. They demonstrated that breast cancer cells expressing mutated galectin-7 were equally or even more resistant to drug-induced apoptosis, compared to cells expressing wt galectin-7 [60]. In addition, galectin-7 proved to accelerate tumor progression in one of the most aggressive forms of breast cancer (HER-2 positive) as was published in a subsequent study, using genetically engineered galectin-7–deficient mice [61].

The observation that galectin-7 may have immunosuppressive properties was made by Labrie and co-workers while investigating the expression of galectin-7 in epithelial ovarian cancer (EOC). It was found that galectin-7 increased the invasive behavior of ovarian cancer cells by inducing MMP-9 and increasing cell motility. EOC cells can also secrete galectin-7. Recombinant human galectin-7 kills Jurkat T cells and human peripheral T cells [62].

In contrast, galectin-7 reduces the invasive behaviors of prostate cancer cells by inhibiting their motility. Galectin-7 is found to be downregulated in prostate cancer cells, and the expression of galectin-7 in prostate cancer cells increases their sensitivity to apoptosis in response to chemotherapeutic agents. The group of St-Pierre showed that the ability of galectin-7 to modulate apoptosis was independent of its CRD activity by using a CRD-defective mutant, i.e., galectin-7^{R74S}. However, CRD activity proved to be necessary to inhibit the invasive behaviors of prostate cancer cells. In vivo, galectin-7 overexpression in prostate cancer cells led to a significant reduction in tumor size, while its CRD-defective mutant form significantly increased tumor growth [63].

The group of Lo demonstrated that human tumorous imaginal disc (Tid1), a heat shock protein (Hsp40), reduces head and neck squamous cell carcinoma (HNSCC) malignancy. It was found that galectin-7 was one of the proteins that interact with Tid1 and the levels of expression of both proteins were measured in HNSCC patients. Low Tid1 and high galectin-7 expression predicted poor overall survival in HNSCC. The interaction between Tid1 and galectin-7 was bridged by N-linked glycosylated Tid1. It is believed that N-linked glycosylation of Tid1 is required to interact with galectin-7 to downregulate galectin-7, which in turn can attenuate cancer progression and metastasis. Galectin-7 played a critical role in promoting tumorigenesis and metastatic progression by enhancing the transcriptional activity of TCF3 transcription factor through elevating MMP-9 expression [64].

Evidence was provided for the pro-invasive activity of galectin-7 in oral squamous cell carcinoma (OSCC) by inducing the expression of not only MMP-9, but also MMP-2. It was observed that galectin-7 overexpression resulted in significant upregulation of MMP-2 and MMP-9. On the other hand, silencing MMP-2 or MMP-9 significantly impaired the invasiveness of OSCC cells that overexpressed galectin-7. In order to explain these results, the signaling pathways involved were investigated. It was concluded that increasing galectin-7 expression significantly enhanced the phosphorylation and activation of extracellular signal-related kinase (ERK) and c-Jun N-terminal kinase (JNK). Moreover, the pharmacological inhibition of ERK or JNK activity significantly reduced OSCC cell invasiveness induced by galectin-7 overexpression [65]. The signaling pathways which direct hypersensitized carcinoma cells to apoptosis was also earlier observed in malignant peripheral nerve sheath tumor cells [66].

The proapoptotic activity of galectin-7 was also attributed to activation of the JNK pathway in cervical and colon cancer [67,68]. Zhu and colleagues confirmed these results by revealing a role for galectin-7 in sensitizing cervical squamous cancer cells to paclitaxel treatment. A galectin-7 knockdown in the cancer cells showed increased viability against paclitaxel-induced apoptosis [69]. As galectin-7 is negatively regulated in cervical cancer, Higareda-Almaraz and co-workers demonstrated the link between the pro-apoptotic response triggered by cancer and the anti-tumoral activity of the immune system. Galectin-7 re-expression affects the regulation of molecular networks in cervical cancer that are involved in some of the cancer hallmarks, such as metabolism, growth control, invasion and evasion of apoptosis. The effect of galectin-7 extends to the microenvironment, where the reconstitution of galectin-7 leads to a change of regulation and interaction networks [70].

It was demonstrated by Menkhorst and colleagues that galectin-7 production increased in endometrial cancer with increasing cancer grade; galectin-7 may promote the metastasis of endometrial cancer by reducing cell–cell adhesion and enhancing cell migration. Furthermore, it was also established that galectin-7 had no significant effect on proliferation or apoptosis [71].

Matsui and co-workers showed that bladder cancer cells expressing upregulated galectin-7 tended to respond more sensitively to chemotherapy, compared to urothelial tumor cells having lower levels of galectin-7 [72].

Kopitz et al. demonstrated for human neuroblastoma cells that galectin-7 is a negative growth regulator not by apoptosis, but rather a switch from proliferation to differentiation of the cancer cells [73].

2.3. Role in Pre-Eclampsia, Menstruation and Recurrent Pregnancy Loss

Pre-eclampsia is a hypertensive disorder of pregnancy and causes maternal and fetal morbidity and mortality. It is defined as the presence of hypertension, proteinuria or other end organs, such as liver or brain, damage occurring after 20 weeks of pregnancy. Severe forms of pre-eclampsia can be complicated by renal, cardiac, pulmonary, hepatic, and neurological dysfunction, hematologic disturbances, fetal growth restriction, stillbirth and maternal death [74,75]. Recurrent pregnancy loss is a prevalent and distressing disorder, defined as the spontaneous end of pregnancy before an embryo has reached viability until 20–24 weeks of gestation [76,77].

Members of the galectin family are expressed within the female reproductive tract and have been shown to be involved in multiple biological functions that support the progression and regulation of implantation and pregnancy via cell adhesion and migration, immune cell activation, apoptosis and hormone production to name a few [78,79].

Menkhorst et al. investigated the expression of galectin-7 in the endometrium during the menstrual cycle of normally fertile women and women who have a history of miscarriage to see whether there is an association with tissue/serum levels of galectin-7 and miscarriage. Galectin-7 was immunolocalized to the endometrial luminal and glandular epithelium in normally fertile women. The serum concentration of galectin-7 proved to be significantly elevated at week 6 of gestation in women with a viable fetus with a history of miscarriage, compared to normal healthy pregnancies. Furthermore, galectin-7 was aberrantly expressed in the non-pregnant endometrium of women with a history of miscarriage. These findings suggested that this allows for inappropriate blastocyst implantation. They demonstrated a role for galectin-7 on trophoblast–endometrial epithelial cell adhesion by acting as an adhesion molecule [80]. In a subsequent study, the same group showed that galectin-7 serum concentration was significantly elevated during weeks 10–12 and 17–20 of gestation in women who went on to develop pre-eclampsia, compared to women with normal pregnancies. It was also proposed that the elevated serum galectin-7 associated with pre-eclampsia may be due to placental oxidative stress and/or hypomethylation [81].

Evans and colleagues were able to identify and compare endometrial expression of galectin-7 in women with normal endometrial repair versus women with amenorrhea who do not experience endometrial breakdown and repair. Their study demonstrated the presence of galectin-7 not only within the menstruating endometrium (being produced by the premenstrual and menstrual endometrium), but also in menstrual fluid. They also established that galectin-7 enhances endometrial re-epithelialization and elucidated the mechanism by which galectin-7 mediates endometrial epithelial wound repair. Galectin-7–mediated re-epithelialization is dependent on integrin-mediated signaling and elevates the expression of ECM factors which are involved in repair in other tissues [80].

In order to study the function of (among other) prototype galectins in placental tissue, the expression of galectin-7 in the placenta in cases of spontaneous abortions (SPA) and recurrent abortions (RA) in the first trimester was analyzed. Galectin-7 was found in the syncytiotrophoblast in placentas after induced abortion and with weaker staining in the decidua. In SPA and RA first-trimester placentas, the expression of galectin-7 in the villous trophoblast/syncytiotrophoblast was significantly lower [82].

In order to determine the role of galectin-7 in the placenta, Menkhorst and co-workers demonstrated that elevated galectin-7 during placental formation contributes to abnormal placentation, thus leading to the development of pre-eclampsia. Augmented galectin-7 during the period of placental formation in mice caused hypertension and albuminuria, and the authors hypothesize that in women, galectin-7 acts via the placenta to induce the systemic features of pre-eclampsia via impaired placental formation, placental inflammation and the placental release of anti-angiogenic factors [83].

2.4. Roles in Allergic Inflammatory and Autoimmune Diseases

Inflammatory autoimmune diseases have large numbers of pathologies characterized by various factors that can contribute to a breakdown in self-tolerance or inflammation dysregulation. Immune cells are sensitive to galectins, and they are important regulators of inflammation or autoimmunity, making them therapeutic targets for some inflammatory autoimmune diseases [84].

Galectins control a wide range of cells involved in the allergic inflammatory diseases by modulating the biological activities of the cells. Hence, galectins may influence the development and course of allergic diseases. Evidence for the involvement of galectins in terms of immunoregulatory activities has been gathered in the pathogenesis of allergic conjunctivitis, atopic dermatitis, asthma and food allergy in the past few years [85].

During their study, Niiyama and colleagues assessed whether galectin-7 could be utilized as an indicator (biomarker) of skin barrier disruption and as an index of local skin symptoms in atopic dermatitis (AD) patients. Atopic dermatitis is a chronic, relapsing inflammatory skin disease characterized by pruritic and eczematous skin lesions. Skin barrier disruption is an important contributing factor in the pathogenesis of AD, as the disruption of the skin barrier allows the penetration of allergens into dry skin, inducing an itching sensation. Galectin-7 expression in keratinocytes increased after skin barrier disruption, and an overexpression in the stratum corneum was detected in tape-stripped samples. Measurement of the galectin-7 content in the stratum corneum might be useful for the evaluation of the skin barrier function in dry skin conditions, such as AD [86].

Niiyama's results were confirmed, and the production mechanism and functional role of galectin-7 in AD patients was investigated. A galectin-7 knockdown experiment on a 3D-reconstructed epidermis was performed; it resulted that endogenous galectin-7 protects IL-4/IL-13–induced disruption of cell-to-cell adhesion and/or cell-to-extracellular matrix adhesion. In addition, IL-4/IL-13–induced galectin-7 release from keratinocytes reflects the skin barrier impairment in AD patients [87].

Luo and co-workers showed that galectin-7 promotes activated CD4+ T cell immunity. The modes of action include the promotion, proliferation and polarization of Th1/2 cells balance toward Th1 in activated CD4+ T cells, and the elevation of immune-enhancement factors in the microenvironment by inhibiting the TGFβ/Smad3 pathway. This means that galectin-7 may have anti-inflammation effects, and it can induce autoimmune disease and transplantation rejection [88].

The airway epithelium plays an important role in the development of allergic inflammation, remodeling, and bronchial hyper-responsiveness. Moreover, the bronchial epithelium plays an important role in immune regulation during the initiation of allergic responses. The integrity of airway epithelial layer structure is the key to the airway barrier and local microenvironment homeostasis. Destruction of the integrity of the epithelium leads to depletion of the ordered airway barrier and increases sensitivity to viral infections and allergens. Eventually, this leads to airway inflammations, such as asthma or chronic obstructive pulmonary disease (COPD) [89].

As galectin-7 was identified to be overexpressed and increased apoptosis occurred in bronchial epithelial cells in asthma, Sun and Zhang investigated the effect of galectin-7 on the apoptosis of human bronchial epithelial cells. They were able to demonstrate that galectin-7 silencing inhibited TGF-β1–induced (growth factor that promotes multiple cell apoptosis, also elevated in asthmatic patients) apoptosis in airway epithelial cells via blocking the JNK pathway [90].

Encouraged by their previously obtained results, Tian et al. showed that the expression of galectin-7 mRNA and protein in bronchial epithelial cells of children with asthma were both increased, and the expression of galectin-7 mainly occurred in apoptotic bronchial epithelial cells. The overexpression of galectin-7 in transgenic mice (Tg(+) mice) showed abnormal airway structures in embryos and after birth; a thin and disordered epithelium layer was observed. Galectin-7 was localized in the cytoplasm and nucleus of bronchial epithelial cells. Increased apoptosis was mediated through the mitochondrial release of cytochrome c; upregulated JNK1 activation and expression destroys the airway epithelium barrier, which predisposes the airways to RSV respiratory syncytial virus (RSV), ovalbumin or OVA-induced epithelial apoptosis. Taken together, the aforementioned results suggest that galectin-7 causes airway structural defects, injury, and other asthma responses [89].

Intracellular galectin-7 proved to be involved in bacterial autophagy, as immunoblotting analysis by the group of Lin and co-workers revealed low-level galectin-7 expression in HeLa cells. Examination of HaCaT cells revealed that intracellular galectin-7 clearly colocalized with and surrounded group A streptococcus (GAS), an intracellular bacterium. GAS proliferation was increased following galectin-7 knockdown in HaCaT cells, which indicates that intracellular galectin-7 plays a critical role in intracellular immunity in the response against bacterial infection [91].

2.5. Role in Transplant Rejection

Based on the facts that galectin-7 is related to immune responses in transplantation and increased expression of galectin-7 in serum from renal allograft recipients (compared with normal volunteers) was identified, Luo and colleagues investigated the galectin-7 response to acute rejection of mouse cardiac allografts. More specifically, they showed that the expression of galectin-7 increased with the severity of allograft rejection. Furthermore, they demonstrated that the upregulation of galectin-7 expression in the allografts was directly related to the T cell response. The results showed that infiltrating lymphocytes and endothelial cells in the allografts expressed large amounts of galectin-7 located in the cytoplasm and nucleus of cardiomyocytes, endothelial cells, and infiltrating lymphocytes. This was not observed in native hearts or isografts, and it is believed that galectin-7 plays a crucial role to accelerate allograft rejection [92].

Table 1 summarizes the various pathophysiological roles and mode of action displayed by galectin-7.

Table 1. Various roles of galectin-7 along with its modes of action.

Role	Mode of Action	References
Epidermal homeostasis of skin	Regulation of keratinocyte proliferation, differentiation and migration	[14–25]
Re-epithelialization of corneal wounds	Mediating corneal epithelial cell migration	[27]
Wound healing of PDL fibroblasts	Promoting proliferation, migration and invasion of PDL fibroblasts	[28]
Promalignant activity in gastric cancer	Lower expression levels of galectin-7 cause increase in gastric cancer cell proliferation, migration and invasion	[54]
Promalignant activity in thymic lymphoma + HNSCC	Induce MMP-9 expression	[6,55,56,64]
Increasing invasive behavior of breast cancer cells	Protecting breast cancer cells from apoptosis	[58–61]

Table 1. Cont.

Role	Mode of Action	References
Reducing invasive behavior of prostate cancer cells	Inhibiting motility prostate cancer cells	[63]
Pro-invasive activity in oral squamous cell carcinoma	Induce MMP-2 and MMP-9 expression	[65,66]
Protective effect on the survival of cervical squamous carcinoma patients	Inhibiting MMP-9 expression and cell invasion in cervical squamous carcinoma cells	[67,69,70]
Promoting metastasis of endometrial cancer	Reducing cell–cell adhesion and enhancing cell migration	[71]
Sensitizing bladder cancer cells to chemotherapy	Increase generation of reactive oxygen species	[72]
Negative growth regulator of neuroblastoma cells	Switch from proliferation to differentiation of cancer cells	[73]
Mediation of endometrial epithelial wound repair	Endometrial re-epithelialization is dependent on integrin mediated signaling	[80]
Abnormal placentation hence leading to the development of pre-eclampsia	Acting via the placenta to induce the systemic features of pre-eclampsia via impaired placental formation, placental inflammation and placental release of anti-angiogenic factors	[81,83]
Skin barrier impairment in keratinocytes	Protecting disruption of cell-to-cell adhesion and/or cell-to-extracellular matrix adhesion	[87]
Anti-inflammation effects, inducing autoimmune disease and transplantation rejection	Promotion, proliferation and polarization of Th1/2 cells	[88]
Causing airway structural defects, injury, and other asthma responses	Increased apoptosis occurred in bronchial epithelial cells in asthma	[89,90]
Intracellular immunity in the response against bacterial infection	Colocalizing with and surrounding group A Streptococcus (GAS, intracellular bacterium)	[91]
Accelerating allograft rejection	Up-regulation of galectin-7 expression in the allografts was directly related to T cell response	[92]

Despite of the many findings mentioned in this section, much has to be discovered at the molecular level of several pathophysiological processes. Knocking down or not expressing galectin-7 would not be sufficient in many cases, and hence it may cause any other complications.

3. Drug Potential of Galectin-7 Inhibitors and Galectin-7 as a Biomarker

It may not always be necessary to solely inhibit galectin-7 either intracellularly of extracellularly. Clearly, inhibitors will eventually aid the elucidation of molecular mechanisms/pathways in a variety of biological processes, but it is also of great interest to support the diagnosis and prognosis of several disorders. The second part of this section will deal with the use of galectin-7 as a biomarker in some cases where it is reported to be overexpressed.

3.1. Drug Potential of Galectin-7 Inhibitors

The activity and function of any galectin can be multi-faceted, due to galectin self-association, and/or interactions with cell surface glycans/other biomolecules, both extracellularly and intracellularly [93].

The approach of carbohydrate-derived small-molecule inhibitors to target the CRD of galectins is mainly based on the use of chemically modified natural galectin ligands, such as the disaccharides lactose (Lac) or N-acetyllactosamine (LacNAc). As the development of these inhibitors involves a full understanding of the biochemistry of galectin–glycan interactions, efforts are being made to generate galectin inhibitors that target individual members (particularly galectins-1, -3 and -7) of the family with higher affinity and selectivity [50]. Most of the current inhibitors only block extracellular functions of a given galectin and neglect intracellular functions [36], except for galectin-3 for which Stegmayr and co-workers were able to synthesize and evaluate the roles of intracellular and extracellular galectin-3 inhibitors [94].

It is warranted in impaired diabetic wound healing to identify and elucidate the status of specific galectin-7 regulating molecules in a high glucose environment. Furthermore, elucidating the specific molecular dysfunction in keratinocytes associated with individual diabetic phenotype will likely result in the development of more effective and personalized therapeutic strategies for optimal wound management in patients diagnosed with diabetes [24].

Wan and colleagues concluded in their review that inhibiting the contribution of galectin-7 to allergic inflammation should be achievable by generating antibodies with the proviso that (1) the antibodies do not exhibit cross reactivities to other galectins and (2) the galectin's contribution should go through extracellular actions. If this is not the case, antibodies will not be suitable, and cell-permeable inhibitors are required [85].

As for the resistance to anticancer therapies, intracellular versus extracellular functions of galectins are an important aspect to keep in mind to understand the role of these proteins in anticancer therapy resistance, as well as in the design of galectin-based cancer treatments [48].

Many publications call for inhibitors and methods for targeting galectin-7 and/or modulating its activity. Yet, no specific galectin-7 inhibitor is available. For example, extracellular galectin-7 promotes cancer via binding to cell surface receptors of cancer cells and induce *de novo* transcriptional activation of *LGALS7*, which in turn render cells resistant to pro-apoptotic drugs. Another example is displayed by the binding of extracellular galectin-7 to glycoreceptors expressed in infiltrated immune cells that triggers a cascade of signaling events, leading to the apoptosis of cancer-killing T cells, or alters their regulatory functions, helping tumors evade anti-tumor immunity [95].

In another example, where the expression of galectin-7 in epithelial ovarian cancer (EOC) is evaluated, it was observed that extracellular galectin-7 is released outside the cells. Galectin-7 is believed to have a significant impact on tumor progression by inducing immunosuppression and increasing the invasive behavior of tumor cells that eventually leads to metastasis. Targeting galectin-7 may represent a valuable strategy to overcome cancer-associated immunosuppression and the prevention of metastasis in EOC [62].

Grosset and co-workers confirmed the expression of galectin-7 in the cytosolic and nuclear compartments of breast cancer cells and the ability of galectin-7 to translocate to mitochondria. However, whether the resistance of breast cancer cells to apoptosis is dependent on the intracellular localization of galectin-7 remains unknown [60,96]. Bibens-Laulan and St-Pierre uncovered how galectin-7 traffics between both intracellular and extracellular compartments in ovarian and breast cancer cells. They reported that extracellular galectin-7 plays a central role in controlling intracellular galectin-7 in cells via two mechanisms: firstly, by increasing the transcriptional activation of *LGALS7* gene transcription, and secondly via re-entry into the cells. However, whether re-entry is dependent on the glycan-binding site of galectin-7 is unknown [97]. Girotti and colleagues concluded that it is still not clear whether intracellular or extracellular activities of galectins should be targeted to halt tumor progression [49]. In addition to their extracellular function, the fact that galectins can alter tumor progression through their interaction with intracellular ligands (sometimes even independently of their CRD) calls for a change in the basic assumptions and may force scientists re-design strategies in order to develop galectin antagonists for the treatment of cancer [98].

3.2. Galectin-7 as a Biomarker

The biomarkers field is shifting from tests analyzing single targets to multiplexed analysis of numerous proteins with or without post-translational modifications or exclusively glycans. These improvements are possible, due to the advances in analytical (detection) techniques, such as mass spectrometry for glycan analyses and lectin-antibody array methodologies. Indeed, a more specific (and early) pathological (i.e., cancer) diagnosis will result in earlier disease detection, improved disease monitoring and assistance and eventually successful patient-specific therapies. However, despite all the literature supporting the value of biomarkers for prognostic and monitoring applications, these tests suffer from limited specificity and sensitivity, which makes it a challenge to come up with a useful biomarker [38,41].

Stevens–Johnson syndrome (SJS) and toxic epidermal necrolysis (TEN) are severe cutaneous adverse drug reactions (cADRs) that can cause a life-threatening condition and late sequelae. Galectin-7 was reported to be one of the seven proteins that showed higher concentrations in the samples of SJS/TEN samples than in the non-severe cADR samples. The proteins were quantitated, using selected/multiple reaction monitoring (SRM/MRM) with stable synthetic isotope-labeled peptides as an internal control. The technique might be useful in the search for a potential SJS/TEN biomarker and key candidates involved in SJS/TEN pathogenesis [99].

Although it was proposed that galectin-7 serves as a negative prognostic factor in ovarian cancer by two independent groups [62,100], Schulz and colleagues studied the prognostic value of galectin-7 (among other galectins) in patients with epithelial ovarian cancer. The staining of galectin-7 in tumor cells was mainly observed in the cytoplasm; only a few individual cases showed nuclear staining. In addition, a significantly reduced overall survival was observed for cases with a high galectin-7 expression and a better survival for galectin-7 negative cases. Lower galectin-7 expression was confirmed as an independent prognostic factor for overall survival in ovarian cancer [101].

Trebo et al. suggested that galectin-7 might be an independent negative prognostic factor in breast cancer and a therapeutic target, especially in HER2-positive breast cancer. The expression of galectin-7 was observed in the cytoplasm as well as in the nucleus of breast cancer cells. Galectin-7 expression in the cytoplasm as well as in the nucleus was significantly higher in no special type (NST) tumors, compared to non-NST tumors. In addition, galectin-7 was also present in macrophages next to the tumor cells. These macrophages might also provide a source of extracellular galectin-7 for tumor cells and might regulate the intracellular galectin-7 pool. Combining the results suggested that galectin-7 might be an independent negative prognostic factor in breast cancer and a therapeutic target, especially in HER2-positive breast cancer [102].

Matsukawa and co-workers aimed to identify predictors of tumor sensitivity to preoperative radiotherapy/chemotherapy for oral squamous cell carcinoma (OSCC) in order to allow oncologists to determine optimum therapeutic strategies. They identified galectin-7 as a potential predictive marker of chemotherapy and/or radiotherapy resistance, as in vitro overexpression of galectin-7 significantly decreased cell viability after chemotherapy (most likely due to growth arrest rather than apoptosis) in the OSCC cell line [103].

Kim et al. indicated that, given the fact that the expression of galectin-7 in gastric cancer is regulated by DNA hypermethylation (as discussed previously in Section 2), the DNA methylation of galectin-7 is a promising candidate biomarker for application in gastric cancer [54].

In order to develop new inhibitors for galectin-7, one must gain knowledge regarding structural information of the binding pocket and to have a understanding of the preferred interactions between the target protein and small molecules. The following section will cover these aspects.

4. Structural Features

Being involved in a variety of physiological processes, many of which are directly linked to immunity and disease, deciphering the complex structures of galectins and their interactions with carbohydrates is of fundamental relevance to gain a deeper understanding of the underlying biological processes involved, the different affinities for different carbohydrates and non-carbohydrate ligands and to develop potential therapeutic interventions [104]. The crystal structures of most of the galectins, also in complexes with glycan ligands, are known. The CRD (consisting of ~130–140 residues) of most galectins is comprised of five- and six-stranded anti-parallel β-sheets arranged in a β-sandwich (sometimes referred to as "jelly roll") configuration that lacks an α-helix. The subunits in the dimeric galectin-7 are related by a twofold rotational axis perpendicular to the plane of the β-sheets [1].

The first crystal structures of human galectin-7, in free form and in the presence of galactose, galactosamine, lactose, and N-acetyl-lactosamine, were published by Leonidas et al. The structure of galectin-7 shows a fold similar to that of prototypes galectin-1 and -2, but has a greater similarity to the related galectin-10. Unlike galectin-1 and -2 that are both dimeric galectins with a single CRD and both known for their multivalent carbohydrate recognition due to their structural organization, the homodimer arrangement of galectin-7 is considerably different because this galectin recognizes carbohydrates in its monomeric form and does not possess multivalency. The dimer interface involves the association of the β-strands, F1–F5, from the two protomers which are held together by hydrogen bonding interactions. These H-bonds involve five residues from molecule (subunit) A, eight residues from molecule (subunit) B, and an extensive set of van der Waals interactions. The dimer interface of galectin-7 is relatively large, 1484 Å^2, compared to areas of 1093 Å^2 (galectin-1) and 1179 Å^2 (galectin-2) [8].

Detailed analysis of the aforementioned galectin-7–carbohydrate complex structures show that His49, Asn51, Arg53, Asn62 and Glu72 are the key residues involved in carbohydrate recognition through hydrogen bond interactions. The highly conserved residues His49, Asn 51, and Arg53 make hydrogen bonds with the galactose O4 in all four complexes. The galactose O5 makes two hydrogen bonds with Arg53 and Glu72, while O6 is engaged in interactions with Asn62 and Glu72. Tryptophan 69 is involved in stacking interactions with the galactose moiety in a manner analogous to that seen in Gal-1 and Gal-2 structures. Residues Arg 53, Thr56, Glu58, Glu72, and Arg74 form a network of ionic interactions. In the galactose and galactosamine complex structures, the O1 (involved in hydrogen bond formation with Pro85 and Ser8), O2, and O3 atoms of the carbohydrate are involved in water-mediated interactions and contribute to the strength of carbohydrate binding. Moreover, the Arg31 residue in galectin-7 could form part of the carbohydrate-binding region, as it was observed that Arg31 in galectin-7 occupies the position of His52 in Gal-1, which is located about 3.1 Å away from the carbohydrate moiety [8]. Figure 1 displays the dimeric structure of galectin-7 as well as its binding to N-Ac-LacNac.

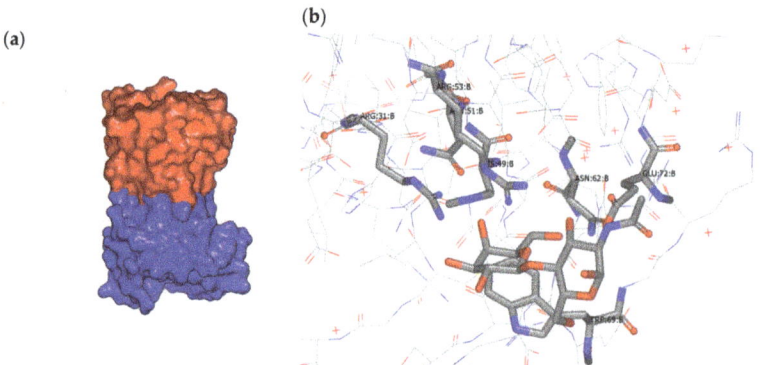

Figure 1. (a) Dimeric structure of galectin-7 (pdb 1BKZ); (b) N-Ac-LacNAc binding to galectin-7 (pdb 5GAL).

By combining nuclear magnetic resonance (NMR) and circular dichroism spectroscopies and molecular dynamics (MD) simulations, Ermakova et al. provided complementary structural information on the binding of lactose to galectin-7 and its impact on protein thermodynamics and conformational dynamics. They were able to show that there is positive cooperativity when lactose binds. Binding of the first ligand enhances binding of the second. Analyzing MD simulations indicated that significant changes occur in the ligand-free subunit (A) when lactose is bound to the other subunit (B). Increased conformational entropy was reflected by an overall increased internal motion in galectin-7. Based on molecular mechanics and MD simulations, an increase in the ligand binding–induced dimer stability (of galectin-7) was observed. This increase was validated experimentally in several assays: gel filtration fast protein liquid chromatography (FPLC), CD-based thermal denaturation studies, fluorescence resonance energy transfer (FRET) and STD NMR. Furthermore, it was observed that the binding of lactose to galectin-7 (K_d = 0.465 mM averaged over two K_a values) alters the lectin conformation and dynamics within the ligand-binding site, as well as through an internal gradient from the ligand-binding site to the dimer interface. The greatest effects were observed in the residues that interact directly with the ligand (the 50–58 and 62–70 loops), the 5-stranded β-sheet at the backside of the lactose-binding site (including the region involved in dimerization of galectin-7) and loops (residues 9–14 and 110–116) down to the dimer interface [105].

Masuyer and colleagues compared the binding affinities of compounds **1** and **2** (Figure 2) and evaluated the structural information by measuring a high resolution crystal structure of galectin-7 in the complex with **2** [106].

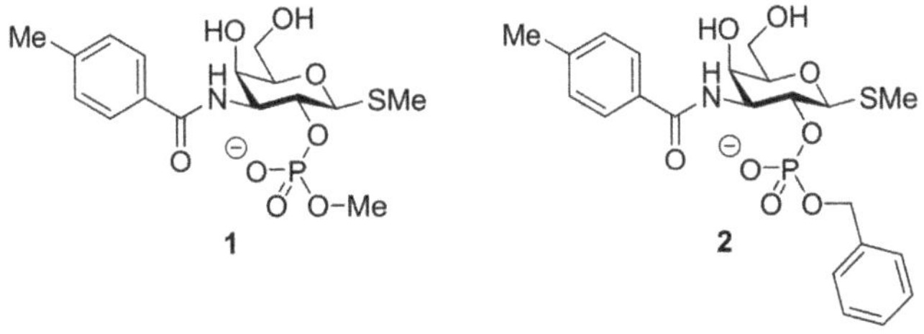

Figure 2. Structures of **1** and **2**: these two molecules differ by the presence of the O-benzylphosphate group in **2**, compared with an O-methylphosphate group in **1**.

They reported that the CRD itself remains unchanged despite a slight movement in the adjacent loop composed of Arg74 and Gly75. The crystal structure highlights stronger binding achieved through the side groups of the 2-O-benzylphosphate ligand **2**, compared to galactose. The phosphate group weakly hydrogen bonds with Arg31 while it is also stabilized by hydrogen-bonded water molecules linked to the same Arg31 and Asn51 of the CRD. The amido group also shows interactions with water molecules linked to Lys64 and Trp69, expanding the binding capacity of the ligand to a region not previously involved in galactoside recognition by galectin-7. The phenyl group does not seem to be involved in the binding of ligand **2** despite being in close proximity to polar residues His33, Glu122 and Asn35. As both His33 and Glu122 are not conserved among galectins, better specificity of inhibition could be achieved by focusing the ligand interaction toward this position. It is also noted that the benzyl moiety of the O-benzylphosphate **2** is not taking part in the inhibitor binding, as it faces away from the CRD. The slightly better affinity of **1** (K_d = 240 µM compared to K_d = 450 µM for **2**) toward galectin-7 reflects this; the smaller methyl group might be able to interact with Arg31, possibly via a different orientation than that of the O-benzyl group, and hence, show a slightly better affinity. The authors proposed to search for a more favorable interaction with Arg31 (and other nearby residues) when the (alkyl)-phosphate groups at the 2-O position is modified for the design of inhibitors [106].

Hsieh and co-workers provided structural evidence of human galectin-7 (hGal7) in complex with Galβ1-3GlcNAc (LN1) and Galβ1-4GlcNAc (LN2) (Figure 3). They compared the results with LN1 and LN2-complexed galectin-1 and (the C-terminal CRD domain of) galectin-3 by means of crystallography [107].

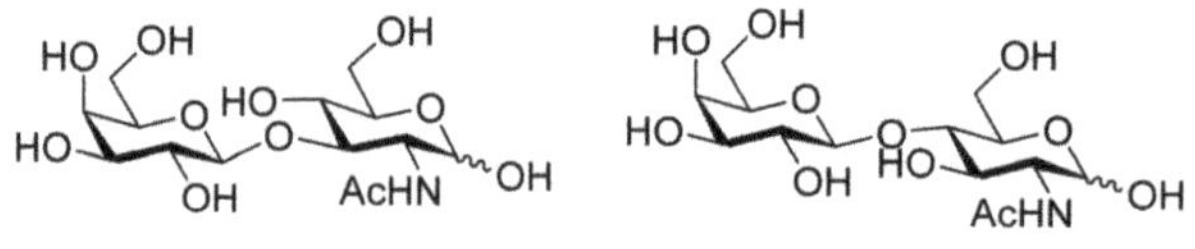

3 (LN1) **4 (LN2)**

Figure 3. Structures of **3** (Galβ1-3GlcNAc, LN1) and **4** (Galβ1-4GlcNAc, LN2).

When complexed to **3**, the crystal structure determination of galectin-7 revealed that the dimer of galectin-7 is present in a back-to-back arrangement. Furthermore, the authors confirmed that the CRD adopts a typical galectin fold, which is composed of two antiparallel β-sheets of six (S-sheets S1-S6) and five (F-sheets F1-F5) strands, jointly forming a β-sheet sandwich structure. The S1–S6 β-strands constitute a concave surface to which β-galactoside-containing glycans are bound. Generally the galactose moiety (Gal) forms more hydrogen bonds with the amino acid residues in the CRD of the galectin than the N-acetylglucosamine moiety (GlcNAc), supporting the idea that Gal serves as the major recognition component [107].

The Gal of **3** (LN1) interacts with the following residues located on S4–S6 β-strands and the loop connecting S4 and S5 strands of galectin-7: His49, Asn53 and Asn62 (through hydrogen bond networks) and Trp69 (via van der Waals contacts). Specifically, the Arg53 residue not only bridges H-bonds to several oxygen atoms of LN1 (C4-OH, O5 of GAL and C4-OH of GlcNAc), but also connects a few carbohydrate-interacting amino acid residues, such as Asn51, Glu58 and Arg74, to form a characteristic interacting network of H-bonds and electrostatic interactions, which are optimal for the carbohydrate orientation. Galectin-7 has more H-bonds to the Gal moiety and a characteristic shorter distance with GlcNAc in **3** (LN1), as compared to those in **4** (LN2) [107].

The electrostatic network consists of Arg53, Glu58, Glu72 and Arg74. Glu58 mediates a unique salt-bridge network by forming two weaker monodentate N–O bridges with Arg53 and Arg74. Neither of the hGal7–LN1 and the hGal7–LN2 complexes contain water-mediated interactions; the main cause is most probably the large distance of Glu58 to the bound sugar. Investigation of the loop L4 between the S4 and S5 β-strands revealed that L4 is shorter, compared to the counterpart in galectins-1 and -3. Glu58 seems to either reside in the end of L4 or the beginning of the S5 β-strand, which makes it impossible for Glu58 to coordinate with the N2 atom of LN2 for additional water-mediated interactions. Based on their results, it was concluded that the length of L4 and the location of the Glu residue (resided in the variable loop L4) are found to influence the geometry of the salt-bridge, which eventually resulted in a higher affinity of galectin-7 toward LN1, compared to LN2 [107].

High-resolution crystal structures of carbohydrate-based dendrons D1, D2 and D3 (**5**, **6** and **7**, respectively, in Figure 4) in complex with human galectin-7 were resolved, as follows. The overall structure of galectin-7 remained unchanged upon ligand binding and appeared as a dimer comparable with that described previously by Leonidas [8]. The dimeric state of galectin-7 did not appear to break down upon ligand binding; however, the interface of dimerization was slightly altered in terms of a decrease in surface contact area. The ligand D1 (**5**) is bound to galectin-7 through its galactose rings interacting with the CRD and a single water-mediated hydrogen bond between the triazole arm and R31. Despite having identical lengths, all three arms do not seem to be long enough to bind to galectin-7 simultaneously. This probably resulted in the disorder and lack of electron density for the third arm in the crystal structure of the D1–hGal7 complex. It was concluded that D1 was able to link two molecules of galectin-7 in a linear fashion as shown in Table 2. Co-crystallization of galectin-7 in complex with D2 (**6**) led to two crystal forms. The first crystal form (D2-1) showed electron density for two of the three arms of the dendrons similar to that observed with D1 (resulting in cross-linking of two hGal7 molecules). In the second crystal form, D2-2, electron density was observed for all three arms of the dendrons with each galactose-terminus bound to one hGal7 molecule; this crystal form has three dimers of hGal7 in the asymmetric unit. Galectin-7 in complex with D3 (7) results in the linking of two molecules of galectin-7 (Table 2). In addition, the D3–hGal7 structure of this complex shows that one terminal galactosyl group binds at the CRD of galectin-7, whereas another galactosyl ring of the adjacent arm interacts with a different CRD of the same galectin-7 molecule [108].

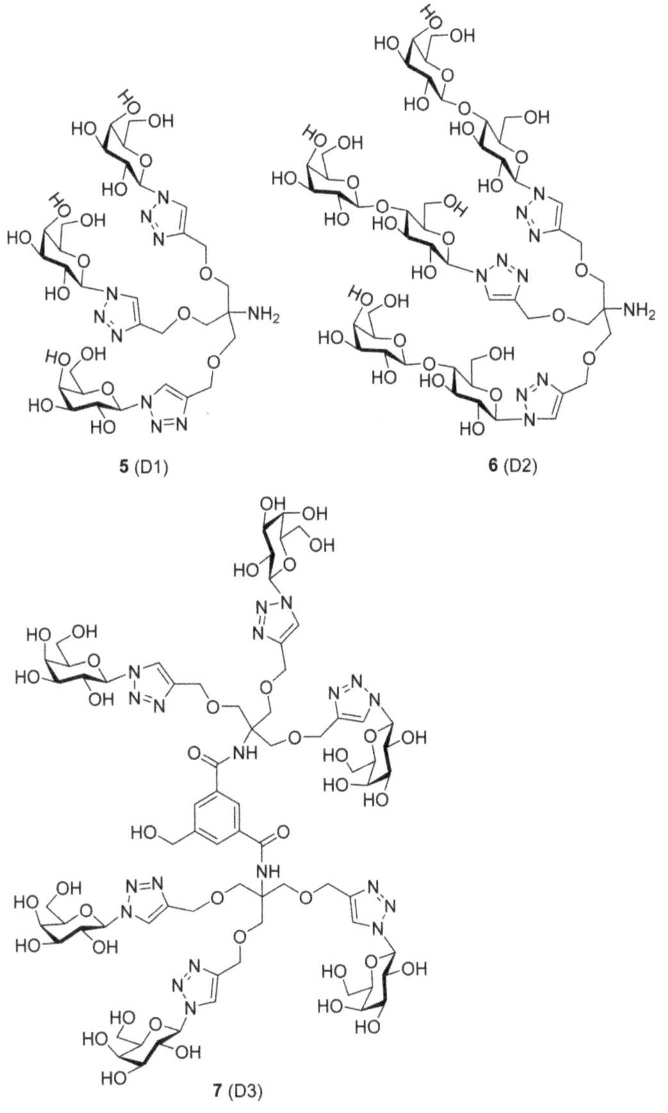

Figure 4. Structures of 5, 6 and 7 (D1, D2 and D3, respectively).

Table 2. Cross-linking of galectin-7 by dendrons D1, D2 and D3. Figures are re-used with permission from the copyright holder.

Compound	Cross-Linked Form
5 (D1)	
6 (D2)	D2-1:
6 (D2)	D2-2:
7 (D3)	

TD139 **8** [109], being in clinical development by the Swedish Galecto Biotech [110], has completed Phase Ib/IIa clinical trials for the treatment of idiopathic pulmonary fibrosis [111]. It displays potent inhibition of galectin-1 and galectin-3, which proved to be increased by a factor up to 200 times, compared to the inhibition of galectin-7 as determined by fluorescence polarization (FP) [112].

Hsieh and co-workers investigated the binding interactions between thio-digalactoside TD139 **8** (Figure 5) with galectin-1, -3 and -7 by means of X-ray crystallography, isothermal titration calorimetry and NMR spectroscopy [113]. The galectin's CRD is described in terms of the subsites A–E in order to facilitate analysis and discussions on ligand binding [114]. According to this model, the best structurally characterized subsites C and D are responsible for recognition of the β-galactoside–containing disaccharides [113].

When the binding affinity of **8** with human galectin-3 (hGal3) was investigated with that of human galectin-7 (hGal7), it became clear that galectin-7 contains Arg31 and His33 at the positions held by Arg144^{hGal3} and Ala146^{hGal3}. Arg31^{hGal7} is placed in subsite B and thus, does not interact with the 4-fluorophenyl substituent of TD139. Likely hindered by the imidazole of His33^{hGal7}, a bulkier residue than the counterpart Ala146^{hGal3}, the 4-fluorophenyl moiety turns ~50° away as compared to that in the galectin-3 complex, having the vacated volume in subsite B of galectin-7 occupied by two water molecules. The orientation in which the 4-fluorophenyl-triazole moiety of TD139 is situated in subsite E (the aromatic substituent interacts with Arg) is a consequence of the previously mentioned salt-bridge in galectin-7, this time involving Glu58, Arg74 and Glu72. Similar tandem arginine–

π interactions between the 4-fluorophenyl-triazole and Arg74[hGal7] were observed, albeit being a weak interaction due to the electron-deficient π system. This π–arginine interaction resides only in subsite E (not in subsite B) as confirmed by ^{19}F-NMR spectroscopy, which led to the conclusion that subsite E of galectin-7 is able to contribute more binding interactions than subsite B [113].

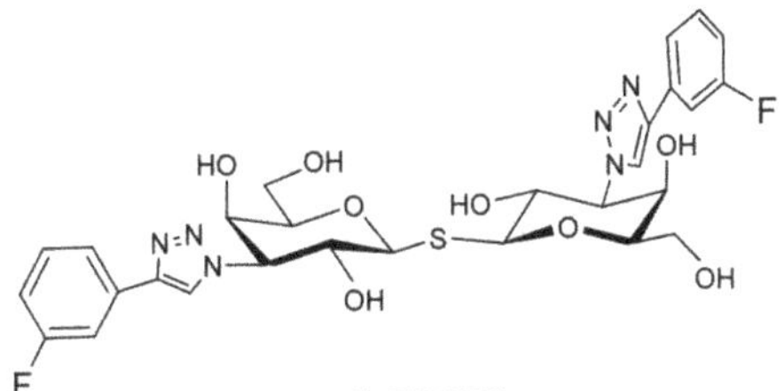

Figure 5. Structure of **8** (TD139).

5. Small-Molecule Carbohydrate and Non-Carbohydrate Galectin-7 Inhibitors

Due to the galectin-7 characteristic that it binds to β-galactosides, most of its small-molecule inhibitors are carbohydrates, or, at least, based on sugar scaffolds. In order to make progress, it is of importance to come up with (glyco)mimetics that are capable of recognizing and blocking galectin-7. These mimetics could be molecules that mimic natural (binding) carbohydrates structurally and functionally. In addition, they should display improved pharmacological properties, have better resistance against glycosidase hydrolysis, and bind more strongly and more selectively to galectin-7 [2]. In particular, the poor selectivity of current small-molecule inhibitors remains an important obstacle to overcome, due to the high similarity of the CRD structures among the different galectins [50]. Hence, developing specific galectin-7 inhibitors that will selectively target the intracellular or extracellular functions of galectin-7 could be a strategy to inhibit not all, but specific galectin-7–mediated processes [13]. Chan and co-workers mentioned in their review that success was achieved in distinguishing between galectin-3 and other galectins. However, having the selectivity be reversed and thus developing inhibitors that are more selective for the weak-binding galectin-7 (than for galectin-3, for example) would certainly be a major breakthrough [115]. In this section, we will briefly discuss the best synthetic inhibitors of galectin-7 based on (non-)carbohydrate scaffolds that were developed in the past.

5.1. Inhibitors Based on a Carbohydrate Scaffold

The first discovery of efficient and selective monosaccharide inhibitors of galectin-7 came from the group of Nilsson during a study in which they synthesized a library of 28 compounds that was tested for binding to galectin-1, -3, -7, -8N and -9N. They demonstrated the potential of 1,5-difluoro-2,4-dinitrobenzene **9** (Figure 6) as a scaffold for the synthesis of combinatorial carbohydrate libraries. Three selective galectin-7 inhibitors (structures **10**, **11** and **12** in Figure 6) were found to have affinities similar to those of the best natural ligands. The K_d values were measured in a competitive fluorescence-polarization assay to be in the range of 0.14–0.18 mM for galectin-7, whereas no inhibition was observed for galectin-1, -3, -8N and -9N [116].

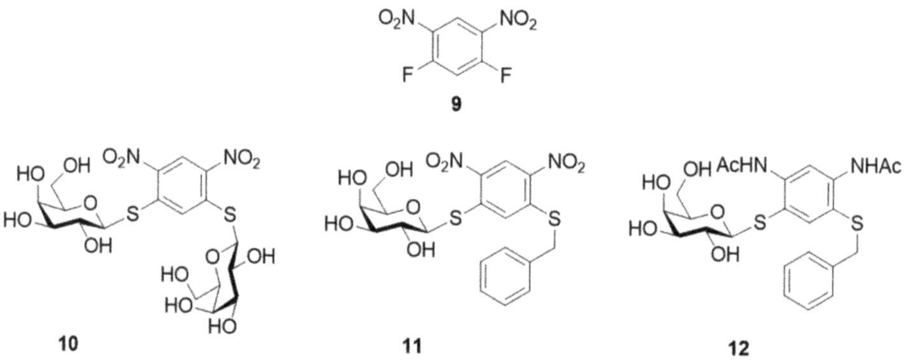

Figure 6. Structures of the scaffold 9 and the inhibitors 10 (K_d = 0.17 mM), 11 (K_d = 0.18 mM) and 12 (K_d = 0.14 mM).

One year later, Bergh and co-workers from the same group published syntheses of galactosides carrying 3- or 4-substituted alkyne benzyl ethers. The group developed a method using a solid phase variant of the Nicholas reaction to provide inhibitors that have alkynyl benzyl ethers. Their approach simplified the purification steps and enabled the use of unprotected carbohydrates in the formation of the *para*/*meta*-substituted products. They found two of them to be the simple straight-chain allyl- and hydroxymethyl-substituted alkynes 13 and 14, which suggests that the binding pocket of galectin-7 close to galactose O-3 is relatively small and does not allow larger cyclic structures to bind. The K_d (mM) values against galectin-1, -3, -7, -8N and -9N were measured in a competitive fluorescence polarization assay and listed in Table 3 [117]:

Table 3. K_d (mM) values for inhibitors 13, 14 and 15 against galectins-1, -3, -7, -8N and -9N as measured in a competitive fluorescence-polarization assay.

Compound	Galectin-1	Galectin-3	Galectin-7	Galectin-8N	Galectin-9N
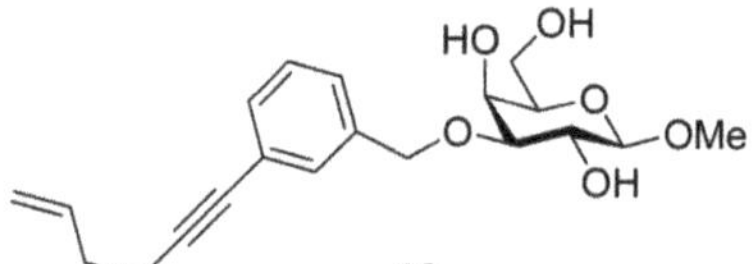 13	27	2.4	0.39	1.0	1.0
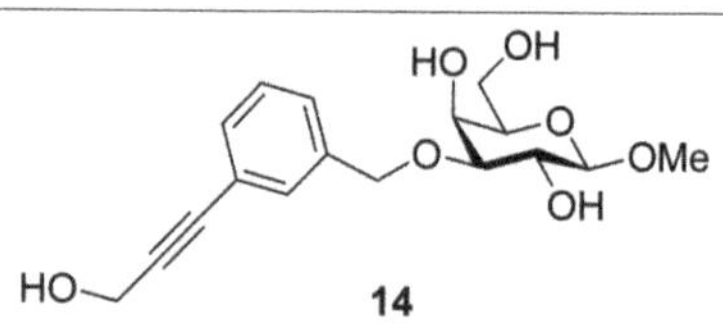 14	6.9	2.9	0.65	3.8	1.9
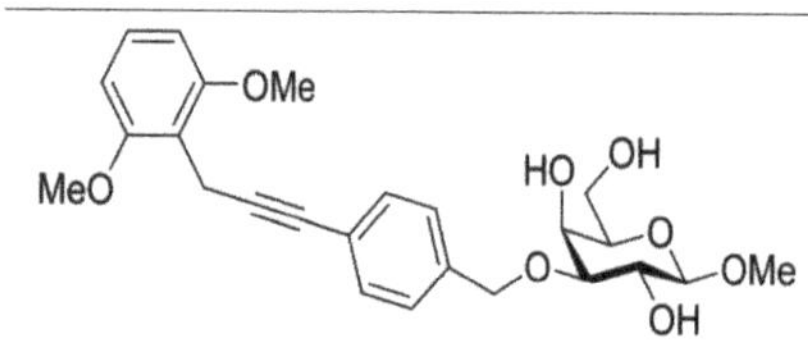 15	n.i. [a]	5.4	0.74	2.4	2.0

[a] n.i. = non-inhibitory.

Compound **13** proved to be the most interesting inhibitor, due to its lowest K_d value and its selectivity. Compared to affinities for other members of the galectin family, preference for galectin-7 is increased by a factor of up to 100 [117].

Salameh and colleagues came up with derivatives of N-acetyl lactosamine carrying diverse thiourea groups at galactose C3. The thioureas obtained upon reaction of the isothiocyanate with amines are known to form strong hydrogen bonds, which makes them suitable for improving the affinity of ligands for proteins. In case of **16** (Figure 7), a K_d value of 23 µM was measured by a fluorescence polarization assay, which makes **16** the best galectin-7 ligand. It is, however, not the most selective, as it binds in the range of 35–47 µM to galectin-1, -3, -8N and -9N [118].

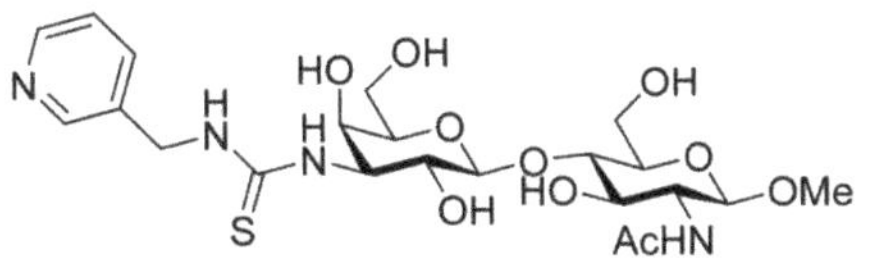

Figure 7. Thioureido N-acetyllactosamine derivative **16**.

In a more recent paper, Delaine et al. continued developing galectin-1 and -3 antagonists with selectivity and therefore, synthesized ditriazolylthio-digalactosides (compounds **8** and **17–26** in Figure 8) [112]:

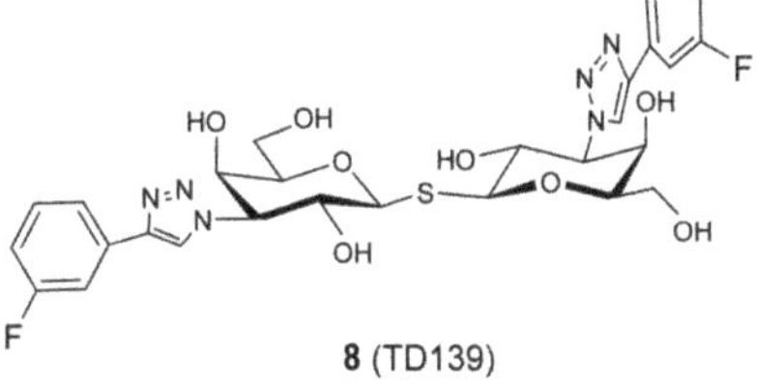

8 (TD139)

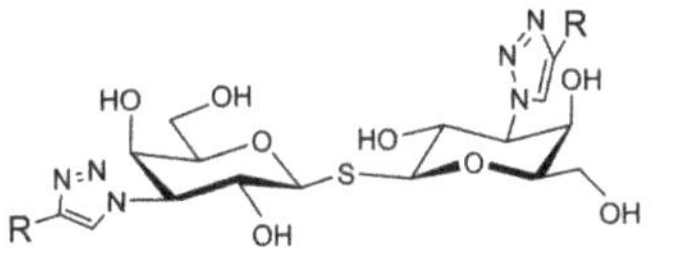

17: R = phenyl
18: R = 2-fluorophenyl
19: R = 4-fluorophenyl
20: R = 4-chlorophenyl
21: R = 2-trifluoromethylphenyl
22: R = 3-trifluoromethylphenyl
23: R = 4-trifluoromethylphenyl
24: R = thiophen-3-yl
25: R = 4-phenoxyphenyl
26: R = 4-biphenyl

Figure 8. Ditriazolylthio-digalactosides developed by Delaine et al.

It was observed from the dissociation constants that, regarding the affinity of these ligands toward galectin-7, the binding is enhanced by the 4-aryltriazolyl groups in **17–24**. The dissociation constants are in the range of 1–10 µM for galectin-7, which are close to those for galectin-2, -4N, -4C, -9N and -9C. The sterically more demanding compounds **25** and **26** did not significantly bind to galectin-7 [112].

5.2. Inhibitors Based on a Non-Carbohydrate Scaffold

Vladoiu and colleagues reported a peptide-based galectin inhibitor that was specifically designed to disrupt the formation of galectin-7 dimers from the monomers and its pro-apoptotic function. They identified critical residues possibly involved in the formation of the dimer interface based on their tendency to form hydrogen bonding, hydrophobic, or van der Waals interactions [8]. In addition, structural analyses of the dimeric interface published by Ermakova and co-workers [105] was also used in their design of a peptide-based inhibitor. Two peptides were designed to rationally mimic and disrupt the galectin-7 segment between residues 13–25 and 129–135 since those residues appear to be directly involved in the stabilization of the dimeric structure: hGal7$_{(13-25)}$ (H-Ile-Arg-Pro-Gly-Thr-Val-Leu-Arg-Ile-Arg-Gly-Leu-Val-NH$_2$) and hGal7$_{(129-135)}$ (H-Leu-Asp-Ser-Val-Arg-Ile-Pro-NH$_2$). Human galectin-7$_{129-135}$ proved to be more potent than hGal7$_{(13-25)}$ in disrupting hGal7 homodimers as measured by mild denaturing native gel electrophoresis. There is an interaction between hGal7$_{(129-135)}$ and galectin-7 through a classical solid-phase binding assay. The decrease in hGal7 homodimers is observed at a concentration range of 100–500 µM of peptide hGal7$_{(129-135)}$. Moreover, an increase in galectin-7 binding on the surface of Jurkat T cells and an apoptotic response were observed in the presence of hGal7$_{(129-135)}$. A reduction in the ability of the protein to induce apoptosis of Jurkat T cells was observed [119]. More recently it was demonstrated that *meso*-tetrakis(*p*-sulfonatophenyl)porphyrin **27** (TpSPPH$_2$, Figure 9) significantly reduced the level of (galectin-7-induced) apoptosis of human Jurkat T cells [120].

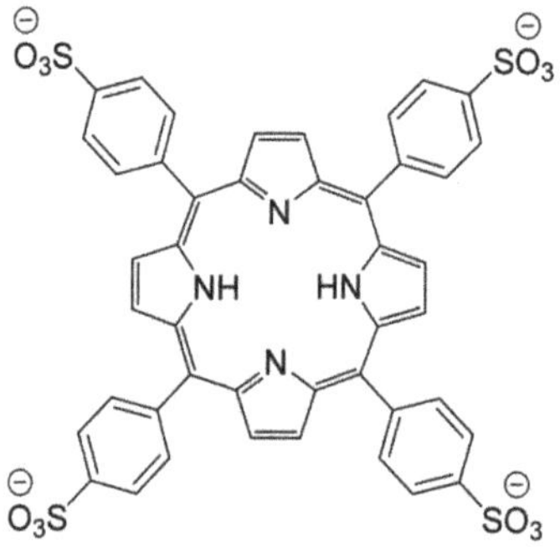

Figure 9. Structure of a novel non-carbohydrate galectin-7 inhibitor **27** (TpSPPH$_2$).

A binding affinity of **27** for galectin-7 was measured by fluorescence quenching to be 9.5 ± 1.6 µM. In addition, TpSPPH$_2$-bridged oligomers of galectin-7 were observed by small-angle X-ray scattering (SAXS) and ^{1}H–^{15}N HSQC NMR of galectin-7–TpSPPH$_2$ complexes. Docking simulations on galectin-7 showed that the TpSPPH$_2$ moiety preferentially binds to three main subsites at the dimer interface. [120].

6. Conclusions

In the world of physiology, pathology and glycobiology, galectin-7 is one of many proteins that require special attention, due to its striking biological properties. Galectin-7 is a member of the prototype galectin family, which is mainly expressed in stratified epithelia of several tissues. While it is known for having multiple biological functions in the human body, much of its molecular mode of action has to be elucidated. Although several strategies were developed to knockout galectin-7 or suppress its translation, we believe the field of cellular pathophysiology would benefit from small-molecule inhibitors which can be administered to evaluate its effect on cellular disorders and even diseases such as cancer. Moreover, the use of small molecules that bind strongly and specifically to galectin-7 may also be deployed for prognosis and monitoring disorders/diseases in search of better and personalized medical treatment. Molecules were synthesized, but both the potency and specificity need to be improved. It is impossible to hit a target when our eyes are closed; therefore, with this mini review, we wish to elaborate on an emerging target within glycobiology, galectin-7.

Author Contributions: Conceptualization, N.V.S. and R.J.P.; literature research, S.A. and N.V.S.; writing—original draft preparation, N.V.S.; writing—review and editing, N.V.S. and R.J.P. All authors have read and agreed to the published version of the manuscript.

Funding: This publication is part of the project 'Synthesis of carbohydrate-based multivalent galectin inhibitors' (023.014.029) of the research programme Promotiebeurs voor leraren, which is financed by the Dutch Research Council (NWO).

Institutional Review Board Statement: Not applicable.

Informed Consent Statement: Not applicable.

Data Availability Statement: Not applicable.

Conflicts of Interest: The author declares no conflict of interest.

References

1. Varki, A.; Cummings, R.D.; Esko, J.D.; Stanley, P.; Hart, G.W.; Aebi, M.; Darvill, A.G.; Kinoshita, T.; Packer, N.H.; Prestegard, J.H.; et al. *Essentials of Glycobiology*, 3rd ed.; Cold Spring Harbor Laboratory Press: New York, NY, USA, 2017.
2. Bertuzzi, S.; Quintana, J.I.; Ardá, A.; Gimeno, A.; Jiménez-Barbero, J. Targeting Galectins With Glycomimetics. *Front. Chem.* **2020**, *8*, 1–17. [CrossRef] [PubMed]
3. Si, Y.; Yao, Y.; Jaramillo Ayala, G.; Li, X.; Han, Q.; Zhang, W.; Xu, X.; Tai, G.; Mayo, K.H.; Zhou, Y.; et al. Human galectin-16 has a pseudo ligand binding site and plays a role in regulating c-Rel-mediated lymphocyte activity. *Biochim. Biophys. Acta Gen. Subj.* **2021**, *1865*, 129755. [CrossRef] [PubMed]
4. Madsen, P.; Rasmussen, H.H.; Flint, T.; Gromov, P.; Kruse, T.A.; Honore, B.; Vorum, H.; Celis, J.E. Cloning, expression, and chromosome mapping of human galectin-7. *J. Biol. Chem.* **1995**, *270*, 5823–5829. [CrossRef] [PubMed]
5. Magnaldo, T.; Bernerd, F.; Darmon, M. Galectin-7, a Human 14 kDa S-lectin, Specifically Expressed in keratinocytes and Sensitive to Retinoic acid. *Dev. Biol.* **1995**, *168*, 259–271. [CrossRef]
6. St-pierre, Y. Towards a Better Understanding of the Relationships between Galectin-7, p53 and MMP-9 during Cancer Progression. *Biomolecules* **2021**, *11*, 879. [CrossRef]
7. Johannes, L.; Jacob, R.; Leffler, H. Galectins at a glance. *J. Cell Sci.* **2018**, *131*, 1–9. [CrossRef]
8. Leonidas, D.D.; Vatzaki, E.H.; Vorum, H.; Celis, J.E.; Madsen, P.; Acharya, K.R. Structural basis for the recognition of carbohydrates by human galectin- 7. *Biochemistry* **1998**, *37*, 13930–13940. [CrossRef]
9. Cooper, D.N.W.; Barondes, S.H. God must love galectins; he made so many of them. *Glycobiology* **1999**, *9*, 979–984. [CrossRef]
10. Morris, S.; Ahmad, N.; Andre, S.; Kaltner, H.; Gabius, H.J.; Brenowitz, M.; Brewer, F. Quaternary solution structures of galectins-1, -3, and -7. *Glycobiology* **2004**, *14*, 293–300. [CrossRef]
11. Nesmelova, I.V.; Berbís, M.Á.; Miller, M.C.; Cañada, F.J.; André, S.; Jiménez-Barbero, J.; Gabius, H.J.; Mayo, K.H. 1H, 13C, and 15N backbone and side-chain chemical shift assignments for the 31 kDa human galectin-7 (p53-induced gene 1) homodimer, a pro-apoptotic lectin. *Biomol. NMR Assign.* **2012**, *6*, 127–129. [CrossRef]
12. Saussez, S.; Kiss, R. Galectin-7. *Cell. Mol. Life Sci.* **2006**, *63*, 686–697. [CrossRef]
13. Advedissian, T.; Deshayes, F.; Viguier, M. Galectin-7 in epithelial homeostasis and carcinomas. *Int. J. Mol. Sci.* **2017**, *18*, 2760. [CrossRef]
14. Bernerd, F.; Sarasin, A.; Magnaldo, T. Galectin-7 overexpression is associated with the apoptotic process in UVB-induced sunburn keratinocytes. *Proc. Natl. Acad. Sci. USA* **1999**, *96*, 11329–11334. [CrossRef]
15. Yamaguchi, T.; Hiromasa, K.; Kabashima-Kubo, R.; Yoshioka, M.; Nakamura, M. Galectin-7, induced by cis-urocanic acid and ultraviolet B irradiation, down-modulates cytokine production by T lymphocytes. *Exp. Dermatol.* **2013**, *22*, 840–842. [CrossRef]
16. Gendronneau, G.; Sidhu, S.S.; Delacour, D.; Dang, T.; Calonne, C.; Houzelstein, D.; Magnaldo, T.; Poirier, F. Galectin-7 in the Control of Epidermal Homeostasis after Injury. *Mol. Biol. Cell* **2008**, *19*, 5541–5549. [CrossRef]
17. Gendronneau, G.; Sanii, S.; Dang, T.; Deshayes, F.; Delacour, D.; Pichard, E.; Advedissian, T.; Sidhu, S.S.; Viguier, M.; Magnaldo, T.; et al. Overexpression of galectin-7 in mouse epidermis leads to loss of cell junctions and defective skin repair. *PLoS ONE* **2015**, *10*, e0119031. [CrossRef]
18. Advedissian, T.; Proux-Gillardeaux, V.; Nkosi, R.; Peyret, G.; Nguyen, T.; Poirier, F.; Viguier, M.; Deshayes, F. E-cadherin dynamics is regulated by galectin-7 at epithelial cell surface. *Sci. Rep.* **2017**, *7*, 1–14. [CrossRef]
19. Chen, H.L.; Chiang, P.C.; Lo, C.H.; Lo, Y.H.; Hsu, D.K.; Chen, H.Y.; Liu, F.T. Galectin-7 regulates keratinocyte proliferation and differentiation through JNK-miR-203-p63 signaling. *J. Investig. Dermatol.* **2016**, *136*, 182–191. [CrossRef]
20. Chen, H.L.; Lo, C.H.; Huang, C.C.; Lu, M.P.; Hu, P.Y.; Chen, C.S.; Chueh, D.Y.; Chen, P.; Lin, T.N.; Lo, Y.H.; et al. Galectin-7 downregulation in lesional keratinocytes contributes to enhanced IL-17A signaling and skin pathology in psoriasis. *J. Clin. Investig.* **2021**, *131*, e130740. [CrossRef]
21. Asano, Y.; Sato, S. Vasculopathy in scleroderma. *Semin. Immunopathol.* **2015**, *37*, 489–500. [CrossRef]
22. Saigusa, R.; Yamashita, T.; Miura, S.; Hirabayashi, M.; Nakamura, K.; Miyagawa, T.; Fukui, Y.; Yoshizaki, A.; Sato, S.; Asano, Y. A potential contribution of decreased galectin-7 expression in stratified epithelia to the development of cutaneous and oesophageal manifestations in systemic sclerosis. *Exp. Dermatol.* **2019**, *28*, 536–542. [CrossRef]
23. Blair, M. Diabetes Mellitus Review. *Urol. Nurs.* **2016**, *36*, 27–36. [CrossRef]

24. Huang, S.M.; Wu, C.S.; Chiu, M.H.; Yang, H.J.; Chen, G.S.; Lan, C.C.E. High-glucose environment induced intracellular O-GlcNAc glycosylation and reduced galectin-7 expression in keratinocytes: Implications on impaired diabetic wound healing. *J. Dermatol. Sci.* **2017**, *87*, 168–175. [CrossRef]
25. Tazhitdinova, R.; Timoshenko, A.V. The Emerging Role of Galectins and O-GlcNAc Homeostasis in Processes of Cellular Differentiation. *Cells* **2020**, *9*, 1792. [CrossRef]
26. Rousselle, P.; Braye, F.; Dayan, G. Re-epithelialization of adult skin wounds: Cellular mechanisms and therapeutic strategies. *Adv. Drug Deliv. Rev.* **2019**, *146*, 344–365. [CrossRef]
27. Cao, Z.; Said, N.; Wu, H.K.; Kuwabara, I.; Liu, F.T.; Panjwani, N. Galectin-7 as a potential mediator of corneal epithelial cell migration. *Arch. Ophthalmol.* **2003**, *121*, 82–86. [CrossRef]
28. Lin, T.; Yu, C.C.; Liu, C.M.; Hsieh, P.L.; Liao, Y.W.; Yu, C.H.; Chen, C.J. Er:YAG laser promotes proliferation and wound healing capacity of human periodontal ligament fibroblasts through Galectin-7 induction. *J. Formos. Med. Assoc.* **2021**, *120*, 388–394. [CrossRef]
29. Picorino, L. *Molecular Biology of Cancer - Mechanisms, Targets and Therapeutics*, 4th ed.; Oxford University Press: Oxford, UK, 2016.
30. Kaur, M.; Kaur, T.; Kamboj, S.S.; Singh, J. Roles of galectin-7 in cancer. *Asian Pacific J. Cancer Prev.* **2016**, *17*, 455–461. [CrossRef]
31. Hanahan, D.; Weinberg, R.A. Hallmarks of cancer: The next generation. *Cell* **2011**, *144*, 646–674. [CrossRef]
32. Cagnoni, A.J.; Pérez Sáez, J.M.; Rabinovich, G.A.; Mariño, K.V. Turning-off signaling by siglecs, selectins, and galectins: Chemical inhibition of glycan-dependent interactions in cancer. *Front. Oncol.* **2016**, *6*, 1–21. [CrossRef]
33. Méndez-Huergo, S.P.; Blidner, A.G.; Rabinovich, G.A. Galectins: Emerging regulatory checkpoints linking tumor immunity and angiogenesis. *Curr. Opin. Immunol.* **2017**, *45*, 8–15. [CrossRef] [PubMed]
34. Kaltner, H.; Toegel, S.; Caballero, G.G.; Manning, J.C.; Ledeen, R.W.; Gabius, H.J. Galectins: Their network and roles in immunity/tumor growth control. *Histochem. Cell Biol.* **2017**, *147*, 239–256. [CrossRef] [PubMed]
35. Chang, W.A.; Tsai, M.J.; Kuo, P.L.; Hung, J.Y. Role of galectins in lung cancer (Review). *Oncol. Lett.* **2017**, *14*, 5077–5084. [CrossRef] [PubMed]
36. Wdowiak, K.; Francuz, T.; Gallego-Colon, E.; Ruiz-Agamez, N.; Kubeczko, M.; Grochoła, I.; Wojnar, J. Galectin Targeted Therapy in Oncology: Current Knowledge and Perspectives. *Int. J. Mol. Sci.* **2018**, *19*, 210. [CrossRef]
37. Rodríguez, E.; Schetters, S.T.T.; Van Kooyk, Y. The tumour glyco-code as a novel immune checkpoint for immunotherapy. *Nat. Rev. Immunol.* **2018**, *18*, 204–211. [CrossRef]
38. Rodrigues, J.G.; Balmaña, M.; Macedo, J.A.; Poças, J.; Fernandes, Â.; de Freitas Junior, J.C.M.; Pinho, S.S.; Gomes, J.; Magalhães, A.; Gomes, C.; et al. Glycosylation in cancer: Selected roles in tumour progression, immune modulation and metastasis. *Cell. Immunol.* **2018**, *333*, 46–57. [CrossRef]
39. Chetry, M.; Thapa, S.; Hu, X.; Song, Y.; Zhang, J.; Zhu, H.; Zhu, X. The role of galectins in tumor progression, treatment and prognosis of gynecological cancers. *J. Cancer* **2018**, *9*, 4742–4755. [CrossRef]
40. Wang, L.; Zhao, Y.; Wang, Y.; Wu, X. The role of galectins in cervical cancer biology and progression. *Biomed Res. Int.* **2018**, *2018*, 2175927. [CrossRef]
41. Dubé-Delarosbil, C.; St-Pierre, Y. The emerging role of galectins in high-fatality cancers. *Cell. Mol. Life Sci.* **2018**, *75*, 1215–1226. [CrossRef]
42. Bartolazzi, A. Galectins in cancer and translational medicine: From bench to bedside. *Int. J. Mol. Sci.* **2018**, *19*, 2934. [CrossRef]
43. Mereiter, S.; Balmaña, M.; Campos, D.; Gomes, J.; Reis, C.A. Glycosylation in the Era of Cancer-Targeted Therapy: Where Are We Heading? *Cancer Cell* **2019**, *36*, 6–16. [CrossRef]
44. Shimada, C.; Xu, R.; Al-Alem, L.; Stasenko, M.; Spriggs, D.R.; Rueda, B.R. Galectins and ovarian cancer. *Cancers* **2020**, *12*, 1421. [CrossRef]
45. Pergialiotis, V.; Papoutsi, E.; Androutsou, A.; Tzortzis, A.S.; Frountzas, M.; Papapanagiotou, A.; Kontzoglou, K. Galectins-1, -3, -7, -8 and -9 as prognostic markers for survival in epithelial ovarian cancer: A systematic review and meta-analysis. *Int. J. Gynecol. Obstet.* **2020**, 299–307. [CrossRef]
46. Manero-Rupérez, N.; Martínez-Bosch, N.; Barranco, L.E.; Visa, L.; Navarro, P. The Galectin Family as Molecular Targets: Hopes for Defeating Pancreatic Cancer. *Cells* **2020**, *9*, 689. [CrossRef]
47. Hisrich, B.V.; Young, R.B.; Sansone, A.M.; Bowens, Z.; Green, L.J.; Lessey, B.A.; Blenda, A.V. Role of human galectins in inflammation and cancers associated with endometriosis. *Biomolecules* **2020**, *10*, 230. [CrossRef]
48. Navarro, P.; Martínez-Bosch, N.; Blidner, A.G.; Rabinovich, G.A. Impact of Galectins in Resistance to Anticancer Therapies. *Clin. Cancer Res.* **2020**, *26*, 6086–6101. [CrossRef]
49. Girotti, M.R.; Salatino, M.; Dalotto-Moreno, T.; Rabinovich, G.A. Sweetening the hallmarks of cancer: Galectins as multifunctional mediators of tumor progression. *J. Exp. Med.* **2020**, *217*, 1–14. [CrossRef]
50. Perrotta, R.M.; Bach, C.A.; Salatino, M.; Rabinovich, G.A. Reprogramming the tumor metastasis cascade by targeting galectin-driven networks. *Biochem. J.* **2021**, *478*, 597–617. [CrossRef]
51. Rorive, S.; Eddafali, B.; Fernandez, S.; Decaestecker, C.; André, S.; Kaltner, H.; Kuwabara, I.; Liu, F.T.; Gabius, H.J.; Kiss, R.; et al. Changes in galectin-7 and cytokeratin-19 expression during the progression of malignancy in thyroid tumors: Diagnostic and biological implications. *Mod. Pathol.* **2002**, *15*, 1294–1301. [CrossRef]
52. Moisan, S.; Demers, M.; Mercier, J.; Magnaldo, T.; Potworowski, E.F.; St-Pierre, Y. Upregulation of galectin-7 in murine lymphoma cells is associated with progression toward an aggressive phenotype. *Leukemia* **2003**, *17*, 751–759. [CrossRef]

53. Demers, M.; Couillard, J.; Giglia-Mari, G.; Magnaldo, T.; St-Pierre, Y. Increased galectin-7 gene expression in lymphoma cells is under the control of DNA methylation. *Biochem. Biophys. Res. Commun.* **2009**, *387*, 425–429. [CrossRef] [PubMed]
54. Kim, S.J.; Hwang, J.A.; Ro, J.Y.; Lee, Y.S.; Chun, K.H. Galectin-7 is epigenetically-regulated tumor suppressor in gastric cancer. *Oncotarget* **2013**, *4*, 1461–1471. [CrossRef] [PubMed]
55. Demers, M.; Magnaldo, T.; St-Pierre, Y. A novel function for galectin-7: Promoting tumorigenesis by up-regulating MMP-9 gene expression. *Cancer Res.* **2005**, *65*, 5205–5210. [CrossRef] [PubMed]
56. Demers, M.; Biron-Pain, K.; Hébert, J.; Lamarre, A.; Magnaldo, T.; St-Pierre, Y. Galectin-7 in lymphoma: Elevated expression in human lymphoid malignancies and decreased lymphoma dissemination by antisense strategies in experimental model. *Cancer Res.* **2007**, *67*, 2824–2829. [CrossRef]
57. Zhu, X.; Ding, M.; Yu, M.L.; Feng, M.X.; Tan, L.J.; Zhao, F.K. Identification of galectin-7 as a potential biomarker for esophageal squamous cell carcinoma by proteomic analysis. *BMC Cancer* **2010**, *10*, 290. [CrossRef]
58. Demers, M.; Rose, A.A.N.; Grosset, A.A.; Biron-Pain, K.; Gaboury, L.; Siegel, P.M.; St-Pierre, Y. Overexpression of galectin-7, a myoepithelial cell marker, enhances spontaneous metastasis of breast cancer cells. *Am. J. Pathol.* **2010**, *176*, 3023–3031. [CrossRef]
59. Campion, C.G.; Labrie, M.; Grosset, A.A.; St-Pierre, Y. The CCAAT/enhancer-binding protein beta-2 isoform (CEBPβ-2) upregulates galectin-7 expression in human breast cancer cells. *PLoS ONE* **2014**, *9*, e95087. [CrossRef]
60. Grosset, A.A.; Labrie, M.; Gagné, D.; Vladoiu, M.C.; Gaboury, L.; Doucet, N.; St-Pierre, Y. Cytosolic galectin-7 impairs p53 functions and induces chemoresistance in breast cancer cells. *BMC Cancer* **2014**, *14*, 1–10. [CrossRef]
61. Grosset, A.A.; Poirier, F.; Gaboury, L.; St-Pierre, Y. Galectin-7 expression potentiates HER-2-Positive phenotype in breast cancer. *PLoS ONE* **2016**, *11*, 1–12. [CrossRef]
62. Labrie, M.; Vladoiu, M.C.; Grosset, A.A.; Gaboury, L.; St-Pierre, Y. Expression and functions of galectin-7 in ovarian cancer. *Oncotarget* **2014**, *5*, 7705–7721. [CrossRef]
63. Labrie, M.; Vladoiu, M.; Leclerc, B.G.; Grosset, A.A.; Gaboury, L.; Stagg, J.; St-Pierre, Y. A mutation in the carbohydrate recognition domain drives a phenotypic switch in the role of galectin-7 in prostate cancer. *PLoS ONE* **2015**, *10*, 1–19. [CrossRef]
64. Chen, Y.S.; Chang, C.W.; Tsay, Y.G.; Huang, L.Y.; Wu, Y.C.; Cheng, L.H.; Yang, C.C.; Wu, C.H.; Teo, W.H.; Hung, K.F.; et al. HSP40 co-chaperone protein Tid1 suppresses metastasis of head and neck cancer by inhibiting Galectin-7-TCF3-MMP9 axis signaling. *Theranostics* **2018**, *8*, 3841–3855. [CrossRef]
65. Guo, J.P.; Li, X.G. Galectin-7 promotes the invasiveness of human oral squamous cell carcinoma cells via activation of ERK and JNK signaling. *Oncol. Lett.* **2017**, *13*, 1919–1924. [CrossRef]
66. Barkan, B.; Cox, A.D.; Kloog, Y. Ras inhibition boosts galectin-7 at the expense of galectin-1 to sensitize cells to apoptosis. *Oncotarget* **2013**, *4*, 256–268. [CrossRef]
67. Kuwabara, I.; Kuwabara, Y.; Yang, R.Y.; Schuler, M.; Green, D.R.; Zuraw, B.L.; Hsu, D.K.; Liu, F.T. Galectin-7 (PIG1) exhibits pro-apoptotic function through JNK activation and mitochondrial cytochrome c release. *J. Biol. Chem.* **2002**, *277*, 3487–3497. [CrossRef]
68. Ueda, S.; Kuwabara, I.; Liu, F.-T. Suppression of tumor growth by the β4-galactosyltransferase gene. *Cancer Res.* **2004**, *64*, 5672–5676. [CrossRef]
69. Zhu, H.; Wu, T.C.; Chen, W.Q.; Zhou, L.J.; Wu, Y.; Zeng, L.; Pei, H.P. Roles of galectin-7 and S100A9 in cervical squamous carcinoma: Clinicopathological and in vitro evidence. *Int. J. Cancer* **2013**, *132*, 1051–1059. [CrossRef]
70. Higareda-Almaraz, J.C.; Ruiz-Moreno, J.S.; Klimentova, J.; Barbieri, D.; Salvador-Gallego, R.; Ly, R.; Valtierra-Gutierrez, I.A.; Dinsart, C.; Rabinovich, G.A.; Stulik, J.; et al. Systems-level effects of ectopic galectin-7 reconstitution in cervical cancer and its microenvironment. *BMC Cancer* **2016**, *16*, 1–22. [CrossRef]
71. Menkhorst, E.; Griffiths, M.; van Sinderen, M.; Rainczuk, K.; Niven, K.; Dimitriadis, E. Galectin-7 is elevated in endometrioid (type I) endometrial cancer and promotes cell migration. *Oncol. Lett.* **2018**, *16*, 4721–4728. [CrossRef]
72. Matsui, Y.; Ueda, S.; Watanabe, J.; Kuwabara, I.; Ogawa, O.; Nishiyama, H. Sensitizing effect of galectin-7 in urothelial cancer to cisplatin through the accumulation of intracellular reactive oxygen species. *Cancer Res.* **2007**, *67*, 1212–1220. [CrossRef]
73. Kopitz, J.; André, M.; Von Reitzenstein, C.; Versluis, K.; Kaltner, H.; Pieters, R.J.; Wasano, K.; Kuwabara, I.; Liu, F.T.; Cantz, M.; et al. Homodimeric galectin-7 (p53-induced gene 1) is a negative growth regulator for human neuroblastoma cells. *Oncogene* **2003**, *22*, 6277–6288. [CrossRef]
74. Ives, C.W.; Sinkey, R.; Rajapreyar, I.; Tita, A.T.N.; Oparil, S. Preeclampsia—Pathophysiology and Clinical Presentations: JACC State-of-the-Art Review. *J. Am. Coll. Cardiol.* **2020**, *76*, 1690–1702. [CrossRef]
75. Phipps, E.A.; Thadhani, R.; Benzing, T.; Karumanchi, S.A. Pre-eclampsia: Pathogenesis, novel diagnostics and therapies. *Nat. Rev. Nephrol.* **2019**, *15*, 275–289. [CrossRef]
76. Larsen, E.C.; Christiansen, O.B.; Kolte, A.M.; Macklon, N. New insights into mechanisms behind miscarriage. *BMC Med.* **2013**, *11*, 154. [CrossRef]
77. Dimitriadis, E.; Menkhorst, E.; Saito, S.; Kutteh, W.H.; Brosens, J.J. Recurrent pregnancy loss. *Nat. Rev. Dis. Prim.* **2020**, *6*, 98. [CrossRef] [PubMed]
78. Jeschke, U.; Hutter, S.; Heublein, S.; Vrekoussis, T.; Andergassen, U.; Unverdorben, L.; Papadakis, G.; Makrigiannakis, A. Expression and function of galectins in the endometrium and at the human feto-maternal interface. *Placenta* **2013**, *34*, 863–872. [CrossRef]

79. Blois, S.M.; Barrientos, G. Galectin signature in normal pregnancy and preeclampsia. *J. Reprod. Immunol.* **2014**, *101–102*, 127–134. [CrossRef]
80. Evans, J.; Yap, J.; Gamage, T.; Salamonsen, L.; Dimitriadis, E.; Menkhorst, E. Galectin-7 is important for normal uterine repair following menstruation. *Mol. Hum. Reprod.* **2014**, *20*, 787–798. [CrossRef] [PubMed]
81. Menkhorst, E.; Koga, K.; Van Sinderen, M.; Dimitriadis, E. Galectin-7 serum levels are altered prior to the onset of pre-eclampsia. *Placenta* **2014**, *35*, 281–285. [CrossRef] [PubMed]
82. Unverdorben, L.; Haufe, T.; Santoso, L.; Hofmann, S.; Jeschke, U.; Hutter, S. Prototype and chimera-type galectins in placentas with spontaneous and recurrent miscarriages. *Int. J. Mol. Sci.* **2016**, *17*, 644. [CrossRef]
83. Menkhorst, E.; Zhou, W.; Santos, L.L.; Delforce, S.; So, T.; Rainczuk, K.; Loke, H.; Syngelaki, A.; Varshney, S.; Williamson, N.; et al. Galectin-7 impairs placentation and causes preeclampsia features in mice. *Hypertension* **2020**, *76*, 1185–1194. [CrossRef]
84. Xu, W.D.; Huang, Q.; Huang, A.F. Emerging role of galectin family in inflammatory autoimmune diseases. *Autoimmun. Rev.* **2021**, *20*, 102847. [CrossRef]
85. Wan, L.; Hsu, Y.A.; Wei, C.C.; Liu, F.T. Galectins in allergic inflammatory diseases. *Mol. Aspects Med.* **2021**, *79*, 100925. [CrossRef]
86. Niiyama, S.; Yoshino, T.; Yasuda, C.; Yu, X.; Izumi, R.; Ishiwatari, S.; Matsukuma, S.; Mukai, H. Galectin-7 in the stratum corneum: A biomarker of the skin barrier function. *Int. J. Cosmet. Sci.* **2016**, *38*, 487–495. [CrossRef]
87. Umayahara, T.; Shimauchi, T.; Iwasaki, M.; Sakabe, J.i.; Aoshima, M.; Nakazawa, S.; Yatagai, T.; Yamaguchi, H.; Phadungsaksawasdi, P.; Kurihara, K.; et al. Protective role of Galectin-7 for skin barrier impairment in atopic dermatitis. *Clin. Exp. Allergy* **2020**, 922–931. [CrossRef]
88. Luo, Z.; Ji, Y.; Tian, D.; Zhang, Y.; Chang, S.; Yang, C.; Zhou, H.; Chen, Z.K. Galectin-7 promotes proliferation and Th1/2 cells polarization toward Th1 in activated CD4+ T cells by inhibiting The TGFβ/Smad3 pathway. *Mol. Immunol.* **2018**, *101*, 80–85. [CrossRef]
89. Tian, J.; He, R.; Fan, Y.; Zhang, Q.; Tian, B.; Zhou, C.; LIU, C.; Song, M.; Zhao, S. Galectin-7 overexpression destroy airway epithelial barrier in transgenic mice. *Integr. Zool.* **2020**, *16*, 270–279. [CrossRef] [PubMed]
90. Sun, X.; Zhang, W. Silencing of Gal-7 inhibits TGF-β1-induced apoptosis of human airway epithelial cells through JNK signaling pathway. *Exp. Cell Res.* **2019**, *375*, 100–105. [CrossRef] [PubMed]
91. Lin, C.Y.; Nozawa, T.; Minowa-Nozawa, A.; Toh, H.; Hikichi, M.; Iibushi, J.; Nakagawa, I. Autophagy Receptor Tollip Facilitates Bacterial Autophagy by Recruiting Galectin-7 in Response to Group A Streptococcus Infection. *Front. Cell. Infect. Microbiol.* **2020**, *10*, 583137. [CrossRef] [PubMed]
92. Luo, Z.; Ji, Y.; Zhou, H.; Huang, X.; Fang, J.; Guo, H.; Pan, T.; Chen, Z.K. Galectin-7 in cardiac allografts in mice: Increased expression compared with isografts and localization in infiltrating lymphocytes and vascular endothelial cells. *Transplant. Proc.* **2013**, *45*, 630–634. [CrossRef] [PubMed]
93. Dings, R.P.M.; Miller, M.C.; Griffin, R.J.; Mayo, K.H. Galectins as molecular targets for therapeutic intervention. *Int. J. Mol. Sci.* **2018**, *19*, 905. [CrossRef]
94. Stegmayr, J.; Zetterberg, F.; Carlsson, M.C.; Huang, X.; Sharma, G.; Kahl-Knutson, B.; Schambye, H.; Nilsson, U.J.; Oredsson, S.; Leffler, H. Extracellular and intracellular small-molecule galectin-3 inhibitors. *Sci. Rep.* **2019**, *9*, 1–12. [CrossRef]
95. Chatenet, D.; Doucet, N.S.; Pierre, Y. Galectin-7-specific monovalent antibodies and uses thereof. International Application No. PCT/CA2020/050024, 1 September 2020.
96. Grosset, A.A.; Labrie, M.; Vladoiu, M.C.; Yousef, E.M.; Gaboury, L.; St-Pierre, Y. Galectin signatures contribute to the heterogeneity of breast cancer and provide new prognostic information and therapeutic targets. *Oncotarget* **2016**, *7*, 18183–18203. [CrossRef]
97. Bibens-Laulan, N.; St-Pierre, Y. Intracellular galectin-7 expression in cancer cells results from an autocrine transcriptional mechanism and endocytosis of extracellular galectin-7. *PLoS ONE* **2017**, *12*, 1–13. [CrossRef]
98. Vladoiu, M.C.; Labrie, M.; St-Pierre, Y. Intracellular galectins in cancer cells: Potential new targets for therapy (review). *Int. J. Oncol.* **2014**, *44*, 1001–1014. [CrossRef]
99. Hama, N.; Nishimura, K.; Hasegawa, A.; Yuki, A.; Kume, H.; Adachi, J.; Kinoshita, M.; Ogawa, Y.; Nakajima, S.; Nomura, T.; et al. Galectin-7 as a potential biomarker of Stevens-Johnson syndrome/toxic epidermal necrolysis: Identification by targeted proteomics using causative drug-exposed peripheral blood cells. *J. Allergy Clin. Immunol. Pract.* **2019**, *7*, 2894–2897.e7. [CrossRef]
100. Kim, H.J.; Jeon, H.K.; Lee, J.K.; Sung, C.O.; Do, I.G.; Choi, C.H.; Kim, T.J.; Kim, B.G.; Bae, D.S.; Lee, J.W. Clinical significance of galectin-7 in epithelial ovarian cancer. *Anticancer Res.* **2013**, *33*, 1555–1562.
101. Schulz, H.; Schmoeckel, E.; Kuhn, C.; Hofmann, S.; Mayr, D.; Mahner, S.; Jeschke, U. Galectins-1, -3, and -7 are prognostic markers for survival of ovarian cancer patients. *Int. J. Mol. Sci.* **2017**, *18*, 1230. [CrossRef]
102. Trebo, A.; Ditsch, N.; Kuhn, C.; Heidegger, H.H.; Zeder-Goess, C.; Kolben, T.; Czogalla, B.; Schmoeckel, E.; Mahner, S.; Jeschke, U.; et al. High galectin-7 and low galectin-8 expression and the combination of both are negative prognosticators for breast cancer patients. *Cancers* **2020**, *12*, 953. [CrossRef]
103. Matsukawa, S.; Morita, K.i.; Negishi, A.; Harada, H.; Nakajima, Y.; Shimamoto, H.; Tomioka, H.; Tanaka, K.; Ono, M.; Yamada, T.; et al. Galectin-7 as a potential predictive marker of chemo- and/or radio-therapy resistance in oral squamous cell carcinoma. *Cancer Med.* **2014**, *3*, 349–361. [CrossRef]
104. Modenutti, C.P.; Capurro, J.I.B.; Di Lella, S.; Martí, M.A. The Structural Biology of Galectin-Ligand Recognition: Current Advances in Modeling Tools, Protein Engineering, and Inhibitor Design. *Front. Chem.* **2019**, *7*, 823. [CrossRef]

105. Ermakova, E.; Miller, M.C.; Nesmelova, I.V.; López-Merino, L.; Berbís, M.A.; Nesmelov, Y.; Tkachev, Y.V.; Lagartera, L.; Daragan, V.A.; André, S.; et al. Lactose binding to human galectin-7 (p53-induced gene 1) induces long-range effects through the protein resulting in increased dimer stability and evidence for positive cooperativity. *Glycobiology* **2013**, *23*, 508–523. [CrossRef]
106. Masuyer, G.; Jabeen, T.; Öberg, C.T.; Leffler, H.; Nilsson, U.J.; Acharya, K.R. Inhibition mechanism of human galectin-7 by a novel galactose- benzylphosphate inhibitor. *FEBS J.* **2012**, *279*, 193–202. [CrossRef]
107. Hsieh, T.J.; Lin, H.Y.; Tu, Z.; Huang, B.S.; Wu, S.C.; Lin, C.H. Structural basis underlying the binding preference of human galectins-1, -3 and -7 for Galβ1-3/4GlcNAc. *PLoS ONE* **2015**, *10*, 1–19. [CrossRef]
108. Ramaswamy, S.; Sleiman, M.H.; Masuyer, G.; Arbez-Gindre, C.; Micha-Screttas, M.; Calogeropoulou, T.; Steele, B.R.; Acharya, K.R. Structural basis of multivalent galactose-based dendrimer recognition by human galectin-7. *FEBS J.* **2015**, *282*, 372–387. [CrossRef]
109. MacKinnon, A.C.; Gibbons, M.A.; Farnworth, S.L.; Leffler, H.; Nilsson, U.J.; Delaine, T.; Simpson, A.J.; Forbes, S.J.; Hirani, N.; Gauldie, J.; et al. Regulation of transforming growth factor-β1-driven lung fibrosis by galectin-3. *Am. J. Respir. Crit. Care Med.* **2012**, *185*, 537–546. [CrossRef]
110. Garber, K. Galecto Biotech. *Nat. Biotechnol.* **2013**, *31*, 481–482. [CrossRef]
111. Hirani, N.; MacKinnon, A.C.; Nicol, L.; Ford, P.; Schambye, H.; Pedersen, A.; Nilsson, U.J.; Leffler, H.; Sethi, T.; Tantawi, S.; et al. Target inhibition of galectin-3 by inhaled TD139 in patients with idiopathic pulmonary fibrosis. *Eur. Respir. J.* **2021**, *57*, 1–13. [CrossRef]
112. Delaine, T.; Collins, P.; Mackinnon, A.; Sharma, G.; Stegmayr, J.; Rajput, V.K.; Mandal, S.; Cumpstey, I.; Larumbe, A.; Salameh, B.A.; et al. Galectin-3-Binding Glycomimetics that Strongly Reduce Bleomycin-Induced Lung Fibrosis and Modulate Intracellular Glycan Recognition. *ChemBioChem* **2016**, *17*, 1759–1770. [CrossRef]
113. Hsieh, T.J.; Lin, H.Y.; Tu, Z.; Lin, T.C.; Wu, S.C.; Tseng, Y.Y.; Liu, F.T.; Hsu, S.T.D.; Lin, C.H. Dual thio-digalactoside-binding modes of human galectins as the structural basis for the design of potent and selective inhibitors. *Sci. Rep.* **2016**, *6*, 1–9. [CrossRef]
114. MacKinnon, A.; Chen, W.-S.; Leffler, H.; Panjwani, N.; Schambye, H.; Sethi, T.; Nilsson, U.J. *Design, Synthesis, and Applications of Galectins Modulators in Human Health*; Seeberger, P.H., Rademacher, C., Eds.; Springer International Publishing: Cham, Switzerland, 2014; Volume 12, ISBN 978-3-319-08674-3.
115. Chan, Y.C.; Lin, H.Y.; Tu, Z.; Kuo, Y.H.; Hsu, S.T.D.; Lin, C.H. Dissecting the structure–Activity relationship of galectin–Ligand interactions. *Int. J. Mol. Sci.* **2018**, *19*, 392. [CrossRef] [PubMed]
116. Cumpstey, I.; Carlsson, S.; Leffler, H.; Nilsson, U.J. Synthesis of a phenyl thio-√ü-D-galactopyranoside library from 1,5-difluoro-2,4-dinitrobenzene: Discovery of efficient and selective monosaccharide inhibitors of galectin-7. *Org. Biomol. Chem.* **2005**, *3*, 1922–1932. [CrossRef] [PubMed]
117. Bergh, A.; Leffler, H.; Sundin, A.; Nilsson, U.J.; Kann, N. Cobalt-mediated solid phase synthesis of 3-O-alkynylbenzyl galactosides and their evaluation as galectin inhibitors. *Tetrahedron* **2006**, *62*, 8309–8317. [CrossRef]
118. Salameh, B.A.; Sundin, A.; Leffler, H.; Nilsson, U.J. Thioureido N-acetyllactosamine derivatives as potent galectin-7 and 9N inhibitors. *Bioorganic Med. Chem.* **2006**, *14*, 1215–1220. [CrossRef]
119. Vladoiu, M.C.; Labrie, M.; Létourneau, M.; Egesborg, P.; Gagné, D.; Billard, É.; Grosset, A.A.; Doucet, N.; Chatenet, D.; St-Pierre, Y. Design of a peptidic inhibitor that targets the dimer interface of a prototypic galectin. *Oncotarget* **2015**, *6*, 40970–40980. [CrossRef]
120. López De Los Santos, Y.; Bernard, D.N.; Egesborg, P.; Létourneau, M.; Lafortune, C.; Cuneo, M.J.; Urvoas, A.; Chatenet, D.; Mahy, J.P.; St-Pierre, Y.; et al. Binding of a Soluble meso-Tetraarylporphyrin to Human Galectin-7 Induces Oligomerization and Modulates Its Pro-Apoptotic Activity. *Biochemistry* **2020**, *59*, 4591–4600. [CrossRef]

MDPI
St. Alban-Anlage 66
4052 Basel
Switzerland
Tel. +41 61 683 77 34
Fax +41 61 302 89 18
www.mdpi.com

Biomolecules Editorial Office
E-mail: biomolecules@mdpi.com
www.mdpi.com/journal/biomolecules

www.ingramcontent.com/pod-product-compliance
Lightning Source LLC
LaVergne TN
LVHW070724100526
838202LV00013B/1166